METHODS
OF
DISASTER RESEARCH

METHODS OF DISASTER RESEARCH

Edited by
Robert A. Stallings

Copyright © 2002 by International Research Committee on Disasters.

Library of Congress Number: 2002095375
ISBN : Hardcover 1-4010-7971-7
       Softcover 1-4010-7970-9

Cover photo of the Hotel Continental, Mexico City 1985, courtesy of Christopher Arnold, Building Systems Development, Inc., Palo Alto, CA.

All rights reserved. No part of this book may be reproduced or transmitted in any form or by any means, electronic or mechanical, including photocopying, recording, or by any information storage and retrieval system, without permission in writing from the copyright owner.

This book was printed in the United States of America.

To order additional copies of this book, contact:
Xlibris Corporation
1-888-795-4274
www.Xlibris.com
Orders@Xlibris.com

16862

CONTENTS

CONTRIBUTORS 11

FORWARD 15

ACKNOWLEDGEMENTS 19

1 METHODS OF DISASTER RESEARCH:
   Unique or not? 21
   Robert A. Stallings

PART I
CONTEXT

2 PREFACE 47
   Lewis M. Killian

3 AN INTRODUCTION TO METHODOLOGICAL
   PROBLEMS OF FIELD STUDIES IN DISASTERS 49
   Lewis M. Killian

4 THE DISASTER RESEARCH CENTER (DRC) FIELD
   STUDIES OF ORGANIZED BEHAVIOR IN THE
   CRISIS TIME PERIOD OF DISASTERS 94
   E. L. Quarantelli

5 FOLLOWING SOME DREAMS:
   Recognizing opportunities, posing interesting questions,
   and implementing alternative methods 127
   Thomas E. Drabek

PART II
CONTINUITIES

6   SURVEY RESEARCH 157
    Linda B. Bourque, Kimberley I. Shoaf, and Loc H. Nguyen

7   QUALITATIVE METHODS AND
    DISASTER RESEARCH 194
    Brenda Phillips

8   THE ECONOMICS OF NATURAL DISASTERS 212
    Anthony M. J. Yezer

9   CROSS-NATIONAL AND COMPARATIVE
    DISASTER RESEARCH 235
    Walter Gillis Peacock

10   MEDIA STUDIES 251
     Marco Lombardi

11   REWRITING A LIVING LEGEND:
     Researching the 1917 Halifax explosion 266
     T. Joseph Scanlon

PART III
PROSPECTS

12   METHODOLOGICAL CHANGES AND CHALLENGES
     IN DISASTER RESEARCH: Electronic media and the
     globalization of data collection 305
     Wolf R. Dombrowsky

13   THE USE OF GEOGRAPHIC INFORMATION
     SYSTEMS IN DISASTER RESEARCH 320
     Nicole Dash

14 PROBLEMS AND PROSECTS OF DISASTER
   RESEARCH IN THE DEVELOLPING WORLD:
   A case study of Bangladesh 334
   Habibul Haque Khondker

15 THE FIELD TURNS FIFTY:
   Social change and the practice of disaster fieldwork 349
   Kathleen J. Tierney

PART IV
POSTSCRIPT

16 FUTURE DISASTER RESEARCH:
   A practitioner's viewpoint on
   public-private partnerships 377
   Ollie Davidson

PART V
APPENDIX

SELECTED INTERNET RESOURCES ON
NATURAL HAZARDS AND DISASTERS 389
David L. Butler

BIBLIOGRAPHY 465

INDEX 503

LIST OF TABLES

Table 1. Comparison of Response Rates in Earthquake Studies to Response Rates in the Los Angeles County Social Surveys, 1993-1996. 165

Table 2. Earthquake Preparedness in California, 1976-1994 (in Percentages). 170

Table 3. Loss of Utilities, by Earthquake. 179

Table 4. Examination of Memory Decay Across Three Waves of Data Collection Following the Northridge Earthquake. 181

Table 5. Examining Dose Response by Modified Mercalli Intensity (MMI) and Earthquake. 184

Table 6. Comparison of Specialized Population to Probability Sample of Los Angeles County. 189

Dedicated to the memory of

Charles E. Fritz
(1921-2000)

*A pioneer of disaster research,
a friend to those who followed*

CONTRIBUTORS

Linda B. Bourque is Professor of Public Health in the School of Public Health, University of California—Los Angeles, Los Angeles, California 90024-1772, USA. [lbourque@ucla.edu]

David L. Butler is Senior Editor and in charge of Web sites and other computer resources at the Natural Hazards Research and Applications Information Center, University of Colorado, Boulder, Colorado 80309-0482, USA. [butler@spot.colorado.edu]

Nicole Dash is a Research Associate at the International Hurricane Center and is a doctoral candidate in the Department of Sociology and Anthropology, Florida International University, Miami, Florida 33199, USA. [dashn@fiu.edu]

Ollie Davidson is Director of Emergency Management Programs at Counterpart International, USA. [PriPubPart@aol.com]

Wolf R. Dombrowsky is Director of the Katastrophenforschungsstelle (KFS) [Disaster Research Unit], Christian-Albrechts-Universität zu Kiel, Olshausenstraße 40, Kiel D-24098, Germany. [wdombro@soziologie.uni-kiel.de]

Thomas E. Drabek is Professor of Sociology in the Department of Sociology, University of Denver, Denver, Colorado 80208-2948, USA. [ZTED@aol.com]

Habibul Haque Khondker is Senior Lecturer in Sociology in the Department of Sociology, National University of Singapore, Singapore. [Habib@nus.edu.sg]

Lewis M. Killian is Professor Emeritus of Sociology at the University of Massachusetts and Faculty Associate in the Department of Sociology at the University of West Florida, Pensacola, Florida 32514, USA.

Marco Lombardi is Professor of Sociology at the Catholic University, Largo Gemelli 1, 20123 Milan, Italy. [marlom@mi.unicatt.it]

Loc H. Nguyen is Program Coordinator at the Center for Public Health and Disaster Relief and a doctoral candidate in the School of Public Health, University of California—Los Angeles, Los Angeles, California 90024-1772, USA. [locn@ucla.edu]

Walter Gillis Peacock is Associate Director for Research at the International Hurricane Center and is Associate Professor of Sociology and Anthropology, Florida International University, University Park Campus, Miami, Florida 33199, USA. [peacock@fiu.edu]

Brenda D. Phillips is Professor at the Institute for Disaster Preparedness, Jacksonville State University, Jacksonvlle, Alabama 36265, USA. [brenda@jsucc.jsu.edu]

E. L. Quarantelli is Research Professor at the Disaster Research Center, University of Delaware, Newark, Delaware 19716, USA. [elqdrc@udel.edu]

T. Joseph Scanlon is Director of the Emergency Communications Research Unit, Carleton University, Ottawa, Ontario K1S 5B6, Canada. [jscanlon@ccs.carleton.ca]

Kimberley I. Shoaf is Research Director of the Center for Public Health and Disaster Relief and Adjunct Assistant Professor of Community Health Services at the School of Public Health, University of California—Los Angeles, Los Angeles, California 90024-1772, USA. [kshoaf@ucla.edu]

Robert A. Stallings is Professor of Public Policy and Sociology, Program in Public Policy, School of Policy, Planning, and Development, University of Southern California, Los Angeles, California 90089-0626, USA. [rstallin@usc.edu]

Kathleen J. Tierney is Professor of Sociology, Department of Sociology and Criminal Justice, and Codirector of the Disaster Research Center, University of Delaware, Newark, Delaware 19716, USA. [tierney@UDel.edu]

Anthony M. J. Yezer is Professor of Economics in the Department of Economics, George Washington University, Washington, D.C. 20052, USA. [yezer@gwis2.circ.gwu.edu]

FORWARD

When I became President of the International Research Committee on Disasters in 1994, *What is a Disaster?*, the first in what was to be series of books sponsored by the IRCD had just been published. The second volume was still at the discussion stage but after a number of meetings in London, it was decided that it would be on methodology and that other volumes on various topics would follow.

Even though Henry Quarantelli and I paid a number of visits to our publisher nothing more materialized and a full manuscript of the second volume seemed to disappear. I found a new publisher and after two meetings—one in Boulder, one in London—worked out an agreement that the new publisher would take over the series. A number of potential titles were approved and, once again, it was agreed that the first would be on methodology.

As was the case for the theory book, methodology began as a special issue of the *International Journal of Mass Emergencies and Disasters* then was turned into a book with some material left intact, some revised and some added. The resultant manuscript was turned over to the second publisher, seemingly found close to acceptable then, despite revisions aimed at meeting every concern, turned down. At that time, a number of persons decided that, perhaps, the route to go was on-line publishing. It was also then that some persons offered financial support to get the on line project off the ground. That whole process has taken roughly seven years.

Throughout that entire period one person has remained calm and patient despite the seemingly endless promises and indications that the project was near an end. That person was Bob Stallings, editor of this book. Perhaps I should say I say almost always calm and patient because I recall one occasion when Bob became exasperated with the seemingly endless problems. But his determination overrode his frustration.

Bob Stallings is one of the many DRC graduates who have made a lasting contribution to the field of disasters, emergency management and emergency decision-making. I think I first met Bob at the World Congress of Sociology in Mexico City. Certainly my first clear memory of him is listening with awe to the thorough way he performed as a discussant, gently devastating a paper that suffered from inconsistency and inadequate methodology. I got to know him and respect him as a scholar at a number of other meetings, then remembering his expertise in methodology, turned to him when the idea of a special journal issue on methodology was conceived. I was so impressed by the way that he dealt with the issue that, with Ron Perry's approval and support, I asked Bob to take over from Ron as editor of Mass Emergencies a task he has performed with competence and dedication. That means that for the past couple of years, Bob has been teaching, doing administration, editing *Mass Emergencies*, and doing the revisions of this book.

Although Bob does not remember this, the book itself arose out of my research into Samuel Henry Prince and the 1917 Halifax explosion. Henry Quarantelli suggested that I should write a journal article on how I did that historical research and with Bob's help, I did so. In the meantime, I had countered Henry's suggestion with the argument that it was time Henry wrote something about how the Disaster Research Center operated. Many of us, including Bob Stallings and myself, have spent various lengths of time at DRC but no one had ever published a description of its methodology. With these two articles as starting points, Bob searched successfully for additional material both from some of the old hands like Lewis Killian, from other established scholars and from relative newcomers. I will leave it to him as editor to talk about that but his ability to meld such a group into producing a coherent end product reflects his skill as an editor.

Now that this volume is finally out I think we can realistically look forward to more volumes. One on gender issues, for example, is well underway, and I am sure others, such as book on popular culture and disaster, will emerge. All these volumes will do far more than add to the knowledge base in disasters. They will use disasters as a reference point and cover topics that should be and will be of interest to a much broader audience. The years of labor by Bob Stallings that brought forward this book thus mark a major contribution not only to the IRDC and to Sociology of Disaster but to social science generally. Because from the start I have watched the time and effort Bob has put into this project more than anyone else I can say a heartfelt thanks from all of us. May I add my own thanks to Benigno Aguirre my successor as President of the Research Committee for allowing me to continue as general editor and, thus, to be able to write these comments.

Joseph Scanlon
General Editor
International Research Committee on Disasters
Book Series

ACKNOWLEDGEMENTS

My personal thanks to all the contributors to this volume who so willingly shared their research experiences and who have waited patiently for this book to appear.

Thanks also to T. Joseph Scanlon and E. L. Quarantelli for their suggestions and comments in helping to structure this volume.

Thanks to the National Academy Press for permission to reprint the report by Lewis M. Killian, originally released in 1956 as *An Introduction to Methodological Problems of Field Studies in Disasters* (Washington, D.C.: Disaster Study Number 8, Committee on Disaster Studies, Division of Anthropology and Psychology, National Academy of Sciences-National Research Council Publication 465).

Thanks also to Charles Perrow for permission to quote from some personal correspondence.

Finally, special thanks to Ms. Artimese Porter for her support on this and numerous other projects and to Ms. Lena Le for preparing the Killian report for this volume.

Robert A. Stallings
Covina, California

1

METHODS OF DISASTER RESEARCH:

Unique or not?

Robert A. Stallings

More than a decade ago at the end of his review of research methods in the sociology of disasters, Dennis Mileti wrote: " . . . from a methodological viewpoint, disaster research is hardly distinguishable from the general sociological enterprise" (Mileti 1987: 69). The topics covered in the chapters in this volume—survey research, historical methods, qualitative research, etc.—support this conclusion. The types of methods used in social science research on disasters are not unique.

Yet, people well-trained and with experience in survey research or qualitative methods will find that the study of disasters is different. The difference does not lie in knowledge of the subject matter; such knowledge is no more important in the field of disaster studies than in any other. What makes disaster research unique is the *circumstances* in which otherwise conventional methods are employed. Put differently, it is the *context* of research not the methods of research that makes disaster research unique.

It follows then that the uniqueness of the circumstances of research varies as a function of the phase of the disaster process one is studying. Research on responses just before, during, and right after impact occurs in a different context that does research on long-term trends in governmental expenditures for disaster relief, for example. It further follows that the collection and analysis of primary data are more affected by the disaster context than are the collection and analysis of secondary data, when disaster phase is taken into account.

Disaster researchers, therefore, need two types of training: first, they need training in research methods in general (e.g., survey research or qualitative methods); and, second, they need training in how, specifically, the circumstances surrounding disaster affect the application of these methods. The following chapters, written by some of the leading practitioners of the several methods used in disaster research, provide materials to meet the second of these needs.

PREVIOUS STATEMENTS ON THE METHODS OF DISASTER RESEARCH

There has been very little written expressly on the topic of methods of disaster research. Although empirical studies of disaster phenomena customarily include a discussion of data collection (and usually also of data analysis), only a handful of previous works have been devoted to a discussion of methods *per se*. Probably the earliest was an article by Harry Williams, at the time a staff member of the Committee on Disaster Studies at the U.S. National Research Council (NRC). The committee assigned high priority to exploratory research:

> "In a field so complex and so little understood, it is felt that exploratory studies should be made in many different disasters, to define the major variables and discover the repetitive phenomena." (Williams 1954: 8-9)

However, the committee simultaneously recognized the importance of hypothesis-testing methods such as existed in the social sciences at the time:

> "There will probably be value for some time to come in general, descriptive studies of disasters. But the time has also come, in the Committee's view, when research can be more rigorously designed to test well-formulated hypotheses." (ibid.: 9)

The article proceeds to give examples of work in progress at the time including economic and demographic analyses as well as experimental and clinical studies.

Most important of the handful of previous statements is the short monograph by Lewis Killian (1956), one of the first sociologists to conduct field studies of disaster. His report, written when he was a member of the NRC committee, is reprinted in its entirety here (Chapter 3); more will be said about it in the next section. Two other early discussions of the methods of disaster research appeared in an anthology edited by George Baker and Dwight Chapman (1962). The better remembered of the two is the chapter by Ira Cisin and Walter Clark (1962); the more valuable of the two is the chapter toward the end of the book by Harold Guetzkow (1962). Cisin and Clark begin with a familiar theme:

> "Strictly, we cannot speak of the methods of disaster research; there are no special methods unique to this field. Its methods are the methods of social research, . . ." (Cisin and Clark 1962: 23)

The "challenge" of disaster research for them is the lack of time between the occurrence of a disaster and the fielding of research: lack of time to develop theory and hypotheses; lack of time to develop research instruments; lack of time even to decide which events are worthy of study; etc. This lack of time is most problematic in studying phenomena associated with what Quarantelli (this volume, Chapter 4) calls the crisis time period of disasters. Cisin and Clark seem troubled that the necessary compromises in disaster research will produce an "inferior product" as judged from the standpoint of "good" social science research. As a consequence, they emphasize the distinction and the complementary relationship between descriptive and explanatory studies. For them, the future of disaster research involves the transition from descriptive studies to "... explanatory studies [that] try to make sense out of the relationships observed in [descriptive] analytical studies..." (Cisin and Clark 1962: 41). Only knowledge based upon explanatory research will be sufficient to achieve the aims of Cold War civil defense policies: "Only such hypotheses can lead toward the shift from explanation to control" (Cisin and Clark 1962: 42).

The distinction between descriptive and explanatory studies of disasters leads to the more valuable chapter in the Baker and Chapman anthology by Guetzkow, who tries to show how to "join field and laboratory work" in the study of disasters. Guetzkow wants to move even further, beyond description and explanation, to prediction. He lists three "fundamental characteristics" of the state of disaster research at the end of the 1950s, the third of which is: "*The soundness of the formulations of disaster research has not yet been subjected to the basic scientific test—the prediction of behavior in disaster*" (Guetzkow 1962: 340; italics in the original). After discussing the built-in limitations of laboratory simulations for disaster research, he argues that quasi-experimental research designs applied in the field would advance the state of the art, especially studies in which advance warning allows for collection of pre-event data. Short of this, he suggests that field and laboratory research should be more closely coordinated so as to take advantage of the strengths of each type of research design. Guetzkow's recommendation to carry out quasi-experimental field studies has largely been ignored; the same recommendation is made, four decades later, by Thomas Drabek in his contribution to this collection (Chapter 5). This model of disaster research as requiring both exploratory and confirmatory studies realized through field and laboratory designs, respectively, was clearly an influence on the early configuration of the Disaster Research Center (DRC; see Chapters 4 and 5 in the next section).

Another, now largely forgotten, work addressing the issues of the timing of research and the possibilities of obtaining pre-event data on "unscheduled events" is the chapter in the Baker and Chapman volume by Albert Biderman (1966). Prepared with funding from the U.S. National Aeronautics and Space Administration (NASA) and geared toward evaluating the impact of unexpected events in the exploration of space, Biderman's chapter reviews previous efforts to carry out systematic research on unanticipated events with the aim of fielding studies as soon as possible after if not in advance of them. The importance of a stand-by capability—and the critical need for funding in advance of any such events—are carefully documented and argued. Although the DRC developed its field research capabilities independently from Biderman's work, his chapter remains a useful supplement to the paper by Quarantelli published here.

An early if very brief description of how the DRC initially incorporated what was required in the study of disasters is contained in Dynes, Haas, and Quarantelli (1967). The separate roles of laboratory and field studies are noted, with the center's use of laboratory simulation briefly discussed. The center's field research techniques with special emphasis on their application outside the U.S. is then described. Drabek (1970), writing at about the same time, looks more to the future from the perspective of the initial work at the DRC. Drawing on his experience with a major simulation of disaster response carried out in the laboratory (see Chapter 5 in this collection), he argues for increased use of experimental and quasi-experimental designs.

One does not belittle the contributions of these early statements on the methods of disaster research by concluding that it is time for a new assessment of the field—where it has been, where it is now, and especially where it should be headed. The authors who have contributed to this anthology were asked to do exactly that. Their papers reflect the twin characteristics of disaster research: its methods are indistinguishable from those in general use in the social sciences; and the circumstances under which these methods are used differ, to a greater or lesser extent, from the conditions encountered in other research fields.

IN THIS VOLUME

The first section of the book contains three papers describing the context from which current and future research methods used in the study of disasters have and will emerge. As indicated above, the most important of the early statements on disaster research methods is the report by Lewis Killian, *An Introduction to Methodological Problems of Field Studies in Disaster* (reprinted here as Chapter 3). After completing his graduate studies in sociology at the University of Chicago—the same program in which E. L. Quarantelli received his graduate training—Killian conducted studies of such well-known disasters as the explosion of two ships in the harbor at Texas City, Texas (Logan, Killian, and Marrs 1952), a series of deadly tornadoes in the Southwest U.S. (ibid.), and a devastating tornado in Warner Robbins, Georgia (Killian and Rayner 1953). His 1956 report on the unique methodological problems posed by conducting social science research in the aftermath of disaster echoes the theme of this book—that even those trained in social research methods, particularly in hypothesis-testing approaches, do not possess the wherewithal to effectively carry out research on disasters. Reinforcing the point made above about the uniqueness of the postimpact setting, Killian concludes that the "methodological problems" of disaster research vary:

> "Some types of research related to disaster present no unusual methodological problems.... It is in the analysis of significant psychological and sociological variables as they affect human behavior *during the course and the immediate aftermath* of a disaster that special methodological problems arise." (Killian 1956: 3; italics added)

It is the unique aspects of the disaster context and their implications for research that Killian examines in detail. He covers all elements of the research process from selecting events for study, research design, populations and sampling (including the selection of control cases), data collection, and data analysis to the writing of reports.

Constraints on conventional hypothesis-testing methods explain why he discusses descriptive and exploratory designs so extensively. Yet Killian continually shows how traditional verification-oriented methods—designs that control for potentially spurious variables and that use probability sampling, for instance—should and can be incorporated into disaster research. Thus his observations are equally relevant for qualitative and quantitative researchers.

Although written nearly fifty years ago, Killian's monograph is required reading for both current and future disaster researchers. It is included here in its entirety for two reasons: the report remains as relevant at the beginning of the twenty-first century as it was in the middle of the twentieth century; and, even though it is frequently cited, few copies remain in existence of what was essentially a committee report. Killian's report is prefaced by his reflection on change and continuity in the methods of disaster research since its original publication (Chapter 2).

Adjustments called for in order to carry out field studies in the immediate postimpact phase of disaster are detailed in the long-awaited description of methods used by the Disaster Research Center (DRC) authored by one of its cofounders and long-time codirectors, E. L. Quarantelli. In Chapter 4, Quarantelli not only describes the methods and techniques employed by the DRC but also explains the theoretical and practical reasons for the selection of those methods and techniques. The center's commitment to exploratory research can be seen in historical context. Forty years ago, this literally was "pioneering" research; the empirical base that made possible later propositional inventories (Mileti, Drabek, and Haas 1975; Drabek 1986) was virtually nonexistent. While the many practical lessons to be gleaned from Quarantelli's chapter will be self-evident, one that deserves highlighting is the importance of the timing of data gathering. The experience of the DRC underscores the importance of arriving on scene as soon as possible after—or before, if possible—disaster strikes. My sense is that barriers faced by later-arriving researchers are increasing, as agency officials become more wary of public reaction (and its potential organizational and career implications), possible litigation, and the like. (Tierney discusses these issues in Chapter 15 of this volume.)

There are a few things that Quarantelli does not say here. Having been one of the graduate research assistants (GRAs) alluded to in his chapter (from 1967 through 1971, in my case), I would like to add two things to the public record about the DRC. First, codification of methods was driven not only by the volume of work undertaken by the center; it was necessitated also by the need for continuity, given the inevitable turn-over of graduate research assistants (who had an annoying tendency to complete their degrees and move on to academic positions elsewhere). This turn-over was made possible by the continuity of the center itself, a research unit that has been in continuous operation for forty years. The continuity of the center came at a price: the hours spent on research administration (getting the air conditioning fixed, straightening out mistakes in billing, hosting site visitors, reassuring agitated representatives of funding agencies, etc.), writing proposals, and preparing interim and final reports. The research contributions of the DRC have been bought and paid for by the unrecognized administrative contributions of its directors, past and present.

Second, Quarantelli describes the participation of GRAs in all phases of the research process, from spelling out the design implications of newly-funded projects to data gathering, analysis, and report writing. This was consistent with the training mission of the center and provided an unequaled opportunity for GRAs to learn disaster research methods in "hands-on" fashion. However, a more subtle form of research training was provided, one that was even more important to those DRC research assistants who moved on to academic careers. Publication was encouraged and supported, both materially and intellectually, but never demanded. Both codirectors during my time as a graduate research assistant (Quarantelli and Russell Dynes) read draft manuscripts, suggested appropriate conferences and panels at which they might be presented, and provided insights into how to "market" manuscripts by matching their content with the editorial traditions of various journals. (Thankfully, they continue to do so.) Writing at the DRC was always driven by the desire to share research findings with those who were most likely to be interested in or to benefit from them. All of this took place without any fear of exploitation. Our writing was our own; no one's name was added automatically to everything that was written. The high ethical and intellectual standards of the DRC, which I naively once assumed were normal in the academic world, are both part of its "story" and part of the reason for its success.

If the expression, "Been there, done that," applies to anyone doing research in this area, it describes Thomas Drabek. His autobiographical account of the many noteworthy studies in which he has participated (Chapter 5) rounds out the chapters in the first section. Like most researchers in this field, Drabek never dreamed that he would make a career of studying disasters. Unlike most researchers in this or any other field, however, Drabek learned and applied many different methods when they seemed best suited to provide answers to interesting questions. His advice—to let research problems determine research methods rather than the reverse—is almost as hard to follow as it is to argue against. His description of the laboratory simulation involving a police communications unit reinforces the point that the Disaster Research Center, where this early study was carried out, was established with a clear understanding of the difference between generating hypotheses (through exploratory designs and fieldwork) and testing hypotheses (under controlled laboratory conditions). Drabek concludes by urging young scholars to commit themselves to good research with practical benefits rather than to a specific research method.

The second section contains six chapters which illustrate the continuity between past, present, and future methods of disaster research. Those who believe that disaster research and qualitative field studies are synonymous will be surprised at the frequency with which survey research has been performed in the study of disaster-related phenomena. Linda Bourque, Kimberly Shoaf, and Loc Nguyen (Chapter 6) do something quite unique. Rather than simply arguing for the relevance of survey research in the study of disasters, they empirically assess strengths and weaknesses of this method with data generated from six surveys conducted in the aftermath of recent California earthquakes. Their assessment challenges some of the alleged weaknesses of survey methods in disaster research. For example, their data show that the size of the population likely to be missed by conducting telephone interviews in the aftermath of disaster is small and its characteristics estimable. Also, there is no evidence that people are less willing to participate in telephone interviews following disaster than at any other time. Further, their assessment challenges the belief that data are so perishable that survey research inevitably misses relevant information. Several of the advantages of population based survey research are also demonstrated with these postearthquake datasets. Bourque and her colleagues have written an important statement that will encourage those interested in conducting survey research on disasters.

Brenda Phillips (Chapter 7) is someone with both extensive personal experience in conducting qualitative research in the aftermath of disasters and extensive classroom experience in teaching qualitative methods. Her chapter brings the reader up to date on recent trends and developments in the application of qualitative methods in the social sciences. Included is a discussion of how some of these recent developments blunt criticism of qualitative research by its positivistic critics. She documents the "fit" between qualitative methods and disaster research—past and present—and makes several specific recommendations for future qualitative disaster research.

Phillips' and Quarantelli's chapters contain somewhat different messages about the role of field studies in disaster research. They differ in a commitment to exploratory research on the one hand (Quarantelli) versus a commitment to qualitative methods *per se* on the other (Phillips). This difference illustrates the point that field studies represent a large umbrella covering a variety of methods. Drabek's chapter—and career—are cases in point. I argued several years ago (Stallings 1986) that the methods many of us were using in conducting field studies of disastrous events were being applied to a type of theory that was not well served by those methods. Qualitative research seems especially well-suited to describe the subjective experience of disasters at the individual level, yet we have most often used it to document "objective" macro-level features of disasters (e.g., changes in interorganizational relationships). I do not mean that qualitative research has been wrongly applied in the study of disasters in the past, only that much of its potential has yet to be realized. Phillips suggests ways of reaching that potential.

Anthony Yezer (Chapter 8) reviews a surprisingly modest—given its political and policy implications—but diverse literature of the economics of natural disasters. As a first step toward future research, he organizes this disparate literature into a small number of common topics and shows various connections among them which future researchers should make. These topics include: the economic impact of disasters; estimating disaster effects using equilibrium models of regional economic development; and research on markets for natural hazard insurance, including individuals' decisions to purchase insurance and invest in mitigation, the interaction among insurance, mitigation, and postdisaster government aid, and instruments for capitalizing private insurance in the face of potentially catastrophic disaster losses. His review of the literature on each topic identifies the indicators and proxy variables that have been or should be used in economic research on natural disasters. In addition to providing this blueprint for future researchers, Yezer shows how standard economic models and their assumptions need to be modified in order to study the economics of disasters.

Cross-national comparative research is a type of disaster study that nearly everyone says is necessary to advance the field (e.g., Taylor 1978) but one that hardly anyone has actually carried out. One who has is Walter Gillis Peacock. Peacock (Chapter 9) points out that *all* research is comparative but not all research is cross-national. As someone long engaged in cross-national research, he recommends multi-national teams of researchers rather than single researchers for the execution of cross-national studies. From his perspective, prospects for successful cross-national disaster research have never been better. Disaster management is increasingly cross-national, the number of disaster researchers outside North America has increased significantly, more nations are concerned about disasters, and more cross-national databases have come into existence. A major impediment remains: the problem of equivalence. This problem encompasses lack of comparability among secondary statistical data, estimates of disaster losses, and even units of analysis such as the family. To date, identical measures of the same variable in different national contexts remain a goal rather than a reality. Peacock's personal suggestion for advancing cross-national disaster research is to examine the complex linkage between disasters and development. The process of development clearly affects disaster vulnerability. It is equally true that natural disasters affect the development potential of developing nations. For researchers interested in cross-national disaster research, Peacock's chapter is a "must read."

News reports have always been valuable data sources in disaster research. Harry Moore, in his comprehensive study of two tornado-stricken communities, provided one of the earliest systematic analyses of newspaper content (Moore 1958: 194-205). In the present collection, Marco Lombardi (Chapter 10) discusses the methodological implications of a constructionist approach to the study of news media. He recommends that researchers think of risk and the events associated with risk not as objective "facts" that are described (more or less accurately) in media reports. Rather, he argues that researchers should think of risk as the *process* of communication about events and threatened events. Thus, to understand risk is to understand the interaction among experts, policy-makers, and the public-at-large. Both the nature and level of risk is negotiated in a contentious process linking these actors. This directs research attention away from a focus on the content of information about risk and onto the interactions occurring in discourse about risk. Against this conceptual background, Lombardi identifies four types of future research on the media and the variables that need to be measured in each.

Joseph Scanlon's paper (Chapter 11) describing his long-time involvement in study of the Halifax, Nova Scotia, munitions ship explosion in 1917 is an invaluable invitation to historical disaster research. With the passing of time, the relationship between researchers and data providers changes. What at the time of disasters people may be unwilling to talk about, many in later life are no longer reluctant to share, especially with a serious scholar who is obviously as passionate—and as knowledgeable—about the topic as they are. Scanlon's chapter, containing numerous tips and suggestions, is one that will inspire both young researchers and old hands alike to devote more attention to historically important events.

There are some things Scanlon does not say that those desiring to pursue this line of inquiry should reflect upon. First, Scanlon's research successes are related in no small way to his skills, "instincts," and experience as a successful *journalist*. While the suggestions contained in his paper can be understood and applied by anyone, not everyone has the personality—the persuasiveness, the persistence, the sheer delight in discovering interesting new details, no matter how minute—to make these suggestions work to produce the same kind of results. (Quarantelli, in his paper, alludes to a relationship between personality and the affinity for research tasks. The same hypothesis can be restated for the relationship between personality and the affinity for certain methods of research. Some people are better suited for analyzing quantitative data with computer software, others to "mucking about" in the world of real human beings.)

Second, Scanlon's investigative pursuits are those of a *senior* scholar, someone who has learned how to "piggy-back" one effort on top of another—and someone with sufficient resources to take risks. Younger researchers should realize that there are risks here; risks related to time, and hence risks related to money—and to one's career. Not every step taken in piecing together the Halifax story could have been planned in advance or described in a research proposal. Not every step resulted in a successful outcome. Scanlon's journey is a relentless, reflective, and exciting one carried out over a long period of time with many different types of support. For those with the temperament and the resources, the historical study of disasters is both an exciting and an important undertaking. Scanlon's chapter is a valuable tutorial. (For another excellent example of historical disaster research, see Dynes' [1998, 2000] work on the Lisbon earthquake of 1755).

The four chapters in Part III describe resources and situations that were generally absent in disaster research even a few decades ago but which will undoubtedly shape the course of that research during the next several decades. Wolf Dombrowsky (Chapter 12) assesses both advantages and disadvantages of the Internet in particular and computer technology in general as resources for future disaster research. While these tools have increased the quantity of our output, Dombrowsky questions whether they have similarly improved the quality of our work. In short he asks: Does the availability of more information on disasters mean that we have more knowledge about disasters? To illustrate his concern, he reports on his own experience in locating nearly a quarter of a million Internet references to a disaster-related topic, "panic," only to discover that fully 90 percent of them were useless for research purposes. More alarming are Web sites advertising the services of "disaster professionals" with unknown—and possibly questionable—credentials.

In his essay Dombrowsky records some of the ironies of computer usage for research purposes, chief among them the obvious fact that modern computers use a less universal language than European scholars used during the so-called "Dark Ages" (i.e., Latin). He reminds us that the earliest statisticians made more progress in using available data for human betterment than we have since the development of computer technology. Furthermore, the massive electronic databases currently available can be used either for good or ill. However, Dombrowsky remains optimistic about the future of disaster research. With current technology, he argues, we have the capability to put together the most accurate assessment of future disaster vulnerability to date.

Dombrowsky's reservations about the Internet and computers as tools for disaster research can be supplemented with an example involving a dataset available via the Internet which posed numerous logistical problems in actually rendering it into usable form. After locating precisely the data he needed on emergency responses to hazardous materials sites, Kline (1995) discovered that the entire database was so large as to overwhelm all of the local computer workstations to which he had access. Yet, subsetting the dataset before file transfer was not possible. Part of this and other problems like it will undoubtedly disappear with on-going technological changes, but a disparity between what the "latest and greatest" computer systems are capable of doing and the capacity of the majority of computers at the disposal of disaster researchers will remain.

Use to date of the Internet as a tool for disaster and hazards research suggests some additional limitations. Most evident is the uneven posting of material. Some organizations—including some national governments—have been quicker and more systematic than others to put documents up on their Web sites (Hwang, Sanderson, and Lindell 2001). Those interested in cross-national research have experienced the thrill of finding an incredible number of documents provided by one government only to experience the disappointment of finding that comparable documents from a neighboring nation-state are unavailable. Presumably this unevenness will recede as organizations and governments discover the cost-savings of distribution via the Internet. However, as Dombrowsky notes, proprietary interests may make certain material available but inaccessible for financial reasons.

Nicole Dash (Chapter 13) offers the reader a first look at Geographic Information Systems (GIS)—a collection of computer hardware, software, and procedures for manipulating and displaying a variety of spatially-referenced data—as both future targets of and tools for disaster research. Her chapter has the benefit of her first-hand experience with the way GIS is actually used in various phases of the disaster process. She is thus keenly aware of both its practical limitations and its strengths. After reviewing some of the recent history of GIS use in hazards management, she provides a case study of the application of a GIS following Hurricane Andrew. Her case study drives home the point that a GIS is most effective if it is created *before* disaster strikes. Dash concludes with five recommendations for basic and applied research involving GIS and disaster. Her suggested research agenda shares with Drabek (Chapter 5) an urging to link theory-based research with practitioner needs, recognized as well as unrecognized.

Habibul Haque Khondker (Chapter 14) notes an ironic feature of existing disaster research: while the greatest toll from disasters is in countries in the developing world, the vast bulk of disaster research has not been conducted there. One reason may be that researchers are unable to distinguish between sudden disasters and the everyday "disaster" of mass poverty and deprivation in the developing world. Another reason for the paucity of research is the relative under-development of sociology in general and disaster sociology in particular in these countries. This under-development is partly explained by the political sensitivity of disaster-related topics, especially famine. Whatever the reasons, much of the current research on disasters in the developing world has been conducted and for the foreseeable future will continue to be conducted by outside (i.e., foreign) researchers.

A continuing barrier to successful disaster research in developing countries is that (secondary) statistical data are spotty and often unreliable. This means that fieldwork will remain the most appropriate method of disaster research in such settings for the foreseeable future. Khondker illustrates some of the difficulties of doing fieldwork with a description of his study of the effects of disaster on women in two Bangladesh villages. This case serves to identify factors that make for successful fieldwork not only in Bangladesh but in all countries in the developing world: the importance of proper "sponsorship"; the role of "networking" in securing entrée; gaining the support and confidence of local elites; the importance of hiring data collectors acceptable to local peoples; and the need to be sensitive to local norms and customs in the gathering of field data.

The final chapter in this section updates the status of many of the methodological issues involved in fieldwork that were discussed in some of the early chapters, especially those by Killian and

Quarantelli. Kathleen Tierney (Chapter 15) documents how social and cultural changes during the second half of the last century have affected the conduct of disaster research. Some of these changes have created challenges, others have made life in the field easier for researchers. Making field studies more challenging are such trends as an increasing emphasis on protecting human subjects, especially those whose status seems to render them especially "vulnerable" when participating in research. For those who know little or nothing about research on the effects of disaster, disaster "victims" seem to be in a state of special vulnerability. University institutional review boards, those groups of administrators and faculty charged with ensuring that researchers do not abuse their subjects, can have a real impact on the success of disaster fieldwork, as Tierney documents. Similarly, the omnipresent threat of litigation can have a chilling effect on disaster research—and on disaster researchers. Organizations more worried about "impression management" and information control are another contemporary challenge in the field. Tierney notes the increasingly tendency for organizations to lump disaster researchers together with representatives of the news media and describes the complications that this entails for researchers. She identifies one ray of hope in this regard, however. As the professionalization of emergency managers increases, they may be more likely to recognize the positive role that research can have in their work and the unique needs of disaster researchers. The growth in the number of disaster researchers and in funding available to them can sometimes create competition among researchers in the field, especially following major disasters. The irony is that such *convergence of researchers* was unthinkable in the 1950s when their predecessors first documented the phenomenon of convergence behavior following disaster (e.g., Fritz and Mathewson 1957).

Another change that has had significant consequences for disaster research is the gradually increasing number of women conducting field studies. Tierney, among the earliest female GRAs at the Disaster Research Center, describes some of the problems and frustrations experienced by women in the field. More than the demographics of the disaster research community has changed since the pioneering days; the increasing involvement of women has changed the way in which disasters are viewed, resulting in changes in the content of knowledge about disasters. On the other hand, Tierney notes that this field has been less successful in attracting people of color, with the consequence that another perspective on disasters is still largely not to be found.

The postscript (Chapter 16) contains an essay by Ollie Davidson, a long-time disaster practitioner. Davidson has been at the forefront in the U.S. in creating partnerships between public and private organizations that can lead to the mitigation of losses and hasten recovery following disaster. Here he makes an explicit appeal to disaster researchers for increased use of evaluation designs. Evaluation research is especially needed, he argues, to determine the effectiveness of existing public-private partnerships and to communicate their successes to other, prospective partners. Davidson's essay suggests how a variety of methods could be employed to benefit practitioners in the future: empirical analyses of business losses and computer modeling of potential future losses; interorganizational analyses of the relationships among organizations from the public, private, and nonprofit sectors; cultural analyses of public—and private-sector organizations aimed at identifying the stereotypes that each have of the other, perhaps across organizational levels, age, and gender lines; the use of GIS to forecast all-hazards damage potential; and studies along the lines previously described in chapters by Drabek, Dombrowsky, and Dash on the utilization of the Global Disaster Information Network.

David Butler has created for the reader an appendix consisting of an annotated list of Internet sites on the World Wide Web that can provide data useful for disaster and hazards research. This listing is presented here in an appendix, not because it is an afterthought but because the Internet is a "dynamic" resource sure to be different soon after readers begin using this list. They may find the home page that Butler manages for the Natural Hazards Research and Applications Information Center more useful in the long-run. The address is: **http://www.colorado.edu/hazards/sites/sites.html**.

INVOLVEMENT VERSUS DETACHMENT IN DISASTER RESEARCH

Prospective disaster researchers should be aware of an important issue that they will confront no matter which methods they use in their work. The issue involves both external and internal pressures to help solve the problems of disaster victims, practitioners, and policy-makers. The issue is more than the usual call for the social sciences to be "relevant." It extends to the worldview of the disaster researcher, creating a potential dilemma between involvement on the one hand and detachment on the other.

Disaster research has always had and will always have an applied side. The sources of funding, from the Department of the Army in the earliest Cold War days of disaster research to current social science funding under the National Earthquake Hazards Reduction Program (NEHRP), are part of the explanation for this. (This emphasis on the applied aspects of disaster research is dramatized by the funding patterns of the U.S. National Science Foundation whose normal program for funding sociological research has provided very little support for disaster research; the vast bulk of NSF support over the years has come from one or another program element in its applied research directorate.)

Perhaps because disasters in the U.S. are less contentious than other types of events making the news (who, after all, is a proponent of more and bigger disasters?), the external pressure on disaster researchers on the side of involvement is probably less than in other research specialties. The experience of Charles Perrow in producing a book about organizations and AIDS in the U.S. (Perrow and Guillén 1990) provides an instructive contrast. Many people refused to be interviewed by Perrow and his colleague; others who had granted interviews later denied the authors permission to use them in their book. Even after they had completed their manuscript, the authors continued to do battle with AIDS partisans. Five publishers rejected the manuscript, first because of hostile evaluations from the public health community, which was criticized in the manuscript, and second because of hostile evaluations from gay activists. Even the editor at the publisher which finally produced the book (Yale University Press) " . . . had to write to some 12-15 people before she could get somebody to even review it because apparently the word was out that people on all sides were unhappy" (Perrow 1993).

To be sure, pressures on researchers from those involved with disaster management are not absent. Tierney notes organizational efforts aimed at "impression management" in her chapter; Dash describes in hers being asked to assist in setting up a GIS in the field; Quarantelli's instructions to DRC field team members alerts them to the possibility of being asked to become involved in disaster tasks and outlines an appropriate response; and in his chapter Drabek recounts how the DRC codirectors resisted more than one attempt by a funding agency to prevent the publication of his monograph on the Indianapolis coliseum explosion, *Disaster in Aisle 13*.

Pressures to "take sides"—for involvement rather than detachment—also exist inside the disaster research community. Quarantelli's recent efforts to create a debate about the meaning of the term "disaster" (1995b, 1998) generated some critical comments about which "side" disaster researchers have been and should be on. Hewitt (1995, 1998) charged that disaster researchers historically have adopted the perspective of government disaster control managers. For the future, he argues for taking the side of disaster victims. Kroll-Smith and Gunter (1998a, 1998b) made a similar argument, calling for a future disaster program that pays attention to the various everyday "disasters" in the lives of ordinary people. The dilemma raised in this exchange is not between involvement and detachment but rather between two different types of involvement, one advocating the side of those in power and the other advocating the side of the powerless (see also Jäger 1977).

Before future disaster researchers conclude that there are only two sides available in thinking about the problems occasioned by disasters, they should know that there is an important counterargument to the side-taking advanced by Howard Becker (1967) in his widely cited article, "Whose Side Are We On?" The counterargument is best laid out by Joseph Gusfield (1984). Gusfield reminds us that there is a third "side" to be taken—that of the uninvolved observer (i.e., detachment). The advantages of detachment in the social sciences have been most systematically explored by Norbert Elias in his long essay on involvement and detachment (1987). Two advantages are seldom mentioned: detachment invites a longer time perspective, making it more likely that the immediate situation will be seen against the background of its institutional origins (ibid.: xv); and detachment raises a barrier against one's emotions, making it less likely that they will color one's professional judgment (ibid.: xxi). Elias uses the story of tool-making to illustrate his basic thesis. Making a tool requires detachment, such as by ignoring one's feelings of hunger long enough to finish making the tool; when it is finished, the tool can be used to satisfy one's hunger more completely than otherwise would have been the possible (ibid.: xxv). For Elias, social order depends on the impulses of involvement and detachment keeping each other in check (ibid.: 3-4). For the researcher, this takes the form of separating the role of participant from that of inquirer and establishing the dominance of detachment over involvement in one's work (ibid.: 16). (Elias' thesis is consist with the logic of ambivalence in social life described by Smelser [1998]; ethnographers discuss this issue in regard to the dangers of "going native.")

Reflecting the intellectual tradition of German neo-Kantian philosophy, most sociologists in the past have embraced the assumption that involvement is acceptable—indeed, inevitable and even desirable—in the selection of topics for research, whereas detachment is required in the conduct of research. Max Weber (1949) outlined this position on the role of values and value neutrality in research in his 1904 policy statement for prospective authors in the *Archiv für Sozialwissenschaft*. He wrote: " . . . [T]hose highest 'values' underlying the practical interest are and always will be decisively significant for determining the focus of attention of analytical activity . . . in the sphere of the cultural sciences" (Weber 1949: 58). Further, "An *attitude of moral indifference* has no connection with *scientific* 'objectivity' " (ibid.: 60) and " . . . in [the] social sciences the stimulus to the posing of scientific problems is in actuality always given by *practical* 'questions.' Hence the very recognition of the existence of a scientific problem coincides, personally, with the possession of specifically oriented motives and values" (ibid.: 61). But " . . . whenever the man of science [*sic*] introduces his personal value judgment, a full understanding of the facts *ceases*" (Gerth and Mills 1958 [1946]: 146; italics in the original). For readers unfamiliar with the issues here, the general discussion by Kaplan (1964: 370-397) is as good as any currently available.

It is the *methods* of research—that is, the techniques of data acquisition and analysis—that are presumed to prevent personal values and biases from seeping into the conduct of inquiry. Research methods, properly applied, supposedly ensure sufficient detachment during the research process to constrain the corrupting influence of involvement. Critics of the assertion that research methods ensure detachment include Elias himself. He cautions that methods, especially those imported from the physical sciences in part because of their supposed "objectivity," can create a façade of detachment around questions already shaped by involvement.

Whatever the reader's opinion about the appropriate or inappropriate role of values in disaster research—about what I refer to here as involvement—the worlds of social science disaster research and emergency management practice are fundamentally different. One difference, related to the preceding discussion, is a contrast between "facts" and "values." For disaster researchers the most frequent dilemma that this difference creates is practitioners' desire for research that can tell them what they should do. While disaster research can identify what has worked and what has not at other times and in other places, "An empirical science cannot tell anyone what he [*sic*] *should* do—but rather what he *can* do—and under certain circumstances—what he wishes to do" (Weber 1949: 54; italics in the original).

The most important—and the most irreconcilable—difference between the worlds of disaster research and disaster practice is that between the *generalizing* (nomological) interests in the former and the *particularizing* (idiographic) interests in the latter. Research is the acquisition of knowledge which is generalizable, that is, which holds beyond the single case (disaster) under study. Its aim is to identify generalizations (as well as of the "brackets" that represent their limits). By definition, to generalize means to omit from the description of a pattern many rich but idiosyncratic details (such as, in sociological research, the personality quirks of most individuals). Drabek's (1986) propositional inventory is a virtual encyclopedia of such generalizations, each with a different degree of empirical support. Practitioners, in contrast, work in a different world. For them, it is the idiosyncratic details of events—including individuals' personality quirks—that determine their ability to successfully manage in the disaster context. In other words, the strength of the research process—methods that permit detached generalization—can contribute to the solution of practical problems, but these methods cannot solve many of the problems that practitioners face.

Future disaster researchers should realize that they can contribute to the solution of practical problems not by becoming absorbed with the particular but by contributing to the empirical base of a *generalized* knowledge. Put differently, disasters involve more than one type of "calling" (as the term is used by Weber [1958: 79-92]). For some the practice of disaster management is a calling; disaster research is also a calling, but a calling of a different type. Volunteers can help staff a one-stop center following a disaster; researchers can analyze the experiences of *many* one-stop centers, looking to find (i.e., to generalize about) what makes one more effective than another. Both types of calling can leave the individual with a sense of having "made a difference." However, most current disaster researchers would argue that a single researcher studying the structure and functioning of one-stop centers, for example, can do more "good" than a single volunteer. Even within the confines of the "publish or perish" world of academia, young (nontenured) disaster researchers have the opportunity to "translate" their discipline-based articles into papers publishable in practitioner-oriented journals and periodicals, and even research universities expect their faculty to spend at least a few hours per week in "public service," which can involve consulting (whether remunerated or not) with disaster management practitioners. The opportunities for making the contributions that Drabek calls for (Chapter 4) are there.

Still in all, the opportunity for future disaster researchers to make contributions, both practical and theoretical, depends on the quality of their research. Quality of research, in turn, depends upon skillful application of methods appropriate for disaster research. All of the participants in this volume on the methods of disaster research hope that we have provided useful guidance on how to make this possible.

PART I
CONTEXT

The three major papers in this section provide the context for the remainder of the book. Lewis Killian's 1956 report (Chapter 3) on the methodological challenges of disaster research deals with what is unique in carrying out research in the disaster context and on the consequences of this uniqueness for the conduct of disaster research. Killian's new preface, written for this volume (Chapter 2), notes both continuity and change in methods used in research on disasters since the report was written. E. L. Quarantelli's description of the methods and techniques employed during the early history of the Disaster Research Center (Chapter 4) shows how many of the unique constraints of the immediate postimpact period—what Quarantelli here labels the crisis time period—were dealt with by the center in conducting its exploratory studies. The section concludes with a reflective essay in which Thomas Drabek (Chapter 5) looks back over a career during which numerous different methods were applied in pursuit of answers to a variety of questions having both practical and theoretical interest.

2

PREFACE

Lewis M. Killian

This report was written when post-World War II disaster research was young. The field was blossoming, however, so the author was able to draw upon not only his own experiences, beginning at the University of Oklahoma, but also the pioneers such as Charles Fritz and the NORC-Chicago team, William Form and the Michigan State group, A.F.C. Wallace, and the staff of the Committee on Disaster Studies headed by Harry B. Williams. If the methodological introduction which follows has not become obsolete, as I hope, thanks are due them and many other members of the early "disaster research fraternity." The field has expanded greatly since the 1950s, extending to include studies of uncertain threat, evacuation studies, analyses of long-term social and economic effects of disasters, and cross-national research.

The most notable change in methods has been more extensive use of survey research and quantitative analysis. In the 1950s, however, some researchers were obtaining data from random samples and snowball samples to reconstruct the sequence of events. Data gathered included protocols of interviews with individuals describing their own behavior and the formation and functioning of small groups. The logs of formal organizations, such as police and fire departments, and the after-action reports of military units provided data for organizational analysis. Whether studies are qualitative or quantitative, the collection of data through field studies remains a significant component of disaster research.

3

AN INTRODUCTION TO METHODOLOGICAL PROBLEMS OF FIELD STUDIES IN DISASTERS

Lewis M. Killian

The validity of the conclusions drawn from any research rests, in large part, upon the scientific adequacy of the methods by which the data are collected and analyzed. But the use of standard, proven methods does not in itself guarantee the production of valid, significant results. Much social science research has been done which is methodologically impeccable but theoretically and pragmatically insignificant because methods have been used mechanically and indiscriminately. Whatever the subject, methods should be carefully selected and skillfully adapted to meet the requirements of the particular area of research.

Basically, the methodological problems of field studies in disasters are those common to any effort to conduct scientifically valid field studies in the behavioral sciences. The disaster situation itself, however, creates special or aggravated problems for field studies. It is the purpose of this report to introduce the scientist who has not previously done disaster research and the persons who might use the results of his research to these problems. It is intended to be, throughout, an *introduction*. There has been no attempt to present specific, detailed solutions to technical problems. An attempt has been made, however, to point out where problems lie and where adaptations of methods may be required. The competent investigator will take it from there. Indeed, the writer and the Committee on Disaster Studies [Ed.: of the National Academy of Sciences-National Research Council] will be disappointed if the ingenuity and wisdom of new and "veteran" disaster researchers do not make the present discussion meet what Dr. Chapman [Ed.: Dwight W. Chapman, cochair of the committee] has aptly called "the happy fate of rapid obsolescence" (Chapman 1954: 4).

Since this is an introduction, it is worthwhile to sketch briefly some introductory concepts and definitions concerning disaster. Many events are disastrous to the individuals or groups involved. Disaster research, however, has been primarily concerned with events which threaten and disrupt communities or larger social units. This is the kind of disaster to which the present report is addressed. While these events may be defined in many different ways, depending upon the interest of the investigator, any disaster basically involves a disruption of the social context in which the individual functions. Deaths, injuries, destruction of property and disruption of communications all acquire importance principally as departures from the pattern of normal expectations upon which the individual builds up his actions from minute to minute. The central problem for research becomes, "What factors produce what degrees of disruption in this social context, and how do individuals and groups behave in the face of this disruptions?" (Killian 1954: 7).[1]

Several kinds of disasters can be distinguished on a functional basis, according to time and spatial characteristics and according to the type of agent causing the disaster. Most research has been conducted on sudden, single-impact disasters, such as tornadoes, flash floods, or a single atom bomb. The discussions in this report are heavily colored by this fact. Many of the problems pointed out apply most directly to this type of disaster. There are other types, however, notably serial-impact disasters, such as a series of conventional bombings or a series of earthquakes, and prolonged-impact disasters, such as epidemics or biological warfare attacks.

Functional time phases and spatial zones have been identified in most disaster studies, and these concepts prove highly useful in ordering the data. The reader who is interested in disaster, either from a research or an administrative point of view, will do well to acquaint himself fully with these concepts (Powell, Rayner, and Finesinger 1953; Powell 1954; Wallace 1956a, 1956b).[2] In the present report, the following concepts are utilized:

> **Time phases**: (1) *warning*—the period during which information is available about a probable danger, but before the danger has become immediate, personal, and physically perceivable; (2) *impact*—the period during which the destructive agent is actually at work; (3) *emergency*—the postimpact period during which rescue, first-aid, emergency medical care, and other emergency tasks are performed; (4) *recovery*—the period which begins roughly as the emergency crisis passes and during which longer-term activities of reconstruction, rehabilitation, and recovery proceed.
>
> **Spatial zones**: This report makes reference only to the Impact Zone. This, obviously, is the area where the danger and destruction occur. In many disasters, this can be divided into Total and Fringe Impact Zones. In some disasters, such as explosions and tornadoes, it is easy to define; in others, such as epidemics, it is harder to define. Beyond the Impact Zones, it is usually possible to distinguish a Filter Zone, where the flows of persons and goods in and out of the stricken area meet and where first-aid stations, traffic control points, and other functions tend to be located. Beyond this area, one can usually define zones of organized community aid and organized outside (extra-community) aid.

Some types of research related to disaster present no unusual methodological problems. For instance, the use of conventional ecological and demographic methods can lead to significant findings on evacuation problems (see, for example, Iklé and Kincaid 1956). Standard sampling and attitude measurement techniques may require little adaptation for the investigation of attitudes towards the threat of disaster, disaster preparations, the operation of relief agencies, or other problems which can be studied before or after the event. It is in the analysis of significant psychological and sociological variables as they affect human behavior during the course and the immediate aftermath of a disaster that special methodological problems arise. The only way in which the effect of these variables can be analyzed completely is through field studies conducted during disasters or after they have occurred.[3] Most of the nascent body of theory about disaster behavior is based on such studies. Field studies continue to be one of the most fruitful and essential means of gathering disaster data. Even though they face formidable difficulties, these difficulties merely signify that the conditions under which disaster field studies must be conducted should be carefully analyzed and that the methods used should be adapted to these conditions.

This report is addressed specifically to problems of field studies in actual disasters. It does not pertain to laboratory studies, demographic and ecological studies using standard data, public opinion surveys in nondisaster communities, or other types of studies which may be made on problems related to disaster and which are important to the full development of this field of research. The report also does not deal with clinical or psychiatric studies, although such studies might be conducted by field interviewing in disaster-stricken communities.

RESEARCH IN THE DISASTER SETTING

There is no area of social research in which the scientist must operate with less freedom than in the field of disaster study. Controlled experiments, except with small-scale, simulated models, are forbidden to him. Since disasters are highly unpredictable, he rarely has the opportunity to select the locus of his study before the disaster has occurred. Usually the locus is determined for him by the unpredictable forces that produce disasters. Cases must be selected on the basis of the few variables that can be controlled, not in terms of the wide range of variables that it might be desirable to control. Insistence on the control of a large number of variables may lead to no research at all.

By the same token, disaster research is usually entirely post hoc. Time always presses upon the researcher, for the longer he takes to get into the field the more remote the disaster experience becomes for his subjects.

In a post hoc study of a disaster, the population is not in, and will never quite return to, its normal predisaster state. If a study is made soon after the disaster occurs, the spatial distribution of the population will be distorted. Many predisaster occupants of the affected area will have been displaced by destruction of their homes, while the area will be crowded with many people who are there temporarily, only because of the disaster. Later, the composition of the population will not be exactly what it was before the disaster, and many potential subjects who were in the area during the disaster period will have dispersed. The fact that those who did not survive cannot tell their stories automatically makes one gap in data on survival behavior. Furthermore, any analysis of the demographic, sociocultural, and psychological characteristics of the population before the disaster must be made retrospectively. In only a few fortunate cases will the investigator find that studies have previously been made in the right community or on the right subjects which are pertinent to his own research objectives.

The disaster experience leaves the people involved heavily laden with emotion and tensions. The great majority are intensely ego-involved in their experience. It is easy for the researcher to be affected by the drama and the tragedy which so strongly affect his subjects, so that interviewer bias can easily become a problem. It also raises forcefully the question of whether the interview responses of disaster victims may be especially subject to faulty memory and retrospective distortion and reconstruction.

In light of all of these conditions under which disaster field studies are conducted, careful attention must be given to the design of the research, the selection of subjects, the collection of data in the field, and the analysis of the data.

DESIGN OF RESEARCH

Field studies of disasters have been made at various intervals of time after the events, but it is generally desirably that fieldwork begin as soon after the moment of impact as feasible. This means that a specific research design must be crystallized hastily, with limited knowledge of the salient features of the situation.

If valid conclusions are to be produced, however, the process of designing research must start before the occurrence of any specific disaster which is to be studied. Significant additive contributions to knowledge of disaster phenomena are not likely to come from hastily designed, entirely ad hoc field studies. An essential part of designing the research is selecting the disaster that is to be studied.

The problem of scope and specificity

Hypothesis-oriented research designs are desirable in disaster field studies, as they are in other areas of social science research, because they give order and precision to methods of measurement and data collection, because they give power and exactitude to analysis, and because they yield results which are additive and more subject to independent test by other investigators. The degree to which disaster field studies can be designed on the basis of hypotheses—or, more properly, the degree of specificity of the hypotheses which can be formulated—still varies, however, with the research topic and with the field circumstances. For some time to come, disaster studies will consist of two important parts—the general, exploratory work required for getting the history of the disaster, and the testing of well-defined, neatly formulated hypotheses through clearly foreseeable operations. In their haste to reach a high level of quantification and control, disaster researchers should not undervalue descriptive case studies, nor should they neglect the need for exploratory studies in problem areas upon which little empirical research has yet been done. The major portion of this report is devoted to delineating the problems which are encountered in executing both of these types of research.

For many subjects, upon which there are already considerable data and a reasonable conceptualization, such as panic flight, it should be possible to formulate operational, testable hypotheses and to plan in advance the procedures of data collection, measurement, and analysis by which the hypotheses will be tested.

In many areas, existing research data give the prospective investigator an idea of the dimensions of the problem and a pretty good notion of the most significant variables and interrelationships. He finds the data cloudy, however, with respect to just how these variables and relationships ought to be defined and measured, and they leave him with the feeling that there are probably still important variables and relationships which have not been discovered.

This type of situation is still amendable to hypothesis-building, but in such a case the investigations should not and need not be limited to the testing of hypotheses which can be foreseen and operationally stated. Data should be collected and analyzed in such a way that new and unexpected relationships can be discerned and the possibility of identifying unforeseen variables is not excluded. No matter how "open-ended" his research design and his data-collecting procedures, however, the investigator will be selective to some extent, and he must have some general assumptions or hypotheses to guide him in the search for new variables and relationships. These should be explicitly stated, both to protect the investigator against his own biases and preconceptions and to guide him in the search for significant data.[4]

This attention to scope, or relative open-endedness, will usually mean in any specific study, that the investigator does not have the controls and data he needs to test his hypotheses as rigorously as he would like. In all likelihood, resources and means usually being limited, he has sacrificed some degree of precision and control to scope and open-endedness. He will then want to pull out what his data show him to be the most critical hypotheses and design a future project or projects to test them as rigorously as possible.

Exploratory research

When the investigator finds that there is very little information available on the disaster problem he wishes to study, he may well decide that his first step is an exploratory study to secure preliminary data and to increase his understanding of the problem in its disaster context. Even in the most unstructured of exploratory studies, however, the investigator has assumptions, or in a sense hypotheses, about what is relevant. His job in planning the exploratory study, therefore, is to make explicit his assumptions, not in terms of operationally stated hypotheses, but in terms of categories of relevant information which he will seek the types of relationships he thinks he may find and the types of hypotheses he hopes to be able to formulate when his data are analyzed.

One further point should be stated emphatically: No matter how narrow the interest and how well crystallized the design, every disaster field study should make provision for securing an accurate description of the overall situation and sequence of events. This is particularly important in disaster because situational variables are so often important in determining human behavior. The physical facts of the disaster—when it struck, what was destroyed and damaged, casualties, conditions or routes of egress and ingress, and similar details—should be established. The general sequence of events—who was notified of the disaster and when, what rescue and relief forces and supplies arrived and when, what steps were taken to organize a coordinating body or directing authority, and similar factors—should be discovered. Other data which should be secured will be readily apparent from a study of existing disaster research reports.

If this discussion has seemed "old hat" or "textbookish," the reader may wonder why it is repeated here. This has been done in an effort to emphasize what this writer and the Committee on Disaster Studies believe to be the greatest needs in disaster research and the greatest weakness of much of the disaster research which has been done, namely, the needs to (1) select problems or problem areas more consciously and more carefully, with a greater attempt at foresight on their potential payoff as research topics, either pragmatically or theoretically, (2) become much more explicit about theoretical and empirical assumptions, whether the intended research is a highly refined project to test a limited number of hypotheses or a new exploratory study, and (3) formulate specific, testable hypotheses and design research projects specifically to test these hypotheses, but (4) not become exclusively devoted to the testing of hypotheses which can now be formulated operationally, because there is still much territory which is only vaguely charted.

The problem of "controls"

Disaster field studies are beset with one problem probably even more than most social science field studies. This is the difficulty in establishing controls over the variables investigated, either through experimental controls, statistical inference, or exhaustive depth study. In the next section we will mention the difficulty in finding stimulus situations which fit closely with the investigator's expectations. It is also frequently difficult to make the distinctions among responses and motivations of actors which one would desire. For example, a common phenomenon in disaster is the convergence of large numbers of outside persons on the disaster site. These persons come for different reasons—for example, to seek relatives and learn of their welfare, to help in the rescue and relief, to "sightsee." The motivations of convergers change—a person comes to stare and stays to help—and their motivations are mixed—a person has a compelling need to find out for himself whether his relatives have been harmed, but he also hopes that he can help them in some way. It would be of great importance, both practically and theoretically, to measure the changes in motivations and the relative strength of different motivations of convergers.

The flux of population in and around the disaster area makes it difficult for the investigator to apply techniques which he would normally think of to study such a problem. It will be shown in a later section how this problem of population flux and dislocation affects problems of sampling in disaster. Disaster research to date has succeed in identifying motivations such as those mentioned above, but it has not yet generally progressed to the stage of measuring their relative importance. This is one type of methodological problem which is presented, through this report, for solution in future investigations.

The problem of "controls" is implicit in much of the discussion to follow, especially in the sections on sampling and all of the discussion of research design. Here it is sufficient to point out the obvious fact that "controls" should be built into the basic design of the research if valid and convincing inferences are to be made about significant relationships among variables. In order that this fundamental problem may be made clear, one further illustration will be presented at this point.

One of the most useful and desirable results of disaster field studies will be a reliable and valid comparison of what forms of emergency rescue and relief organization are most effective. Such a comparison obviously requires reliable and valid methods of rating the actual effectiveness of different forms of organization. But what criteria can be used to rate or rank effectiveness of performance— number of lives saved?; number of dollars or amount of goods given in relief?; speed of mobilization and performance of prescribed or assumed functions?; effects on the population involved?

When the investigator has decided upon useful, reliable, and valid criteria for rating effectiveness of organizational performance— which take account of the differences of functions prescribed to or assumed by the different organizations—he has to find data which permit him to apply the criteria to the forms of organization being compared. He may find records on some aspects of performances, but not others. Relief agencies will usually have a record of the number of dollars and amount of goods given out, but it is unlikely that there will be reliable records of the number of injured and dead removed and handled by different organizations. In a group that rescues X number of persons, there may very likely be representatives of more than one organization and several volunteers. Who is to say how "credit" is prorated? Painstaking research may measure speed of mobilization of different organizations, but speed of performance of functions after mobilization is even more difficult to measure.

There is no reason to believe that these problems are insuperable. For instance, progress has been made by William H. Form and others at Michigan State University towards developing criteria for rating organizational performance (Form, Nosow, Stone, and Westie 1954). Yet it is quite clear that there is no easily accessible and measurable output of disaster organizations which can be used as a dependent variable—as, for example, can the production rate of industrial workers. As was indicated earlier in the present section, the same difficulties hold for the measurement of independent variables. This, then, is another root of the problem of "controls" and therefore of making valid inferences concerning significant relationships among variables in disaster.

Selecting the event to be studied

Having a research design prepared in advance does not mean that the investigator can rush into the field at the first news that a disaster is impending or has occurred, to apply a particular design mechanically.

One of the difficulties in determining whether a given disaster is the appropriate one to be studied is the difficulty of determining in advance whether the stimulus was sufficiently distinct and meaningful. In the winter of 1954, a large part of London was covered by a very heavy, black layer of smog, bringing sudden darkness. Newspapers reported that people had reacted excitedly. Preliminary investigation seemed to indicate that it was defined by many people as an unusual situation, not just another London fog. It therefore seemed an excellent opportunity to study people's reactions to a strange, ambiguous threat, and a study was launched. The study itself, however, revealed that 52 percent of the people interpreted the situation as a fog or smog. Fortunately, there were enough people who did define the situation as unfamiliar and threatening to make the study worthwhile, but this example does illustrate the difficulty of knowing, until the final results are in, whether the desired stimulus, as perceived by the subjects, did exist in a distinct way. One need have no doubt, of course, that a tornado creates danger and distress, but he may not know, prior to considerable investigation, what situations people were in with respect to entrapment and being cut off from escape. These are situational variables he must be concerned about if he is studying survival and escape behavior. Since it is desirable to commence fieldwork as soon as possible after the disaster, it is sometimes unavoidable that the investigator take a chance that the stimulus situations will be as expected.

A preliminary reconnaissance of the disaster situation should always be made. This reconnaissance may be made by members of the researcher's own staff or by reliable, scientifically trained colleagues who can reach the scene quickly. News reports, particularly early bulletins, cannot be relied on for they often present an incomplete, distorted picture of the situation.[5] Hence they should be supplemented by interviews with a few key personnel who are believed to be in a position to hold authoritative, even if incomplete, information. City officials, the police, directors of relief activities, and hospital directors are among the people likely to be fruitful as original sources. On the basis of this reconnaissance, the research director can decide if the disaster is actually of the type appropriate to his problem and whether it is feasible to follow his design in this particular situation.

Even after a disaster has been selected for study on this basis, it may be found that modifications of the design are necessary. For example, a preliminary reconnaissance of a coastal community which had been threatened by a hurricane indicated that the reactions of beach residents to official orders to evacuate could be studied. By the time fieldwork was launched it was discovered that this population, composed of transients, had changed almost completely since the hurricane threat. As a result, the study had to be done in an area containing a more stable population but one in which the warnings had been issued as *advice* rather than as *orders* to evacuate.

Attempts to adhere closely to a particular design may result also in long periods of waiting without any actual research being done. If the research team goes only into those situations which are appropriate to its plans, it may have to wait many months for the "right" kind of disaster to occur.

This problem can be alleviated if the researcher has several alternative research topics in which he is interested, with an appropriate design for each. With a broad range of preformulated problems, the researcher will more frequently find situations which offer opportunities for meaningful research and will still have some criteria by which to select his cases.

To illustrate, a research group might focus on several problems relating to leadership and social organization, including emergent leadership, the operations of preexisting disaster plans, and the performance of formal leaders in a disaster emergency. On entering the disaster area for research they might find that a disaster plan reported to have been put into operation existed only on paper and that formal leaders, such as civic officials, would not cooperate in the research, so that a study of formal organizations would be unfeasible. They still might find an opportunity to study emergent leadership in rescue and relief activities, using subjects drawn in an area sample of the entire community.

Matching the design and the disaster

It may seem that the preceding discussion has placed the prospective disaster field investigator in a dilemma, first by urging predesigned, hypothesis-oriented research and then by arguing that the occurrence of specific research problems and therefore the applicability of detailed research designs are unpredictable in disaster. It has been suggested as a solution that the investigator should (1) conduct a preliminary reconnaissance in a given disaster to determine if it provides the variables and the subjects he desires to study and to determine if he can secure the necessary cooperation, and (2) formulate alternative topics for field investigation and general research designs appropriate to each, which can be adapted specifically in the field. The dilemma still remains, however, if either of these suggestions is carried too far. If the investigator waits until reconnaissance has revealed the ideal situation, he may never launch the investigation; if he develops too many alternative topics and plans, he may never succeed in developing more than a very general design. In either event—while he conducts reconnaissance or while general plans are adapted—time is lost in commencing the investigation.

In final analysis, the solution must be a practical one for each investigator, in terms of his own interests and resources. In general, it might be along the following line: The investigator should become thoroughly familiar with the existing disaster literature and data. He should think out how *his* research interest can be applied to different kinds of disasters and different situations within disasters. If he has a specific area of interest—e.g., civil defense organization, spontaneous groups, leadership, medical care, communication, community organization, problem-solving behavior under stress, emotional reactions to traumatic experiences, evacuation, social change—he should discover that most community disasters will allow him to study *some* of the variables and interrelationships in which he is interested. Combining competence in his own area of interest with information from previous disaster research, he can perhaps formulate a tentative, general theoretical model. This model will help him to specify the hypotheses which are likely to be testable in different kinds of disaster situations and suggest preliminary research designs to have ready for use. Under any circumstances, however, the investigator must expect to make some adaptation of his design in the field.

This strategy has not yet been tried in disaster research, so it can be illustrated only hypothetically.[6] Let us assume, as a very crude example, that the investigator is generally interested in communication research and theory and that he believes theoretical or pragmatic value should result from the study of communication in disaster. Study of disaster material will show him that communication problems are very different in different time phases of a disaster—warning, impact, emergency, and recovery. It will show him further that communication problems of a given time period—for example, warning—vary with different types of disaster, such as epidemic, hurricane, and tornado. This is because of different time dimensions and because different threats have to be communicated. He will see further that the communication problems vary in different disasters of the same type, according to conditions (for example, the telecommunication system was or was not knocked out by the disaster) and according to conditioning factors such as previous experience (for example, the community has or has not experienced previous hurricanes). If he has a very specific research problem—such as the relative effects of warning type A and warning type B upon populations which have not previously experienced floods—then the kinds of differences between disasters which are cited above will be all-important. If, however, he has a more general model and set of hypotheses about the effects of different kinds of warnings, communicated in different ways to different kinds of populations, he can go into a wide variety of disasters and test *some* of his hypotheses. The same thing should be true for other phases of disaster communication, such as the mobilization and disposition of rescue and relief forces.

THE SELECTION OF SUBJECTS

Having decided that a particular disaster is suitable for the study of some of the variables in which he is interested, the researcher faces the problem of selecting subjects from a vaguely defined, more or less disorganized population. The type of problem he chooses to study has much to do with the method by which the subjects are selected. When the population is small and easily accessible, the universe may be studied. In some instances a probability sample may be used. In yet other cases, subjects may be selected for certain desired characteristics with no randomization.

Selecting the subjects to fit the design

Certain precautions must be observed to insure that the subjects selected are adequate for testing the hypotheses of the study. This requires attention to the physical and ecological situation in the disaster and foresight in sampling the subpopulations which logically can be expected to have been involved in the actions or to have been subject to the variables being studied. These logical expectations, to some extent, can be checked empirically by preliminary interviewing before the final sample is drawn.

One research team set out to investigate the reactions of a town's population to a rumor that an upstream dam had burst. It was decided that a random sample of city residents, drawn from a complete and up-to-date city register, was the best method of sampling in this case. Preliminary interviewing, however, showed that a small proportion of the city's residences had actually been flooded a few days previously and that this previous experience had probably affected their occupants' reactions to the rumor. The sampling rate in the flooded area of the town was therefore increased to ensure sufficient cases for cross-tabulations on this important variable. New hypotheses and new questions for the schedule were also formulated, illustrating the previous point about the need to make adaptations in the field.

Another hypothetical example of this problem, adapted from experience in an actual study, is as follows: It was desired to compare the amount of self-help and help by immediate family members with help by nonfamily members during the emergency period in two communities. Area samples were drawn for the total impact areas and a narrow fringe next to the total impact area in each community. In Community A, the area sampled included approximately 85 percent of the inhabited area and a corresponding proportion of the population; in Community B, approximately 33 percent of the inhabited area and a corresponding proportion of the population were in the sampled area. The results showed clearly that more people received help from nonfamily members in Community B than in Community A. However, before this difference could be attributed to other differences between the communities, it was necessary to know whether it could be accounted for by the different physical situations—i.e., in Community A, only 15 percent of the population was not directly involved in the impact area; in Community B, 67 percent of the population was not involved in the impact. Those not caught in the impact could have been more free from concern for missing family members and more able physically to go to the aid of fellow citizens. The sample, representing total and fringe Impact Zone victims, was of little help in solving this problem. One could find out, on a statistically reliable basis, whom victims reported helping and being helped by, but one could not investigate the opportunities to help and the problems of giving help, as these would be reported by nonvictims.

Sometimes the physical and ecological facts of the disaster may make such problems incapable of any neat solution, but the investigator should always at least consider the possible need to increase his sampling rate in subsamples to correct ecological or situational distortions in the sample and provide sufficient cases for analysis of important variables.

Difficulties in sampling

Probability sampling has been used less frequently in disaster research than in other social science field studies. This is largely because the conditions described above make it difficult to define, locate, and reach the universe to be sampled. In spite of the problems of sampling in a disaster area, however, probability sampling should be used as fully as possible whenever conclusions are to be stated in quantitative terms, for generalization to an entire community or other known population. This is particularly true when the findings relate to the frequency and distribution of various types of individual behavior, attitudes, or emotional reactions. More frequent use of rigorous sampling procedures is needed before the knowledge now existing about the types of behaviors and events which occur in disaster can be translated into knowledge about the frequency with which they occur under specified conditions.

For some problems, such as the reactions of people within the Impact Zone, it may be sufficient to sample within the disaster-stricken community. If the focus is on convergence behavior, participation in rescue activities, or other activities in which persons from zones beyond the Impact Zone are likely to be involved, it may be necessary to sample in surrounding areas. Even then it should not be assumed that the total population involved in such activities is included in the area from which the sample is drawn, for some participants may come from very distant points.

The sample should take into account the differential experiences of subjects in different zones. Subjects in areas outside the Impact Zone have been in a situation very different from that experienced by those within it. Even though they may be residents of the same community, the occupants of the different zones should, in most cases, be treated as separate populations, making it necessary to draw separate samples. In any event, the sampling techniques should always permit the separation of respondents on the basis of spatial zones.

Where physical destruction or evacuation has resulted in displacement of portions of the predisaster population, special problems arise. In this case, it is necessary to reconstruct the composition of the predisaster population from available records. For urban areas, land-use maps obtained from city engineers or from real estate firms provide a fairly accurate indication of where residences were located before impact. On the basis of this information, a sample of blocks may be drawn, and, in turn, a sample of every n^{th} household may be drawn from these blocks. At this point the aid of local informants must be relied upon. Their assistance will be needed in tracing the present, postdisaster location of subjects who have been displaced. Obviously, the time and travel required for actually reaching subjects selected in this manner are great.

For rural areas, county highway maps generally indicate the location of residential structures prior to impact. Populated rural areas may be broken into identifiable segments using roads and other landmarks as boundaries. Systematic sampling of segments, and of households within segments, may follow.

Occasionally it may be possible to obtain a pre-list or even a sample of the predisaster population from commercial or governmental research agencies. There are many areas in the United States for which such agencies already have drawn samples.

When adequate information is available, the sample of blocks may be stratified for severity of destruction, distance from point of impact, and other disaster-related variables. If predisaster studies of the social and demographic characteristics of an area are available, stratification may be introduced for such characteristics as sex or age composition, or socioeconomic class level.

This procedure may be followed when the population to which conclusions are to be generalized is contained entirely within the area sampled and when no other population is included within the area. Sampling becomes more difficult when a special population, not assumed to be randomly distributed throughout some spatially-defined universe, is to be studied. For instance, the problem selected may require that the population be limited to persons who were in the Impact Zone, regardless of where they came from, or to persons who participated in certain phases of disaster operations. Here sampling by standard probability procedures is feasible only if, with the aid of records or informants, a nearly complete pre-list of the relevant population can be made. Sometimes a pre-list can be made up by use of a preliminary questionnaire survey. The questionnaires may be distributed to a larger population believed to include the totality of the desired subject population, the pre-list then being obtained from the returned questionnaires. Completeness of the pre-list obviously depends upon a high rate of return of questionnaires.

Selecting subjects for studies of group processes

Data obtained from a probability sample of the population may reveal the frequency of various types of individual or group reactions, but unless the sample is very large (or unless a very large proportion of the population participated in the types of groups being studied, which is unlikely unless the family is the object of study) such a sample gives insufficient insight into group structure and process. A feature of disasters having great theoretical and practical significance is the emergence of new groups and the transformation of the structure and functions of preexisting groups. To study the dynamics of group formation, structuring, and behavior in a disaster situation requires a different method of selection of subjects.

Whatever the goal of research on disaster groups, the first step is the identification by any means available of single members of various groups of the type to be studied—spontaneous rescue teams, for example. From these initially selected subjects, the names and addresses of other members of their group and of members of other groups may be obtained as "leads" for the selection of additional subjects. Then, if the goal is a case study of the structure and function of one or a few groups, such leads are followed until the entire membership of the group or groups to be studied is located, or until all leads are exhausted. If the goal is to develop a typology of groups or to compare different types, such leads are followed until no new types of groups are identified. In this case, the group members who are located serve as informants reporting on the formation and operation of their groups rather than (or as well as) respondents.

This is, obviously, a time-consuming process. In spite of its limitations, however, it must be followed if the processes of social disorganization and reorganization are to be fully understood.

Selecting informants—sampling "points of observation"

Most field studies include a reconstruction of the natural history of the disaster, or of certain phases of it. In the collection of data for this history, the subject has a dual role. He is a respondent, reporting his own reactions, and also an *informant*, reporting his observations of what went on around him, including the behavior of other people. Informants must be selected as carefully and as systematically as are respondents. Bias in their selection may lead to a grossly inaccurate picture of what took place. In order to maximize accuracy, a number of informants representative of populations which viewed the disaster from different vantage points should be used. Position with reference to the Impact Zone, time of entry into the disaster area, degree of involvement, professional training, and orientation are among the factors which should be considered in selecting informants. In the confusion of a disaster, the portion of the total situation which any one observer witnesses is very small, and there is likely to be too much distortion of his perception. The accounts of many differently-placed informants serve to complete the *gestalt*; they also constitute checks on each other.

In studying formally structured organizations, it is important to secure reports of observations by (1) persons at different levels in the hierarchy of the same organization, (2) persons at corresponding levels in different organizations, (3) persons in any organization which had higher or wider jurisdiction or coordinating functions, (4) persons not in an organization who were (a) in a position to observe its work or (b) recipients of its services.

A matrix of positions, units, or locations appropriate to the particular disaster, community, and organizations involved can be constructed by the investigator to give system and greater surety to his attempts at securing the total picture and "cross-checking." Such a matrix, properly constructed, provides a sample of "points of observation (or participation)," of opinions, or both. This is not intended to imply, of course, that probability samples of members of organizations cannot or should not be used in the study of organizational behavior.

This procedure of constructing a "sample" of "points of observation" must apply to many aspects of the study of group, collective, and community behavior in disaster. Sometimes, for the reasons given above, the investigator cannot define statistically the universe to be sampled. Some of the basic units of investigation are likely to be events or chains of events rather than the distribution of attitudes, emotional reactions, or behaviors in a population. Here his concern is to obtain the most accurate reconstruction of the event possible, and this need may best be served by tracing down and interviewing those persons who were participants in or witnesses to that event. His concern is sometimes with a particular decision—for example, a decision to form a disaster committee representing different agencies—or a particular point in a process—for example, the origin of a rumor. There is no sample of unique events, and the general population does not contain a random sample of witnesses or participants.

Much important data, therefore, has to be gotten through what Wallace has called "jig-saw puzzle" research methods (Wallace 1956a: 3). Sometimes the jig-saw puzzle never falls completely into place—different observers or participants with an equally good *a priori* chance to be correct give irreconcilable reports, or the same informant gives contradictory reports in succeeding interviews. Often, however, patient investigation will yield at least a "most probable" account. The suggestion here is that this type of investigation should be planned as carefully as possible, on the basis of a matrix of likely points of observation and participation. Just because it cannot be guided by the precise rules of probability does not mean it must be pursued willy-nilly. The investigator can first exhaust the possible points of observation and participation logically: What persons (or positions) would one logically expect to have witnessed or participated in an event? This picture can then be modified empirically as he asks: "What other persons were here when it happened?"; "Who else did you talk to about the decision?"; etc.

Reports of discrete events or trends are not only a significant part of the fieldworker's data, they also are one of his problems. Every competent research worker is alert to the need of verifying reports of particular happenings which have an important place in his analysis. It is particularly important to verify such reports in disaster because a certain number of stories always gain wide circulation, credence, and durability. They are dramatic and, if accurate, they are often highly informative about human behavior and the course of events. In articles on the Texas City disaster, one is very likely to find a statement to the effect that many people were packed around the docks, watching the ship burn following the first explosion. They were there, so the statement goes, when the second explosion occurred, and the number of casualties was greatly increased as a result. Although this story has gained wide credence and is frequently cited as an important illustration of human behavior in disaster, later investigation by a trained, objective research team seems to dispute it entirely—there were very few people on the docks at the time of the second explosion. The only available information on an important problem of human behavior in disaster may be an unverified report from a particular disaster. Such an account, if true, might fill a significant gap in existing data. Under these conditions, the temptation to accept and repeat the account without verification is great, even for research workers.

COLLECTION OF DATA

Careful selection of subjects is of little avail unless the data obtained from them are relatively free from bias and are in a form which permits systematic analysis. Experience has shown that the deep emotional involvement of disaster subjects in their experiences and the confusion of the events of the disaster creates special problems of data collection.

Securing cooperation of subjects

Research teams have achieved great success in getting the cooperation of disaster populations. To gain such success, researchers must take cognizance of the emotional state of their subjects. People who have experienced a disaster have many sensitivities about their experiences. Particularly during the period shortly after the disaster, they may object to any study which is perceived as exploiting their tribulations for theoretical, scientific purposes. They want to be convinced that the research will "do somebody some good." Most victims are likely to resent any implication that they have "psychological problems" or, worse, that they might need psychiatric treatment. Leaders, particularly formal leaders, may be defensive about the way in which they played their roles during the disaster. Since many investigations of disasters are made in connection with insurance claims or law suits, people may have an initial suspicion of, and resistance to, any sort of fact-finding. For some, their experiences may have been so disturbing that the interviewer must overcome an initial reluctance to discuss them.

It is easy to get the cooperation of subjects when they are approached tactfully, in a way that will not be offensive or threatening. Extensive explanation of the purpose of the research and explicit identification of the sponsors are sometimes necessary. Cooperation is maximized if subjects can be convinced that the research has an immediate, practical purpose and will contribute to the alleviation of the effects of future disasters. It is particularly important in disaster research to avoid offending any subject or giving rise to misconceptions as to the purpose. Group cohesion tends to be heightened following a disaster, and negative reactions evoked in even one subject are likely to be reflected throughout the community in a very short time. Even if there is no such reluctance, the interviewer has a duty to be especially considerate of the psychological well-being of subjects who have experienced serious injuries, of those who are bereaved, and of children.

Whereas subjects are almost always cooperative when properly approached, officials, especially those who are in a position to be sensitive to public opinion, will sometimes be afraid that interviewing people will "upset them" and impose barriers to interviewing. In a study of an epidemic, officials in three communities out of six communities approached refused to cooperate in the research on the grounds that the subjects would be disturbed. When the study was made, however, only 2.2 percent of the persons in the sample refused to be interviewed, and the experienced field director described the public cooperation as "terrific." This brings up the problem of gaining entrée to the community, which will be discussed in a later section.

Questionnaires

Perhaps it is because of the need for such a careful approach to the subjects that questionnaires, mailed or left with subjects, have structuring his own answers, than can be gotten by holding him to a set pattern of response. Affect may be more freely expressed if he is permitted to "warm up" to his subject and give vent to his feelings. But if answers are permitted to flow in this unstructured fashion, frequent probing questions must be used to insure complete, systematic coverage of the questions in the schedule. While interviewers should improvise some of these probing questions, the types of probing questions that may be required should be identified in advance and included in the schedule whenever possible.

Various devices have been employed to maximize freedom and flexibility in the interview while insuring completeness. Questions of the type described above may be asked directly from a printed schedule on which the answers are to be recorded. If responses are recorded in note form, to be expanded and organized later, a list of questions or even cues to questions may be used for reference by the interviewer. In one disaster study a single "word," the letters of which suggested areas for questioning, was memorized by the interviewers as a mnemonic device.[8]

This sort of interviewing requires careful training of the interviewers. It permits numerous digressions by the subject, so the interviewer must become skillful in bringing him back to the relevant theme without destroying rapport. He must learn to keep himself from being sidetracked into irrelevant areas of discussion. He must be alert to indications that the respondent is on the verge of bringing out emotionally-toned, hard-to-verbalize material. If it seems that this is about to occur, the interviewer should pause and allow the subject to structure his answers in his own way and to formulate them at his own speed. He should be especially wary of asking a new question at such a time, as this may divert the respondent from his difficult struggle to express that which is painful to talk about. He must guard against his own tendency to block or divert the respondent when the latter begins to verbalize material which is painful or anxiety-provoking to him, the interviewer. Review of disaster interview protocols reveals clear-cut instances of this very human mistake by well-trained interviewers.

The interviewer must be prepared to listen to, and even to probe, responses which are gruesome and tragic, in spite of the affect which they arouse in him. At times he may find himself performing a therapeutic function for the subject who needs to ventilate his feelings about traumatic experiences. In such cases he must combine the role of the permissive, sympathetic listener with that of the efficient interrogator. The greatest danger is not that the interviewer will appear unsympathetic to the respondent but that he will become so identified with him that he drops the role of scientific observer. While he should evince signs of sympathy he should maintain within himself the attitude of scientific observer. Few interviewers in field studies will be qualified psychotherapists. The interviewer must not ever allow himself to fall or be drawn into a therapeutic role for which he is not qualified. Amateurs can do harm. This is not to deny that well-conducted interviews with disaster victims do sometimes have therapeutic effects. The thing the interviewer must be trained to do is to recognize when the subject is coming to depend upon him as a therapist and to avoid this happening. Disaster field studies, like many other field studies, have almost always involved just one interview with each respondent. When this is the case, even the interviewer who is a qualified psychotherapist must remember that he will not have the opportunity to help respondents with emotional problems through a series of interviews.

In spite of all these sources of interference, the interviewer still must give sufficient structure and direction to the interview to produce a coherent protocol in which all relevant questions are covered. In extreme cases, interviewers have become so emotionally involved that the interview becomes a medium for the interviewer to ventilate his own feelings. In one interview, the interviewer persisted in asking such questions as: "Everybody has just been wonderful, everybody has just pitched in together, haven't they?" The respondent persisted in giving such answers as "yes" and "no."

Supervision of interviews

This type of interviewing situation demands careful attention to supervision and quality control throughout the study. Even highly trained and skillful interviewers will often need advice and supervision in adapting to the disaster interview situation. If it is at all possible, the initial one or two interviews of each interviewer should be tape recorded. The research supervisor should go over these carefully with the interviewers, correcting faults and advising on pitfalls. With a group inexperienced in disaster interviewing, the supervisor may have to decide that some interviewers cannot interview skillfully in a disaster situation even though they may be very good in other situations. Checking of tapes, or a least protocols or notes, should continue throughout the study. The supervisor should continue the training process, and the group of interviewers should meet to discuss their problems. The need for supervision and quality control seems particularly great in disaster because the dramatic nature of the material and the ease of becoming emotionally involved with it can cause even well-trained interviewers to develop stereotypes and biases in the course of early interviews which they then introduce systematically into all their future interviews.

As was indicated in the discussion of the selection of subjects, people who were in different spatial zones at the time of impact constitute different populations. Interviewers should be aware of the differences they may find in subjects from the various zones. Respondents in nonimpact areas usually are not so highly motivated to "tell their story" as are those from impact areas. It is more difficult to make them see the value of data they may provide. In fact, the disaster is likely to have a meaning for these people which is quite different from that which it holds for those who are in the impact zones. Interviewers must take into account this difference of perspective in justifying the interview, stating the questions, probing, and interpreting the answers.

Recording interviews

This type of interviewing demands careful, skillful recording. No matter how they are recorded, disaster interviews are likely to contain a mass of detailed information out of which must be winnowed the data that are relevant and usable. Unless highly-structured, limited-choice questions are used exclusively, some content analysis is necessary. Since a flexible schedule and open-ended questions permit the respondent to structure his own answers, there is always a problem of organizing the data for systematic analysis. Tape-recording of interviews provides the greatest wealth of faithfully recorded data.

Tape recorders have been used successfully in several field studies. They have not been found to inhibit the subject or interfere with the establishment of rapport. On the other hand, they free the interviewer from note-taking and make possible easy, uninterrupted interaction between him and the respondent. The most serious limitation on the use of tape recorders is the problem of analysis. Transcription, checking, organization, and coding of the mass of data so obtained can be costly and time-consuming. Sometimes budgets and time-schedules do not permit the use of this method. One research organization which tape-recorded several hundred disaster interviews found that it cost an average of about $25 [Ed.: about $160 U.S. in 2000] to transcribe and edit each one-hour interview. Translated into time units, it took six hours of typing and three hours of editing time for each hour of tape.

A much cheaper, generally more feasible method of recording is through note-taking. Usually notes may be taken during the interview, to be expanded soon after its conclusion. If a subject is obviously disturbed or inhibited by note-taking, it may be necessary to reconstruct the interview from memory immediately upon completion of the interview. The schedule or mnemonic device aids in this reconstruction.

Even when notes may be taken during the interview, the free flow of questions and responses is somewhat impeded, and much material is inevitably lost. If interviewers can work in pairs, with one taking notes while the other conducts the interview, more content can be recorded by the pair than by a single interviewer.

Even more serious than the loss of material is the effect of interviewer bias on the retention or loss of content. What is retained or lost is of greater consequence than how much is retained. Only the interviewer himself can guard against this deficiency, since the completed notes provide no way of completely ascertaining the effects of his biases. Interviewers must be cautioned and recautioned against retaining only the data which support the hypotheses of the research design. Supervisors must be alert for indications of the structuring of data in terms of some hunch which may occur to an individual fieldworker. When an interviewer follows such hunches too assiduously, his data not only reflect this bias but also become less comparable to the data of other interviewers. An obvious safeguard is to insist that interviewers write down their own questions and probes.

The complete interview protocol should be written up from the field notes as soon after completion of the interview as is possible, and certainly on the same day. While time schedules do not always permit it, the protocol should be completed before other interviews intervene. Otherwise data may be transposed from one interview to another. If several interviews must be conducted consecutively, it is essential that the field notes from each be edited and organized immediately after each interview.

In most disaster research interviews, it is desirable that as much material be recorded verbatim as is possible. This is especially true, of course, if emotional responses are being studied. It is also advantageous to have the data recorded in the order in which they were obtained, even though this does not correspond to the schedule or the topical outline. The data can be reorganized in terms of a schedule, a topical outline, or a code during analysis, but the analyst should be able to see the context from which each item was drawn.

The use of projective tests

Very little use has been made of projective tests in disaster research, but it is believed that they would be of value in some types of investigations. If the influence of personality variables on individual reactions to disaster is to be explored more thoroughly than it has been in the past, techniques for employing such tests in disaster research must be developed. Projective tests also promise to be useful in the study of children's reactions. They might also disclose extreme emotional reactions of which adult subjects are unaware and which they are unable to verbalize.

There are both methodological and theoretical problems involved in the use of projective tests. A practical limitation is that most such tests require a relatively long time for administration. Careful scheduling of the tests and the whole-hearted cooperation of the subjects are needed. It may be somewhat more difficult to convince subjects that the research has some practical significance than it is when an ordinary interview is to be obtained. Theoretical problems arise from the nature of projective techniques themselves. Research is needed to ascertain to what extent results of tests administered after a disaster reflect stable, preexisting personality characteristics rather than reactions to the disaster experience itself. There is an unresolved question as to whether facets of personality revealed by the tests should be treated as independent or as dependent variables vis-à-vis disaster experiences and reactions.

The study of personality variables in relation to reactions to disaster should not be based on the use of projective techniques alone. These techniques are recommended as a supplement to direct methods of personality study. Gordon Allport has pointed out that projective techniques frequently fail to show dominant motivations in normally balanced persons and that on the basis of projective tests alone the research may be unable to distinguish a well-integrated personality from one that is not well-integrated (Allport 1953: 111).

Sociometric techniques

Little use has been made of formal sociometric techniques in disaster studies. One reason for this is the essentially transitory nature of group structure during the course of a disaster. While there is little chance that sociometric studies can be done both before and after a violent disaster, many opportunities are offered for the study of group structure and interpersonal relations at various periods after impact. It has often been hypothesized that social structures undergo dramatic changes during the early periods of a disaster but return to something closely approximating the predisaster state later. Sociometric techniques are well-fitted for the testing of such an hypothesis, with the group structures revealed by them at various times being used as a dependent variable. In some instances, it might be possible to use sociometric data as an independent variable, relating group structure and lines of influence to such things as rumor transmission or leadership.

The use of documents

Documents constitute an important source of data in the study of disaster, but they must be used with caution. They are most useful when treated as sources of data supplementing first-hand data collected from the subjects themselves.

Newspaper accounts of a disaster are one of the most accessible but often least valid documentary sources. Because of the many uncontrolled biases of reporters and editors, particularly their tendency to overemphasize the dramatic, newspaper stories cannot substitute for systematic, controlled field studies. Even accounts which answer specifically the newsman's questions, "Who, What, When, Where, and Why?" must be checked for accuracy, for many errors are made under the pressure of disaster conditions. Yet news stories are useful, particularly in the early stages of a study, for the general, though tentative, description of the disaster which they provide. Both news and editorial columns are more valuable as sources of data on the role of mass communications. They record the official pronouncements of public officials; they reveal the reactions of a potentially influential opinion leader, the editor, to various stages of the disaster and to specific, disaster-related events. They are an important variable affecting individual and collective reactions during the emergency and recovery periods, and they often reflect or report rumors which arise during the disaster. For instance, early, exaggerated estimates of casualties are often reported in newspapers even though they may be corrected later. A good indication of the accuracy and efficiency of communications at various stages of a disaster may be found in news columns.

Another type of document with great utility is the tape recording of broadcasts during an emergency which a radio station may keep. Content analysis of these recordings may reveal much about the content of communications at various stages. Such records often provide incontrovertible evidence of the official pronouncements, the orders, the warnings, and the requests for aid which are broadcast at the behest of various agencies or officials. These materials are also valuable for revealing the biases and distortions introduced by mass communications agencies themselves.

A similar picture of the development of the disaster through time as perceived by certain people can be obtained from the logs, taped records, or teletype sheets of agencies which are in communications nets, such as the police or the American National Red Cross. There is a danger in the use of official message records, however, in that they are likely to represent only a small segment of the total information communicated in disaster.

Although the bias of the writers must be kept in mind, use may be made of official records and reports of operations prepared by public and private organizations. These reveal something of what the agency did and sometimes even more about its needs for justification and defense of its operations. Comparison of the reports of different agencies may provide important clues to interagency cooperation and conflicts.

Researchers should always make a careful search for documents which provide background data on the disaster population. These may include reports of demographic, economic, or sociological studies conducted before the disaster occurred. Occasionally it may be found that personality studies have been made before the disaster using some portion of the disaster population as subjects. If available, the data from such studies provide an excellent starting point for the analysis of the relation of personality factors to reactions to disaster.

Other documents which may be useful, if available, include:

Tape recorded, on-the-scene interviews conducted by reporters for radio stations.
Personal documents, such as diaries, and memoirs of survivors.
Letters received by agencies and officials, such as the letters which often accompany donations to relief funds.
Police files.
Building permits and other indices of economic activity.
Reports of surveys and investigations made by other agencies, such as the Army. Engineers or insurance underwriters' laboratories.

In conclusion, documents are never adequate alone as sources of data for a comprehensive study of a disaster. They may be adequate for certain types of research, such as the study of mass communications. In all types of disaster research documents are a valuable adjunct to data collected firsthand.

TIMING AND THE PROBLEM OF RETROSPECTIVE INTERVIEWS

Field studies of disasters have been conducted beginning as early as two days and as late as five years after impact. Little is known about the effect of timing on the validity of disaster data, but these effects may be highly significant.

The difficulty of initiating the collection of data immediately after impact has been indicated in the discussion of the planning of research. Furthermore, if researchers enter the field while rescue and clean-up operations are still in full swing, they are likely to find that many people do not have the time to be interviewed. They may also find that there is pressure on them to become participants in these operations, since the role of the scientist may appear callous and unsympathetic.

Later in the emergency phase, or early in the recovery phase, certain advantages accrue to the researcher from a reasonably early entry into the field. By this time, many people may have the time to cooperate as subjects, and the role of the scientist may be more acceptable to them. More important, there are indications that this is the time when it is easiest to gain rapport. Time and again survivors of disasters have shown themselves eager to talk about their experiences at this period.

In view of the present state of development of disaster research there is no reason for delaying fieldwork until more than a year after impact, except when studies are concerned with the long-run aftereffects of disaster. Entry into the field may still be delayed for several months by problems of obtaining funds and personnel. There is a danger that during this interval subjects will have forgotten many details of their experience and that distortion of other recollections will have occurred. Interviewers should be on the alert for signs that a subject is giving stereotyped answers which have become standardized in previous discussions of the disaster. Through the rumor process quite inaccurate versions of what happened may come to be accepted by a large segment of a population. A "group version" of an event can develop during a very short period of discussion. In experiments in which small groups were subjected to a simulated emergency, Dr. Bradford Hudson [Ed.: a psychologist at Rice University who performed experiments and simulations on people's reactions to possible air raids and other threats; e.g., see Hudson (1954)] found it necessary to separate the subjects immediately after exposure to the stimulus. He found that within 30 minutes a group might develop a standardized interpretation of the stimulus which would be accepted and reported by most of the members.

Persons experienced in disaster field studies report that disaster victims usually want to talk, to tell their stories, and there is quantitative evidence to this effect. Interviewers are seldom refused. Willingness to be interviewed, however, does not answer the question of whether serious biases creep into retrospective accounts of disaster experience. Experiences of disaster researchers and fragmentary evidence show only that we do not know with any precision how much bias, of what kinds, occurs among different types of respondents, under different circumstances, and at different periods after the disaster has occurred. Review of some tape-recorded interviews with disaster victims shows that they persisted in telling the story of what happened to them and to others in spite of leading questions and improper probes from the interviewers, which might have been expected to divert the respondent in an ordinary interview. In these cases, the respondent seemed to insist on telling his or her story in spite of the behavior of the interviewer.

In general, the facts that (1) people so often report that they felt confused, were afraid, behaved maladaptively, or even did things of which they are now ashamed,[9] (2) the experienced interviewer can often sense distortions, bring about clarifications, and break through defenses in the respondent's story by skillful probing, and (3) stories concerning overt behavior can often be cross-checked with other respondents' stories of the same behavior, have led many disaster researchers to the subjective conclusion that by and large no gross distortions have been introduced into the findings of disaster research by retrospective distortion of faulty memory on the part of disaster victim respondents. Experienced researchers have learned to be more wary of such biases in the stories of officials or agency representatives who feel the need to represent the performance of their own or their agency's official roles in a favorable light. Amazingly frank stories are often secured from officials, but in reconstructing the story of actions of officials or agencies, extensive cross-checking is important.

In spite of the experience which has been developed and the inferences which are possible from this experience, the fact remains that no precise measurements exist to show us the extent and consequences of retrospective distortions and faulty memories in *post hoc* disaster interviews and to chart specifically the pitfalls which must be avoided in disaster field studies.

Systematic research on this problem is necessary before disaster field studies move to higher levels of scientific validity. Subjects should be reinterviewed at different times after a disaster to determine how much distortion and loss of detail occurs with the passage of time. In one study where subjects were interviewed two or three times, only minor difficulties were encountered in obtaining reinterviews with subjects who had already been interviewed once. The fact that another similar disaster had intervened between the first and subsequent interviews may account, in part, for this success.

ANALYSIS OF THE DATA

The methods of analyzing disaster field study data are the ones a competent investigator would normally select to suit the research design and the data collected. This is not a special problem. A few of the customary problems tend to become exaggerated in the analysis of disaster data, however.

The first point is to urge the investigator to give very careful consideration to situational variables and matters of context in building a code or other analytical scheme. As this report points out in several places, situational variables are sometimes especially important in understanding behavior in disaster. Many times the physical situation permits only a limited number of alternative actions by individuals and groups. An analysis of the perceptions, motivations, and actions of such subjects may be quite misleading if the code does not account for the possibility that some actions or some sources of information were physically unavailable to the actor. A simple example will illustrate the importance of this point. The investigator may wish to know what proportions of persons of different types tried to aid other persons during the emergency. If the disaster is a flood and large numbers of people are isolated on rooftops, the analysis must take account of the people who had no physical possibility of helping others. Thus the Institute for Social Research in the Netherlands had to cross-classify the subjects by "physical circumstances during the flood" in analyzing assistance to others and other behaviors of disaster victims (Lammers 1955).

The investigator will probably be unable to include all of the factors of situation, context, and background in his code and keep it manageable, but he should make as certain as he can that he has included the variables relevant to the particular hypothesis and relationships he wishes to analyze. It seems quite safe to say at this time that *every* code should identify the behaviors or events reported and each individual respondent in terms of time and space. The categories of time and space which are used will, of course, depend upon the research design and the data obtainable. The determination of how fine a time-space breakdown should be made in the sequence of events reported will also depend upon design and data, and upon the time and resources available for analysis. In any case, the events or sequences of behavior or feelings of most significance to the investigation should each be identified in this way. The spatial relationship of the respondent to the events he is describing should also be coded if possible.

Another thing which the analyst may be reminded of is the difficulty involved in ascertaining predisaster background characteristics of respondents from interviews conducted after the disaster. This is not difficult with respect to demographic and similar characteristics, but applies to any perceptions, memories, or personality characteristics of the respondents which may have been influenced by the disaster. It is the *post*disaster individual who is interviewed. The investigator will, of course, deal with this problem as best he can in the planning of the research, but it may be well to remind coders of the dangers involved.

As has been previously pointed out, the field of disaster behavior inherited more than its share of stereotypes and semantical confusions. It has been suggested that interviewers be trained and supervised to avoid falling into these pitfalls; the same training and supervision needs to be given coders and analysts.

PROBLEMS OF ENTRÉE IN FIELD STUDIES

It has been found that even two or three years after a disaster has occurred certain sensitivities of the population must be respected in gaining entrée in a disaster area. Disaster research requires probing into areas which are, for various reasons, regarded as confidential. While victims will discuss their physical injuries, their property losses, and even their bereavements, they are careful with whom they discuss them. Individuals who have insurance claims or lawsuits pending are likely to be suspicious of interviewers. Officials of public agencies or private concerns may fear that research is designed to fix responsibility for disaster events.[10] At a time when the nation is security conscious, researchers are quite likely to be asked to prove the legitimacy of their inquiries into the organization or operation of industries and branches of municipal or state government.

It is essential, therefore, that the research team establish an acceptable identity at the very beginning of a study. Presentation of credentials from a state or federal government agency, a university, or a well-known private foundation or research organization usually serves this purpose. In some cases, however, emphasis on one of these may increase resistance. For example, if a disaster has resulted in claims against the government it may be necessary to "play down" federal sponsorship and emphasize university connections. These conditions vary from one disaster to another and cannot be anticipated for any given disaster. To some extent the investigator must always "play it by ear." The alert investigator will soon discern that it will be to his disadvantage to become identified with a particular agency that happens to be in disrepute in a particular disaster.

It is desirable to secure the cooperation or, at least, the concurrence of city and county officials or of the heads of private concerns, before beginning fieldwork in areas under their jurisdiction. Sometimes they will furnish credentials which are accepted locally without question. The use of such credentials is not always advantageous, however. A political rift in a city may make sponsorship by an incumbent mayor a handicap in getting the cooperation of his political opponents. Company approval makes workers suspicious if they feel the company is trying to evade responsibility for an industrial disaster.

It is always a safe procedure to inform police officials that fieldwork is to be carried on in their bailiwicks. This usually ensures their cooperation and precludes embarrassing incidents such as the detention and questioning of interviewers by police officers.

In many disasters it will be necessary to secure passes in order to gain access to the stricken area. These will usually be secured form the local police, but it may be the national guard, state police, or local civil defense agency that issues passes. In some cases more than one agency issues passes, and, assuming the principle of maximum unhappiness, it could occur that one agency does not honor another agency's passes at first. The question of passes will usually have to be settled in the community, but investigators may sometimes arrange credentials with the national guard or state police which will serve at the locality.

Before embarking on a field study, any team should make a vigorous effort to find out if anyone else is doing disaster research in the same area. The efforts of both teams may be hampered if they are not coordinated. Researchers should make a special attempt to avoid overworking the same subjects and thereby arousing resentment.

It is important in disaster research, as in other areas of social science research, to convince respondents that information they give will be held in strict confidence and that their anonymity will be preserved. As has been stated before, it is also important to persuade them that the data obtained with their cooperation will be put to significant, practical use.

THE REPORTING OF DISASTER FINDINGS

While they have important theoretical implications, field studies of disaster are preeminently applied research. To the extent that a field study is designed to produce significant and useful conclusions, the results will be in demand among lay people who have been, or may be, affected by disaster. With this in mind, reports of disaster studies should be written as if they might be read by at least some of the erstwhile subjects.

This standard makes it imperative that, in the writing, the anonymity of subjects actually be guarded as carefully as they were assured it would be. It also demands that conclusions be stated in a concise, intelligible fashion, with neither their theoretical nor their practical meaning being obscured by professional jargon. Such reporting will not only enhance the reputation of disaster research among its sponsors but will also contribute to the advancement of social science and the acceptance of its findings.

NOTES

1. This language is "sociological," but the writer believes and intends that it embrace the interests of the psychologist, psychiatrist, social psychologist, anthropologist, political scientist, economist, and administrator as fully as those of the sociologist.

2. Wallace's (1956b) *Tornado in Worcester* will give the reader a good picture of the actual situation in a tornado disaster and will help to make the discussions which follow more meaningful.

3. Obviously opportunity to do research *during* the event is much greater in prolonged-impact disasters.

4. There is probably no area of human behavior about which so many untested stereotypes and so much word-magic exist as the area of human behavior in disaster. Another thing that a scientist just beginning work in disaster will notice is a considerable semantic confusion about certain widely used terms. The term "panic," for example, is used by different writers, and sometimes the same writer, to mean a subjective feeling of fear or terror, bad judgment, inefficient behavior, acting too fast, not acting fast enough, blind flight, any kind of flight, paralysis, or a vague, global concept of wild, animal-like behavior. Scientific investigators, though usually more specific in their definitions, also use the term "panic" to mean quite different things. The terms "shock," "looting," and other widely used terms connoting behavioral and emotional responses in disaster are used with different and sometimes undefined meanings. Since the person coming into disaster research inherits

this confusion about terms, and possibly some of the common stereotypes, he needs to exercise special care in his definitions, descriptive categories, and other terminology, and he needs to be doubly careful about making his own assumptions and hidden hypotheses explicit.

5   With respect to the London black smog episode referred to above, a newspaper story in the United States reported that "one of the world's biggest cities experienced a near mass panic." In a sample survey, one percent of the respondents reported themselves as having panicked, and one percent stated that other people with them had panicked (Anon. 1955).

6   One study does provide an enlightening similarity, however. The Department of Anthropology and Sociology of Michigan State University is conducting a long-range study of cultural and social relationships and differences along national boundaries. The Mexico-United States border is one of those being studied. When a disastrous flood swept down the Rio Grande River in June 1954, the Michigan State University group, which had previously done field studies in disaster, immediately saw an unusual opportunity. Mexican and American towns on opposite sides of the river were being subjected to approximately the same disaster stimulus. Would there be differences in response associated with differences in culture and social structure? The investigators hypothesized that there would be such differences, that basic differences in social structure and cultural value orientations would be manifested through differences in the communication media through which persons and families learned of the flood danger, in the nature of groups formed for evacuation, rescue, and rehabilitation, and in other responses to the disaster situation. They designed and carried out a study to investigate these and other hypotheses (see Clifford 1956). While this study does not illustrate our suggestion of preplanned alternatives for disaster research, it does illustrate how a preexisting interest and competence in a special area of research can be readily translated into hypothesis-oriented research in disaster.

7   The discussion of interviewing pertains to a field study in which some type of schedule or some form of structure is used in the interview to seek answers to predetermined questions. The reader will see that we are not

dealing with the very nondirective, clinical type of interview which would be required for some kinds of studies.

8   The mnemonic device was developed by Michigan State University investigators for use in a study of spontaneous rescue groups. The device employed was GLIRCS, representing respectively: (1) spontaneous rescue *G*roupings; (2) *L*oyalties activated by the disaster; (3) consequences of the disaster for the informant's *I*dentification with his community; (4) *R*ole conflicts; (5) consequences of the disaster for *C*ommunity solidarity; and, (6) consequences of the disaster for social *S*tratification (Form, Nosow, Stone, and Westie n.d.).

9   William A. Scott describes this problem and the type of judgment which is often employed in evaluating it in his report of a study of people's reactions to an air raid signal which proved to be a false alarm in Oakland, California. This situation, on the assumption that distortion is encouraged when people are later proved wrong in their interpretation of the situation, would be one in which retrospective distortion might especially be expected to occur. Dr. Scott writes:

> When people are asked to recall what they felt, thought, and did a month or more ago, there is bound to be some omission or distortion unless the particular events involved were extremely striking. While those respondents who took the alert seriously at the time may have been impressed enough by the experience to recall all the details of their thoughts and actions at a considerably later date, it is quite likely that people who more or less ignored the alert would have forgotten how they actually did feel at the time. Moreover, there is a possibility that memory would err in the direction of how the respondent feels he *should* have behaved. A person who feels silly about having become excited over the siren may be reluctant to report, or even remember, his own excited reactions. Another who feels guilty for not having paid more attention may, consciously or unconsciously, make his behavior appear more responsive to the siren than it really was.

latter being what is currently known as the later preparedness and early response phases of disasters). Organized behavior included not only formal organizations but also informal and emergent groups.

The focus on organized behavior was a very conscious one. It is impossible in this chapter to detail the complex of theoretical, professional, and practical reasons for the choice. Suffice it to say here that a major reason was that, while the earliest studies in the area had already established the parameters of the behavior of individuals in crises (see Quarantelli 1988), almost nothing was known about group behavior. Also, field studies of persons undertaken during the *actual* crisis times of disasters always have difficult sampling problems with an unknown universe of participants, whereas the universe of most organized groups in a community is known and finite (e.g., typically there is only one fire and police department, a handful of mass media outlets and hospitals, etc.). Finally, even in the pioneering days of disaster research it was evident that the most effective and efficient planning and managing measures for responses during crises would have to be primarily carried out by organizations and could not be done by individual households.

The basic model DRC employed was set early. It was initially drawn from the pioneering field operations of the National Opinion Research Center (NORC) in its 1949-1954 research on disasters (see Quarantelli 1987c, 1988). However, also influential was the research focus on organizations per se for which conventional survey methods during actual crisis times were not appropriate or difficult. There were minor refinements and augmentations over the years. In the main what we depict is the later rather than the earlier versions of the procedures. Also, what is set forth, using a Russell Dynes characterization, is the "generic" version rather than all the variants used in the different research projects DRC undertook. Finally, while we generally depict what was intended to happen, we do note difficulties in achieving what was wanted or planned with relation to specific matters and near the end of the chapter indicate some problematical aspects of the whole enterprise.

GENERAL BACKGROUND

The possibilities and problems inherent in the kind of fieldwork DRC undertook requires understanding the general context within which it operated. This is not the place for a history of the center, but we note selective aspects since they affected the research planned and done.

From its inception, DRC was administratively nested in and informally part of a department of sociology. However, the center for most of its existence never had any funding from any university source, thus making organizational control by other entities over its operations mostly nominal. Except in very recent years, funds for all DRC activities were from the research grants and contracts it obtained on its own. In the early years most funding came from the predecessors of the Federal Emergency Management Agency (FEMA) (e.g., the Office of Civil Defense—OCD), but that was increasingly replaced in later years by grants from the National Science Foundation (NSF). While the support from OCD came through contracts, the initial relationship that quickly evolved soon allowed DRC, within very broad limits, to do what research it wanted in whatever way it thought best. Overall, DRC had more freedom in its operations than many centers or institutes typically have, a professional advantage the center used in the studies it independently launched, as illustrated later.

As an organization, DRC was never systematically planned or even formally created. The center was never a formal part of The Ohio State University. In many senses, it was an emergent group with decisions on group structures and functions forced by situational contingencies. Thus, its evolution had neither the advantages nor the disadvantages of a structured path. But implicitly at least, DRC from the start was thought of by its founders as primarily a social science research entity. It was never visualized as having teaching functions, providing any formal training, or having a consultative role; and as a group no such activities were ever undertaken. Nevertheless, the sharp focus on research did lead to several auxiliary activities that eventually were important in the center's history, namely the systematic creation of an archive of its own field data, the establishment of the largest specialized library in the world on the human and social aspects of disasters, as well as the development of its own publication program.

From a research perspective, DRC from its inception looked at all kinds of "disasters." The natural versus technological disaster distinction was ignored in its work (contrary to some statements others have made about its focus), although sometimes particular projects because of their funding source focused on one particular kind of disaster rather than another (e.g., the DRC pioneering study of chemical disasters in the early 1980s; see Quarantelli 1984). As an indication of its generic approach, the first ten events studied were a hurricane in Texas, the overflow of a dam in Italy, the Coliseum explosion in Indianapolis, a nuclear plant accident near San Antonio, a nursing home fire in northern Ohio, the dam break in Baldwin Hills (Los Angeles), California, a plant explosion in Massachusetts, a flood in Cincinnati, the Alaskan earthquake, and a student civil disturbance in Columbus, Ohio. Also, as just said, DRC sometimes studied civil disturbance and riot occasions (Warheit and Quarantelli 1969), mostly for comparative purposes since it was assumed such conflict-type episodes are somewhat substantively different from natural and technological disaster occasions. While DRC found it could use the generic field research methodology in all its work, conflict situations did require adjustments that are not discussed in this chapter.

The quantity of fieldwork done also partly dictated the need for a standardized field procedure. From 1963 to 1984, while DRC was at The Ohio State University, it undertook 457 different field studies (in about a third of the cases, this involved more than one actual trip to the site). In several years, field trips were quite numerous, there being 55 different field studies in 1969 alone, 51 in 1972, 50 in 1982, and 46 in 1979. Only in two years did the number of field studies drop below double digits. As such, it was not rare to have two different teams in the field simultaneously, and occasionally three teams were concurrently in the field.

The great majority of work was done in the United States. However, 18 field studies were done between 1963-1973 in 11 foreign countries (five times in Canada, three times in Italy, twice in Japan, and once each in Mexico, Chile, Greece, El Salvadór, Australia, Iran, Curaçao, and Yugoslavia). These foreign studies mostly ended with the emergence of native disaster researchers in many places.

The actual fieldwork, especially after the first few years, was done by graduate students. Depending on the research funding available, these at any given time numbered between four and fifteen (the total DRC personnel once peaked at 59 staff members, but two dozen was the more typical work force). These students, appointed as graduate research assistants (GRAs), were mostly drawn from sociology but in later years were also selected from other areas (political science, journalism, anthropology, nursing, etc.). In principle they were employed only on a half-time basis (20 hours per week), but the informal understanding was that there was no limit in doing fieldwork. While major efforts were made not to disrupt class attendance and examination-taking, it was nonetheless an unambiguous stipulation for employment that all GRAs had to be available for fieldwork on any day at any time of the year, university holidays and vacations to the contrary. In part this was related to the fact that DRC never employed GRAs for a specialized work role (e.g., field interviewer, coder, or any other specific task). Instead, it was an explicit condition of employment that all would work on all research aspects, ranging from the designing of the fieldwork through data gathering to data processing to data analysis and initial report writing. It was also assumed that GRAs would work on all research projects in being during their employment (usually two or three; in rare instances, more).

Most center funding was for specific research topics. As an example, DRC once had a five-year contract with OCD to study the major community organizations involved in disasters; under this contract, particular studies were done of local emergency management agencies, police and fire departments, hospitals and related entities, the public utilities, and the Red Cross. DRC also specifically studied the delivery of mental health services under a grant from the National Institute of Mental Health (NIMH) and the delivery of emergency medical services through a Health Resources Administration grant.

In addition, DRC at times was funded to study very general topics such as "community coordination" or "organizational functioning," labels deliberately vague but which allowed the center to venture into different research areas. For instance, DRC studied the military, religious groups, and—in the early days of its existence—the mass media, which otherwise probably could not have been attempted at that time. This work did at times lead to more specific studies (for example, a later NSF grant to study news reporting by community mass communication systems). In addition, DRC, taking advantage of the professional freedom mentioned earlier, did studies not directly funded by anyone. These were "piggybacked" in various ways on funded projects. Among such research were studies on the handling of the dead in mass casualty situations, disaster-induced long-run organizational changes, the characteristics of disaster subcultures, state-level disaster planning, rumor control centers, nonriot looting behavior, etc.

DRC chose all the topics it studied. It rejected suggestions by funders that were not of professional interest to the directors or that made little sense on the basis of earlier research (e.g., a proposal by a federal agency to fund the center for a field study of the "looting" in Hurricane Camille). Important is that a conscious effort was made to keep moving on continually to new topics for study. In only a few cases did the center replicate or build upon previously studied topics (the major exception was a multi-year restudy for FEMA of an earlier five-year OCD study of emergency organizations). There was a logic to this pioneering. For years DRC was the only research entity of its kind anywhere; given that, studying new topics was seen as the best way to call attention to the importance and significance of social science studies of disasters. At another level, a pioneering effort is more of an intellectual challenge, requires innovation in field design, and is certain to generate unexpected findings. In short, we found it both "fun" and "interesting" to pioneer, and accordingly did so.

DRC research in one sense operated at two levels. At one level, the intent was to understand disaster phenomena per se such as the *conditions* that generated disaster problems, the *characteristics* of organized disaster behavior, the later *consequences* of that, and the *careers* of disasters (this jokingly came to be known as the "C" model). At another level, the goal was to further sociological understanding of emergent groups and organizational behavior. Put another way, the DRC sociologists, true to their disciplinary background, assumed that the better sociology they did, the better would be the research on disasters. Therefore a conscious effort was made in analysis and report writing to interplay the descriptive disaster data with sociological ideas, but a balance was not always achieved. While the center was successful in explicitly resisting the development of a separate field of "disasterology" and had some success in developing the sociology of disasters, the latter was not as much as might have been ideal.

Yet the center never operated with one explicit theoretical orientation. There was in fact a conscious effort to avoid the development of a center orthodoxy or "party line." Nevertheless, certain views about social phenomena were implicitly used more than others. A sociological orientation was always present and not a geographical one, as in the early hazards studies by others. In addition, the implicit social psychology framework used was that of symbolic interactionism. Similarly, ideas from the classical University of Chicago view of collective behavior permeated the approach to emergent behaviors and groups (for the historical background of these two orientations see, Quarantelli 1987b, 1994). Organizations, on the other hand, tended to be viewed, but not always consistently, in an amorphous structural-functional framework. Our point is that the field operations and research procedures were influenced by the indicated theoretical preference, so while a rigid theoretical orthodoxy was avoided, DRC did lean implicitly in certain directions rather than others.

As loose as the theoretical preferences were, the DRC

methodology was even more eclectic and catholic. There was, to be sure, a preference for methods that allowed induction rather than those requiring deduction and which allowed qualitative rather than demanded quantitative analyses. In a general sense, what DRC by trial and error evolved was similar to the "grounded theory methodology" which was being created roughly at the same time by other sociologists with no direct connection to the disaster area (see Glaser and Strauss 1967). However, the center did not explicitly or consciously borrow from the formal literature on grounded theory. In fact, it was quite a while before DRC consciously recognized that it too had gone down the same methodological path developed by grounded-theory scholars.

PRIOR OR PREFIELD PROCEDURES

While there was "trial and error" in training the first cohort of GRAs, from the first prefield training was deemed very crucial for their ability to do well at disaster sites. Therefore, much time and effort were spent on training. Among procedures usually followed were providing all new GRAs with: (1) a general introduction to the history of disaster research and of DRC; (2) the procedures, promises, and problems in qualitative field research; and (3) a detailed introduction to the specific research project(s) in which they were to be involved. Indicative of the coverage of this training for new GRAs are the topics of the 30 sessions listed in the outline for 1987 and given by the DRC directors:

OUTLINE FOR 1987

I. BACKGROUND ON THE DISASTER AREA AND DRC. (1) Nature of disasters and disaster preparedness and response in the U.S.; (2) History of disaster research in general in the social and behavioral sciences; (3) Overall view of substantive disaster findings in general; (4) History and activities of DRC; (5) Past DRC work and resources; and (6) DRC operations, including logistical issues.

II. DRC FIELDWORK. (7) General orientation and policies (including ethical issues); (8) Preparations and entry; (9) Interviewing problems and procedures; (10) Observing problems and procedures; (11) Documenting problems and procedures; (12) Processing of field data; and (13) Report writing.

III. THE FEMA STUDY. (14) FEMA as an organization and its interests; (15) Earlier related DRC work; (16) Last year's work; (17) Projected fieldwork for the coming year; (18) The research designed, new issues and questions that need consideration; (19) Specifics of the research design; and (20) Planning and actually doing the upcoming fieldwork.

IV. THE NSF MASS MEDIA STUDY. (21) NSF as an organization and its interests; (22) Earlier related DRC work; (23) Projected fieldwork for the coming year; (24) The research design: issues and questions that need consideration; (25) Specifics of the research design; and (26) Planning and actually doing the specific fieldwork.

V. THE NSF MEXICO CITY EARTHQUAKE STUDY. (27) The study design used; (28) The survey work undertaken; (29) The organizational data obtained; and (30) Current status of the work and what needs to be done to finish the study.

After the training sessions, new GRAs typically conducted practice field interviews with officials in local emergency-related groups (defined for them as studies DRC was doing of preparedness planning). The tape recorded interviews were then listened to by veteran DRC researchers, with the positive and negative aspects of what had been done individually discussed with each new GRA.

Even more important, at the start of any research project all GRAs were given copies of the funded research proposal. After being told to read the proposal carefully, a series of meetings were held, the first of which started with roughly the following statement: "This formal proposal gives you a vague idea of what we think we are going to do; now we are collectively going to have to work out the details of how we will actually proceed in the real world and what we concretely need to find out." Succeeding sessions particularly focused on developing research design specifics (e.g., the field instruments, early versions of which were often drafted by smaller task forces of GRAs). The basic goal was to involve GRAs in the process so they knew what information was to be sought from whom, why specific questions were asked, and generally how much detail was wanted given the data analysis projected. Fully involving the GRAs in building the research design ensured that they understood the logic of what DRC would do in the field. Interview and other guides were always produced, but by that point GRAs needed no further guidance on their use. GRAs who joined DRC in the middle of research projects were not as heavily involved in building the research design, but there was always overlap of "veteran" and "rookie" GRAs so that the latter could informally learn from the former.

A most important idea conveyed in the training was that DRC mostly wanted the *overall* picture of the occasion, and no one or even several officials or persons together could possibly provide by themselves such a picture. (It was stressed that in disasters perceptions are even more selective and narrower than at routine times.) So, like doing detective work, it was emphasized that the team would have to develop the larger picture by putting together information from a variety of different kinds of field data. Three major mechanisms were used to sensitize the GRAs on this point: (1) emphasizing this process in the prefield training; (2) interviewing at different levels and positions in the groups studied (from administrative heads to liaison personnel); and (3) in-the-field team meetings. There is not space here to discuss how inconsistent or contradictory statements were handled, although it was stressed in the training that, when the perception and social location of informants/respondents is taken into account, there are far more "seeming" rather than actual irreconcilable statements.

The specifics of what social dimensions were to be examined varied somewhat in each study. However, in general information was almost always sought about the structure and function of the group involved, the formal and informal division of labor in being, its interorganizational contacts, the available material resources and facilities as well as the group's prior disaster planning and earlier disaster experience. It was always very explicit that the fieldworkers should make a very conscious effort to obtain a picture of the organization as it operated during normal times and how it acted during the crisis period of the disaster being studied. This contrast was frequently talked about as the differences between Time One (preimpact) and Time Two (impact). In its later work, DRC tended to make more of a differentiation between the preimpact or preparedness phase, the impact or transemergency period, and the immediate postimpact period.

Also stressed in the training was that, while particular officials or workers would be approached in the field, the picture the center wanted was how the group or organization of which they were a part operated. DRC only had a secondary interest in individuals as such. Put another way, the theme was that the focus of the center was on the group level, what the organized group as a social entity did, not what social roles were played by particular people. This point was not always easy to communicate in the training because even professional sociologists do not always understand the difference between the study of a group as such and the study of the members of a group. (The emphasis on using as much participant observation and systematic document collection as possible was also part of the effort to get the GRAs to think in group rather than individual terms; this distinction is easier to see by using these techniques rather than interviewing individuals.) This focus on the group level became even more clear in later DRC work and was therefore made even more explicit in the training.

Related to the point in the last paragraph is that in almost all projects DRC wanted the multiple perspectives of those involved. Thus, it was fairly standard in the research design that certain work positions or occupational roles were automatically part of the listing of those who should be interviewed (or approached for other kinds of data). It was stressed in the training that it should be expected that DRC would obtain different accounts of "what happened" depending on the perspective of those reporting the happening. The center did assume that there was no one "true" story. As such, in the training it was pointed out that in almost all cases there were at least a minimum number of perspectives we wanted to obtain. These usually included the administrative head of the organization, its operational head (which DRC early learned usually knew more specific details about everyday activities of the group, whether in routine or crisis times, than did the administrative head who was often more concerned with policy and political issues), boundary personnel such as liaison persons or secretaries who represented or linked their organization to the outside world, and communication personnel at almost any level. The last role in particular could be rather low in the formal organizational chart but frequently would not only have substantial information on the "what" and "who" in the communication flow, but also often could supply to DRC relevant statistics or at least numbers on the flow. Of course, as was indicated in the training of the GRAs, there was frequently more than one occupant of the indicated roles in organizations, and that should be taken into account on who should be approached in the fieldwork. In emergent or informal groups, the kind of social roles just indicated would be less obvious, but for DRC purposes we would seek out those persons who engaged in equivalent activities. In addition to the somewhat fixed sampling of certain work roles, "snowball" sampling was also used as a very useful procedure given what the center was often studying.

Apart from the training for substantive issues, DRC also paid much attention to logistical matters in the dispatching of field teams. This was important for achieving the major objective of getting to a disaster site as soon as possible, while emergency operations were underway. In the 1964 Alaskan earthquake, DRC had five persons, including all three directors, in Anchorage within the first 24 hours. This process is less complicated than might be believed if routines are developed as DRC did early in its life. This included: (1) training the center's office staff in how travel arrangements were to be handled both within and outside the university context, including educating others that normal travel procedures could not be followed and the need sometime to operate outside of usual university hours; (2) preparing a standardized field kit always in the personal possession of all GRAs; (3) insisting that all GRAs at all times have luggage bags prepared for travel; and (4) putting in the possession of all fieldworkers a master check list. A later version of this list specified checking the following:

MASTER CHECK LIST

1. Tape recorder (test by recording and playing back before leaving). 2. Charger, cord. 3. Blank cassettes and boxes with identifying cards. 4. Field kit folder. 5. Personal DRC business cards. 6. Personal identification letter with photo. 7. Auto rental credit card*. 8. Receipt forms. 9. Stack of DRC leaflets. 10. DRC fact sheets. 11. Forms for reporting expenses. 12. Sheet with office, home addresses and phones. 13. List of current telephone billing numbers. 14. Pens. 15. Pencils. 16. Tablets for note taking. 17. Traveler's checks (as well as enough cash). 18. DRC card for auto sun visor. 19. Interview guides. 20. Special handouts. 21. List of prior contacts in the area*. 22. Data analysis forms. 23. Data inventory forms, if relevant. 24. Video cassette and cameras*. 25. Self addressed stamped envelopes. [The field coordinator is to have the items indicated as * .]

Learning about disasters

All personnel at the center were instructed to be attentive to news reports or weather stories that might have implications for possible fieldwork (although DRC learned early that initial accounts tended to overstate damage and destruction). Usually such information was directed to the field director, who had the authority to dispatch field teams. Generally the judgment on whether a team should be sent was made on whether the reported occasion fitted the existing criteria for inclusion in whatever projects the center had under way. (Sometimes field data could be concurrently obtained for two different projects.) At times a decision had to be made on whether a team could be gotten into a locality before actual impact (e.g., possible in developing hurricanes or floods). Thus, there was more than one occasion where a DRC team managed to arrive on site before the airports in the area were closed. In other cases, where DRC had done previous field studies, knowledgeable officials known to the center were contacted and information obtained from them.

The value of being on the scene at the height of crises cannot be overstated. It is worthwhile to be in such situations for two basic reasons. First, observations can be made and documents collected that cannot be obtained through later interviewing. The social barriers that normally exist to restrict access to high-level officials and key organizations simply do not exist. A second reason for being on the scene early ensures a high degree of access and cooperation. Victims are typically candid, cooperative, and willing to talk in ways far more difficult to get later.

Arriving on the scene

Field teams on arrival did two things as quickly as possible: (1) finding living quarters; and (2) going to Emergency Operations Centers (EOCs). The latter was to learn what was going on as seen from the perspective of community officials, keeping in mind that a "command post" bias would be present. The basic purpose was to obtain an overview of the situation, learn who the key officials and organizations were and also what they were doing, make personal contact with them for later follow-up interviews, gather ephemeral observational and documentary data, and lay the groundwork for a later systematic in-depth study. Contrary to popular imagery, there can be many periods of lull and inactivity even in the most hectic of disasters where interaction with officials can be initiated, provided entry is gained into EOCs or coordinating centers. (Although not discussed here, obtaining entry often requires considerable skill with a special need to be seen as researchers and not "journalists.")

A variety of research techniques were used for data gathering, particularly open-ended interviewing, selective participant observing, and systematic document collecting.

INTERVIEWING

The guides and procedures used by DRC were basically for open-ended and in-depth interviewing. The usual lead question was deliberately very general, such as: "Tell me what happened in X (e.g., the tornado)." Other suggested questions were often also very general, putting a premium on the fieldworker's ability and knowledge of when, where, and what to specifically probe. However, while formal interview guides were always prepared (with questions, probes, and instructions) and taken to the field, they were never thought of as field manuals to read from. Given the involvement of GRAs in their production and thus their presumed understanding of what was wanted, the guides were instead used to ensure that fieldworkers covered all relevant topics.

Certain principles that all fieldworkers were to follow evolved over time. In a later explicit version, these were, to paraphrase the oral and written statements provided in training sessions:

1. Always tell the truth about who we are and what we are doing. Apart from ethical considerations, from a pragmatic point of view it is much easier to proceed in that way since a "cover" story does not have to be remembered. However, you should not volunteer too much information or details, unless asked to do so. Overexplaining often confuses people and may raise unnecessary questions.
2. DRC does not seek publicity, especially in the field and particularly from the mass media. Try to fend off inquiries by using the DRC handouts which indicate who we are, what we do, etc. If such contacts are unavoidable, try to

structure the picture of us we want portrayed. The principle to follow regarding mass media contact for information about the study is to refer the problem back to the home office of DRC.

3. Indicate very explicitly the confidential nature of the information we seek and obtain. Make clear that once in the center's hands we have the responsibility to protect any data, and we take that very seriously. If any problems on this score occur, have them contact the home office of DRC.

4. We should try to have ourselves identified as researchers and from a university. Such labels evoke positive reactions. You must make a conscious effort to avoid being misidentified as "journalists" or "investigators." We are not there to judge, to evaluate; this is why words such as "investigation," "investigator," etc., should never be used. The term "government" also needs careful use since not everyone is positive to it.

5. Adjust to what is going on rather than trying to fit others into our study. In crises (and even outside in many emergency-related groups), there is no 8:00 a.m.—5:00 p.m. weekday schedule. This means that field teams can make contacts literally around the clock and in nonroutine locations (e.g., at EOCs, emergency shelters).

6. Make field decisions on the basis of future consequences. The question of pressing for a particular interview, entry into a specific organization, seeking information about some sensitive topic, etc., is a field team decision. The judgment should be based on the future rather than the immediately present situation. Before pressing regarding something, ask yourself: Would it matter, and in what way, if you or another researcher were to later contact the involved parties again? Since we do follow up and repeat studies, our reputation is important.

7. Under no circumstance should any DRC fieldworker get in the way or interfere with any emergency operation or

personnel. Similarly, never under any circumstance should you participate in any emergency disaster response even though there are times when you may be asked to do so, and the task may be minor.

8. There may be crises where there might be some general personal risk involved for team members. It is understood that no research result is worth any personal risk. Thus, use your common sense and avoid or move away from such situations.

9. There is a danger in hearing more or less the same "story" over and over again; it may be assumed that the overall picture is clear. However, team members should ask the same question to everyone and to probe the answers, because that assumptions should not be made. Beware also of the "collective consensus" and "retrospective redescribing" that often occurs the later from the time of the actual occurrence of any action. Keep placing your interviewee back to the time period being reported on. Ask for a step-by-step temporal and spatial chronology. Stress that our interest is in the interviewee's own perception of happenings.

10. Keep in mind that, for diplomatic reasons, some nonsubstantive interviews may have to be conducted. In complex organizations, it is not wise to interview lower— and middle-level personnel without first obtaining an interview with a high-level official (where the field study and topics to be probed can be noted). Interviews at different levels of an organization ensure getting different perceptions of what occurred, because of the DRC assumption that there is not only one "true" story.

11. Never forget our distinction between informants and respondents as this is understood in sociology. Different kinds of information are obtainable from each perspective. That informants are discussing matters in which they were not personally involved does not mean that the information is not useful. Always disentangle statements

involving the two perspectives because, in any interview, interviewees may wander from one to another.
12. Field teams are not micromanaged. While consultation back to DRC is encouraged, teams should make their own decisions about field problems and questions. You have considerable autonomy; you will not be second-guessed on your decisions.

These instructions seemed to work well. Not only were outright refusals to participate in the study extremely low (in most field studies none ever occurred) in the thousands of contacts made, but the cooperation was such that candor was the norm rather than the exception. We could attribute all of this to the excellence of the DRC planning and training and its implementation in the field, but that would be an overstatement. Our fieldwork was helped by several psychological reasons, primarily that persons under stress are far more willing to talk without reservation than during normal times, and secondarily by the fact that respondents and informants were often very impressed that the team had traveled from afar to try to learn from their experiences on how other communities elsewhere might cope with a disaster.

A member of the field team was usually designated as the field coordinator with final responsibility for field trip decisions. For the educational experience, this temporary position was typically rotated among the GRAs (when the field director was not a member of the team). This choice and the actual composition or mix of the team when choices were available required some tact and sensitivity in the decision-making. As in any work situation, not everyone liked everyone else, various kinds of rivalries and conflicts sometimes existed, and, despite attempts to standardized the role, there were varying views of what constituted responsibility. So while DRC found liking others was not necessary for good field performance, having a professional attitude about the GRA role was crucial. Overall, most team efforts went well, but there were occasions when, because DRC had put together a poor field team mix, there was a negative effect on the data obtained in field operations.

Observing

Although DRC did work out several systematic field observational guides for crowds and civil disturbance situations, it never developed systematic and general parallel ones for disastrous occasions. (In retrospect, this is probably a major shortcoming of the DRC field research methodology.) Nonetheless, field team members were trained to be alert to make relevant observations of unscheduled as well as scheduled happenings (briefings, press conferences, etc.), and if possible to tape record their observations live (including photographing or drawing diagrams of EOCs, etc.). The simple fact is that fieldworkers who do good participant observations can "see" things that cannot or will not be reported on in a later interview.

Let us illustrate through a personal example from Hurricane Betsy which hit New Orleans. At the height of the crisis we were able to walk into the Mayor's office, which then was on the top floor of a high rise building which was informally being used as the major EOC, and sit in his chair without having to go through a single secretary or obtain the permission of anyone. We just walked in. While there, we monitored the radio messages being sent. At one point a police car reported several stores with broken windows might be vulnerable to "looting." To our surprise, a half hour later the dispatcher ordered two other police cars to go to that location to stop the "extensive looting going on." But the initial police car had only reported certain open store fronts were susceptible to looting! It is very doubtful that later interviewing would have evoked the sequence of events that we saw and heard, especially since the informal log used only recorded the later sending of the two cars.

Also, observations were often intermingled at the same time with other procedures for obtaining field data. For instance, in one study the center focus was on outside-of-the-hospital handling of mass casualties. As indicated in the following outline reproduced from what was used by the GRAs, much of the information sought could be obtained by observing particular situations, although important. For instance, a team member in one disaster was told by a police major that he had ordered and then quickly rescinded an evacuation order concerning police cars in an endangered area, and showed the informal logs on it. The DRC research assistant alertly asked for and obtained a copy of the log and brought it back to DRC. Later when the center obtained the official organizational logs, the order and the rescinding of it were not at all noted. This was not a cover up. It turned out that the official logs only listed actions taken; since the police cars never left the area, this was never recorded. Yet it surfaced the problems of organizational officials attempting to follow plans that do not indicate what is to be done when contradictory advice is offered.

Developing a field consensus about the data

Very important is that at the end of the day in the field the team members met and collectively discussed what they had learned in their interviewing and observing. When done every day, this allowed a slow reconstruction of happenings during the disaster and also provided clues on where the research attention ought to be directed the next day (e.g., in studies of emergent groups where "snowball" sampling ought to proceed). Also, as pointed out in training sessions, this meant that after several days the DRC team would collectively have a broader and more comprehensive picture of the disaster than would be known to even the most heavily involved organizational official. The danger of assuming too much because of this knowledge required that team members had to be very alert in asking the same questions, and probing appropriately, later respondents and informants. This did not always occur. DRC early established that later interviews in the field were without good reason often shorter in length than earlier ones.

POSTFIELD PROCEDURES

Upon returning to the center the field team would present a detailed briefing to the staff, make its recommendations on whether to do an in-depth study, and process the material it had collected. Reconnaissance trips led to in-depth studies about 40 percent of the time. These initial trips were not wasted abortive efforts since they provided good feedback for improving field instruments, gave valuable insights on substantive issues, and was realistic field training for new GRAs. The directors always made the final decisions on in-depth studies. It was learned early that the best timing for such work was usually two to three weeks after impact and not while the crisis period was still in being.

A field report of about 3-10 pages was usually written for almost all trips taken. Its purpose was twofold: to produce a quick historical record of the study, and to force the fieldworkers to think substantively about what they had studied. This is illustrated by a 1978 guide for writing field reports that required the following information:

GUIDE

I. Identification Data: Field trip report #, event name, date of report, author of report, field team members. Type of trip (baseline, planning, actual event, follow-up, other), purpose of trip.

II. Disaster Agent Characteristics: Briefly describe disaster agent and community context, agent type (see inventory list), scope of impact—localized or diffuse, speed of onset—sudden or gradual, prior warning or not, length of warning period, scope of disaster planning—community wide, organizations mainly, interorganizational, little or no planning. Previous disaster experience in last five years—if so, what kind. Losses to community—deaths, injuries, residences and businesses destroyed, evacuees. Was there a federal declaration?; if so, date of declaration.

III. The Organized Response. Note organizations involved in disaster planning and/or response. Specify tasks and responsibilities of each. Note if there were any emergent groups.

IV. Problems in Communication, Coordination, and Control. Note in general terms problems encountered by responding groups such as conflicting planning, duplication of effort, etc. Note also adaptive mechanism used.
V. Relevant Comparisons and Contrasts. Tell briefly in what ways the preparations for or the response is similar to or different from the organized response to other events we have studied.
VI. Methodological Notes. What implications does the study of this situation have for future field operations? What implications for data analysis are there in the material gathered? Are there any special problems regarding entrée, confidentiality, etc.? Are there any other groups besides DRC doing research in the area—who and what are they doing? Any contact made with them?

There was also the putting together of a master list of the field data obtained. In one form prepared in 1979 the field coordinator had to indicate if and how much of the following had been obtained:

I. Field interviews obtained for Guide A (preparedness informants), Guide B (organizational informants), Guide C (organizational respondents); interview notes made in person and over phone.
II. Field analysis forms for planning, response, organizational linkages.
III. Participant observer sheets: (1) photos of activities at site, EOCs, etc.; (2) diagrams/charts of EOC layout, etc.; (3) description of physical topography of impact sites; (4) other items.
IV. Documentary data: (1) phone books; (2) maps with disaster site marked; (3) disaster plans—list of organizations; (4) communication tapes; (5) organizational logs; (6) after action reports; (7) minutes of emergency meetings; (8) organizational charts; (9) mutual aid agreements; (10)

radio and TV tapes; (11) local newspapers—before, during, and after impact; (12) other items—leaflets, booklets, news releases, etc.
V. Statistical data and sources of: (1) casualties, (2) property damage, (3) injured—treated at hospital, admitted, DOA; (4) other.
VI. Follow up calls/letters needed as well as anything DRC promised to send back.

In the early days of DRC almost all interviews, the great bulk of which were tape recorded, were fully transcribed. This followed the processing model set up by NORC in its pioneering effort in disaster research (see Bucher, Fritz, and Quarantelli 1956a, 1956b). In time this led to the setting up of a massive transcribing operation, and more than 6,000 interviews were transcribed. However, the time, cost, and effort were too costly to be continued, and no funding could ever be obtained after 1974 for such an operation. This was unfortunate since analyzing from tapes is very tedious. The later making of summaries of tape content by the actual interviewers themselves was a partial but not perfect solution, although it did have the unintended consequence of GRAs spending more time and effort in the field on ensuring the quality of recordings!

Although informal case studies following an outline were often written, very few were ever actually formalized or published. DRC never had a major interest in case studies as such nor in the disaster history of any given social entity. Instead, to the extent there were case studies, they were aggregated or combined to draw more general observations or conclusions. A worthwhile side product of involving GRAs in writing case studies is that they quickly learned, as a result of finding unnecessary gaps in the information, why it was necessary in their own fieldwork to obtain the details specified in the data gathering procedures.

SOME PROBLEMATICAL ASPECTS

Although the system developed worked well, there were some

inherent problems. For one, about the time GRAs acquired substantial field experience, they graduated from the university. While a very few of the best were occasionally kept on for a year or two as full-time field directors, the center never had the funding to do this on a regular or continuous basis. So while veteran workers were overlapped with new recruits on field teams, the value of having much field experience was always eventually lost to the center. Second, it was the rare graduate student who was very good in all phases of the research process. So there was a tendency to let senior research assistants gravitate to what they were best at (e.g., interviewing or report writing). This, however, generated problems of equity in use of staff members. Third, and related to the second problem, not all persons are comfortable and good at working concurrently on several different projects at varying stages of development. Some people work much better on only one sequentially-developing project. Yet the center had the former rather than the latter style. Fourth, research assistants sometime had "overlearned" their classroom training. Thus, some thought scientific research could only be of a deductive, quantitative, and hypothesis-testing nature. Such students had difficulty operating in an inductive, qualitative, and hypothesis-generating framework. Finally, postfield data processing (not data analysis) activities are time consuming and not intrinsically interesting. GRAs tended to delay this work with a compound of memory failures. Unless there is a continuous monitoring effort, this work is often not well done, leading to a compromising of the gathered data.

IMPLICATIONS AND LESSONS FOR FUTURE RESEARCHERS

This chapter has another goal besides providing a detailed historical documentation of a major pioneering social science research effort in the disaster area. It is also to indicate the implications and lessons that the DRC effort has for future field researchers on crisis time disaster behavior. These will be briefly discussed under four general themes.

1. The kind of field research described cannot only be done, but can be done well. When DRC first started its work, other social scientists often expressed doubt that such field research could even be undertaken, or if attempted, could produce any worthwhile research results. The question asked was: How could one do research in the midst of social chaos and extreme personal stress and essentially very difficult working conditions? As late as 1976 a highly placed administrator in the U.S. National Academy of Sciences very strongly objected to a proposal that the National Research Council set up quick-response field research teams to study disasters. His objections were:

> How can the investigative team be alerted and got to the right place in time? It is difficult for me to see how you can select a site and collect the sort of data outlined . . . Officials are going to be so preoccupied with their own immediate problems that I cannot imagine their talking to researchers in advance of a known emergency. How can the monitoring system envisioned . . . actually be put in place in the face of imminent disaster?
>
> (Brooks 1976)

Perhaps because of the now known DRC work, such views are seldom expressed anymore by other researchers. Actually the problem of skepticism may exist more outside the research area, among research funders and users. This is increasingly less so in developed societies, but skepticism still occasionally surfaces in some developing countries. Possibly this chapter will give "ammunition" and professional support to those who still have to convince policy-makers and research funders that such research can be undertaken.

A statement on how well the work was done might appear self-serving. Nevertheless, DRC did obtain thousands of interviews, record innumerable observations, and collect tens of thousands of documents. More important, from those data nearly five hundred publications were produced, including 29 Ph.D. dissertations. Of even more value, the published results are a prominent part of the disaster literature, which in turn shows up in introductory sociology texts, summaries of the literature, and compilations for research users. Here too what was done can be pointed to by researchers and research users who still have to make a case that not only can field research be undertaken but that the end results can be worthwhile.

2. Certain relevant data and assessment of data gathered in other ways can only be obtained by the kind of field study described. There are several aspects to this, depending on whether interviewing, observing, or documenting is the research technique in question. For example, as previously illustrated, certain activities can be noted via observations that could probably not be recalled in later interviewing. However, even the episodic and informal interviewing that can sometimes be done during the midst of a crisis at a central site like an EOC is likely to be more candid and honest than a more formal interview later in a less hectic setting. (As one respondent remarked to us in a crisis setting, "I could tell you I know what I am doing, but you can clearly see I'm wildly guessing in much of what I'm doing.") This is apart from the typical reconstruction that occurs toward what should have happened in later interviewing away from the time of the occurrence. This is a serious problem about formal interview data that to this day has been almost completely ignored by disaster researchers, but it can be partly counterbalanced by having trained researchers present in the crisis setting. It has also always bothered us that the "decision-making" we have observed during actual crises seldom corresponds to the picture evoked in later interviews outside of the actual crisis context, where the process is often depicted as explicit, conscious, individually based and involves the consideration of alternative options. This is why we think that it is very unfortunate that too many current disaster researchers who are the ultimate analysts of data often not only get the information third-hand via first an interviewer and then a coder, but also have absolutely no direct experience in disaster occasions which would give them a larger context for interpreting the data. In part, we are paralleling here

what a famous social scientist we had in our graduate studies used to say in his introductory lecture on doing community studies:

The very first thing you must do is to walk very slowly and several times through the area and observe everything you can. Your interpretation of all the statistics you may later play with will differ depending on your observations. In any case they will certainly be more accurate if you make the walk.

Of course, many possible observations and many gatherable documents at a time of crisis are very ephemeral and if not collected at the time are lost forever. (DRC did try to get its field teams to photograph and/or tape record such material as much as possible, but for unclear reasons the admonition to do so was not consistently followed.) Future field researchers should attempt even more than did the center to collect such kind of fleeting and transitory data.

3. For this kind of field research to be done and done well requires much prior planning as well as continual monitoring of the implemented measures. The notion that research consists of an initial training and pilot period and then a moving on to the actual gathering of data and its later analysis is not an appropriate model/image of what should be in place (although this is the naive view in some guidelines on how to write research proposals or even in some formal courses for students on how to do research generally). The methodology used has to be kept in mind from the start to the end of the work. The quality and quantity of the data obtained will be mostly determined long before the first study in the field. As such, it is important that much time and much effort be explicitly allocated to teaching the methodology involved. However, it is also crucial that a continuous evaluation be undertaken of whatever is done. Even very well trained and experienced personnel will show slippages and regressions back to inappropriate behaviors and actions.

4. Finally, just because something has been done and done well does not mean that whatever the "traditional" approach has been should be blindly followed. In particular, we would strongly urge that even for the kind of disaster research just described it will be necessary to take into account whatever larger social changes

are occurring. These changes can both hinder and help future crisis time field studies that might attempt to follow or build upon what has been described in this chapter. For example, the increasing litigiousness nature of Western type societies as well as the increasing tendency of negative disaster effects to cut across national not to mention community boundaries (see Quarantelli 1996) were not major problems that DRC researchers faced. Yet that is part of the future for field students of disasters, and the methodology used will have to take such difficulties into account. However, there are also some positive aspects of the changing social context of disasters. For instance, as suggested elsewhere (Quarantelli 1995a), the Internet can now be used to directly collect "real time" data on organizational behavior and interorganizational communications during disasters. And of course, as discussed in other chapters in this volume, there are now many very useful technological tools and machines such as camcorders and laptop computers that were not available to DRC in its early years of existence. The newer technological devices ought to be incorporated into the methodology of future field studies, taking into account their negative as well as positive effects.

In the preceding pages the focus has been almost exclusively on the field research undertaken by the center. However, as noted in the initial paragraph of this chapter, many other methodologies as well as other tools besides tape recorders were concurrently used by the center. The research results reported by DRC reflected this larger effort. As such, we favor linking field research to other methodologies or incorporating newer tools in future large-scale and continuing field research. In fact, any study that merely mirrored what DRC once did would be an unnecessarily limited effort. Studies indicate that emergency managers should not plan on the basis of the last disaster they experienced, no matter how efficiently and effectively it was handled. Rather, they should plan on what reasonably can be projected in the future. The same is true of disaster field studies. No matter how well past studies were done, future researchers need to continually improve and to add whatever new or novel relevant methodologies and tools become available. If that is done, then someone later in this century should be able to write from a historical viewpoint a chapter similar to this one but yet an improvement on it.

5

FOLLOWING SOME DREAMS:

Recognizing opportunities, posing interesting questions, and implementing alternative methods[1]

Thomas E. Drabek

My career in disaster research began in September 1963, when I became the first full-time professional hired by the three cofounders of the just-created Disaster Research Center (DRC) at The Ohio State University (OSU). I was a Ph.D. candidate at the time and had spent an interesting summer working for E. L. Quarantelli as a research assistant on his study of emergent conceptions of professionalism among OSU dental students. I learned much from the DRC cofounders—E. L. Quarantelli, Russell R. Dynes, and J. Eugene Haas—but never imagined that decades later I would still be conducting disaster research. This essay has afforded me opportunity to review and reflect on the many dreams I have pursued as a researcher with special emphasis on the methods used. Always, my work has been guided by three goals: (1) to test and extend sociological theory related to human response to disaster; (2) to identify insights relevant to emergency management practitioners; and (3) to communicate the results to both the academic and practitioner communities. Unlike those who prefer using one methodological strategy for all or most of their work, my choices differed. Rather, I have chosen to pursue my dreams. That is, I first have become excited about a question and then have borrowed or invented the methods required to pursue it. In my early years, unique events—certain disasters—stimulated the initial research question. Two decades later, I posed the question and waited for appropriate events. My plea to those who continue in this research tradition is that they adopt my goals, *not* my choice of methods. The profession of emergency management needs your results, your theory, and your methodological rigor. And so do future disaster victims. Helping to prevent or ameliorate human suffering remains the promise of disaster research. As C. Wright Mills (1959) said to those of us who were born professionally in the 1960s, always remember that the potential of the sociological imagination is an increased capacity for human freedom. Start with questions that are worthy of your efforts, and then select or invent whatever methods are required.

I will return to this conclusion at the end of this essay. To get us there, I have organized my analysis into three parts: (1) a historical review of several of my studies with brief comments on the methods used; (2) an assessment of the state of disaster research with emphasis on existing defects; and (3) speculation on future directions, needs, and potentials.

LOOKING INTO THE REAR-VIEW MIRROR

Upon reflection, I outlined nine topics thereby to highlight many of the dreams I have pursued during the past 34 years. My focus will be on the methods used, but both the historical and theoretical contexts will be noted too. Hopefully, this autobiographical approach will help my peers—both young and old—better understand how certain methods can be used. I have learned much by reading similar statements prepared by others who were willing to take us behind the scenes of their research so that we could get a glimpse of them practicing their craft (e.g., Hammond 1964; Whyte 1943).

My story begins in Indianapolis, Indiana, and ends on the

Outer Banks of North Carolina, with a lot of stops in between. So let's start by returning to 1963 when the DRC was being birthed.

An unexpected case study

Hiring and orienting several research assistants, arranging office space, establishing a library of disaster research studies, and conceptualizing a code form for study abstracts were initial priorities that the three DRC codirectors set for "the workers" during September and October 1963. The last night of October—Halloween night at 11:06 p.m.—there was a massive explosion inside the coliseum at the Indiana State Fairgrounds in Indianapolis. Early the next morning, Russell Dynes telephoned me and indicated that the three codirectors had decided this event would make an excellent field training opportunity. Later that day, Dynes and I, along with two research assistants, arrived at the coliseum. A temporary morgue had been established there since an ice show was the attraction that had been disrupted. Over 400 people were injured and 81 died, making this the worst disaster in the history of the state.

Our mission was to: (1) identify which organizations were most involved in the emergency response; (2) conduct interviews with organizational directors so as to determine their role and major activities during the response; (3) ascertain the range and type of disaster planning they had completed; and (4) obtain relevant reports and memoranda. En route to Indianapolis—about a four-hour drive in those days—we designed a brief interview guide that contained a series of basic questions regarding organizational activities, resources, communication, and authority relationships.

This was my first encounter with disaster. Initially I viewed the several days of interviewing as nothing more than good training. But as we shared our experiences on the drive home, it became clear that the event and our database merited more. In addition to detailed descriptive interviews and piles of reports, we obtained copies of the audio recordings of all radio transmissions made by both fire and police departments during the emergency response.

Since citizen telephone calls to the police department were recorded routinely, too, numerous additional tapes were promised. Before these arrived, or at least before I listened to any of them, I had begun organizing the materials the team acquired and outlined a report the codirectors had asked me to prepare. A mental image of organizational response and key operational problems had emerged. But when I played the tapes, I became angry. To some extent that emotion has stayed with me all of these years. As I listened to what citizens—husbands, wives, grandparents—were being told on into the wee hours of the morning after the explosion, my feelings of disappointment intensified. As I reviewed radio interchanges regarding the whereabouts of equipment needed at the scene, I had but one thought: There has to be a better way! This was followed quickly by another: Maybe the explosion couldn't have been avoided, but too many people suffered even more because these emergency workers did not do their jobs. Or more accurately, emergency workers had been swallowed up in a system that failed them despite their best individual efforts. As would be repeated hundreds of times in subsequent years, all of us had interviewed or learned about heroic actions taken by some who were there. Yet, their individual actions were neutralized by a strained system that was minimally integrated.

A little over one month after the explosion, Indianapolis was rocked again. This time, however, it was not leaking propane gas but rather indictments that followed a Grand Jury investigation. The legal system was focused on a search for "the guilty." Yes, the Indianapolis Fire Department, like the State Fire Marshall, did have responsibility for inspections. An existing ordinance did prohibit the gas company that had delivered several tanks containing liquid petroleum gas from doing so, as it did the coliseum management from storing them. None of this activity, however, brought public focus to fundamental defects in the emergency response, and I doubted whether future inspections would be improved by legal actions that were limited to sanctioning so-called guilty individuals. The systems needed fixing, and such actions just distracted from the real problem. Even worse, by learning that a few individuals had been punished, the public was misled into thinking that corrective action had been taken. Henry Quarantelli and I discussed this matter in detail and later decided to document our observations and interpretations (Drabek and Quarantelli 1967).

One year after the explosion, I made a return trip to Indianapolis with three research assistants to determine what changes had been made beyond the search for the guilty. I was pleased that we could document many. These were described in the fourth chapter of the emerging case study I was drafting. After the manuscript was completed, following extensive rewrites—some for clarity and some for tact—publication was delayed. In part, the delay was due to the DRC establishing a monograph series in which *Disaster In Aisle 13* (Drabek 1968) would be the first, but the several year hold-up also reflected sponsor concerns. Fortunately, the three codirectors took a strong stand regarding the autonomy of university-based research, and the monograph was published as written.

In an essay prepared for a volume honoring E. L. Quarantelli (Drabek 1994c: 32), I documented how this case study continued to impact the institutional memory of Indianapolis for years to come. For example, the opening paragraph in the preface of a plan prepared in 1989, over two decades after *Aisle 13* was published, contains a quotation from this case study. Former local directors of emergency management for the community have indicated to me that this descriptive analysis was still used in their training as a point of reference to help others better learn why disaster response requires a community system perspective if it is to be effective. Clearly, the case study, despite its limitations, remains a powerful method that future researchers studying aspects of human responses to disaster should use.

Laboratory simulations

While drafting the Indianapolis case study, I was assisting J. Eugene Haas collect ideas from professors in psychology and political science regarding the requirements for a new small-group laboratory. One of the original research objectives of the DRC was to conduct laboratory studies of disaster-related phenomena. In large part because of the hours I had spent working with materials, interviews, and audio recordings from the police and fire departments in Indianapolis, another dream began to emerge. Instead of bringing groups of students into the laboratory and then asking them to perform various types of stress-inducing activities, why not bring in a real police department? Obviously, the whole department wouldn't fit very well, but then the whole department never is in one room, anyway. After considering several options and the physical constraints of the soon-to-be-operational laboratory, I proposed to the codirectors that we simulate a police communication system under stress. The system to be stressed would be composed of the actual police officers who worked daily as the communication dispatch unit—a four-person team which functions as the linkage between the citizen with a complaint or problem and the cruiser operator who responds. Such a "realistic simulation" (Drabek and Haas 1967) could be developed so that the patterns of decision-making, communication, and authority, like other aspects of their performance structure, could be observed and recorded. Once baseline measures were obtained, say with three two-hour sessions of routine police work, information about a major disaster could be given to the team. How would their performance structure change under that type of stress?

With the help of five research assistants who were trained to talk police radio "lingo," a dozen or so other DRC staff ranging from secretaries and undergraduate assistants to research assistants, and a skilled technical crew, three shifts of the Columbus, Ohio, police department communication unit (four officers each) were introduced to a simulated work environment. Aside from being new and clean, in contrast to their dingy room in the downtown headquarters building, the laboratory equipment resembled their daily work setting. More importantly, it functioned identically. That is, they sat at their respective stations, processed calls from simulated citizens, and dispatched police cruisers. After a prearranged time period, based on a protocol I constructed using their annual statistics, another simulator representing a cruiser operator would report back regarding outcome (see Drabek 1969a and Drabek and Haas 1969 for details). My several-week field study within the Columbus Police Department communication unit, the tapes from the Indianapolis Police Department following the coliseum explosion, and the commitment of the DRC staff, who really got motivated to "sock it to the police," permitted the construction of this simulation. It functioned so realistically that we had some trouble convincing the participants that the whole thing was an illusion. Several indicated that they believed we had switched telephone lines so that they were responding to actual police calls while in the laboratory. For ethical reasons, they were told that all calls and cruiser responses were simulated. They were not told, however, that after the three sessions conducted during a two-week period with each of the three teams a disaster would be simulated. Since the laboratory experience was viewed by their superiors as worthwhile training, this deception was defined as ethically acceptable. If I were to use this method in the future, I would forewarn participants that unusual events might occur during the simulation since I do not believe their behavior would differ.

Although the DRC continued to use laboratory simulations after my departure in 1965, they have not become popular. While local officials routinely engage in highly realistic disaster simulations at the Emergency Management Institute (EMI) in Emmitsburg, Maryland (EMI is a component of the Federal Emergency Management Agency [FEMA]), these are defined as training experiences. No program of research has been developed to evaluate or capitalize on these simulations. Community-based disaster exercises probably occur at least weekly somewhere within the U.S., but researchers have not viewed these field experiments as data collection opportunities, either. That should change in the future.

A victim survey

During June 1965, my wife, daughter, and I left OSU for a quick trip to Colorado to participate in my brother's wedding and to inspect rental houses owned by the University of Denver where I had accepted an assistant professorship for the fall. The laboratory simulation study served as my doctoral dissertation and was nearly complete so as to permit August graduation. About three-fourths of the way back to Columbus, Ohio—in Effingham, Illinois, to be exact—we learned that Denver and several other communities had been flooded severely. This was the most costly flood in the history of the state. Upon our return to the DRC, I was advised that the organizational warning responses to this flooding required documentation. My wife's father had just had serious surgery in Colorado, so we decided to fly back the next day. I did extensive interviewing for a couple of weeks. The database was expanded further by two other DRC staff. Henry Quarantelli suggested that as a new Ph.D. I might secure a small grant from the National Institutes of Mental Health (NIMH) if I identified a relevant topic. "Reactions to Sudden, Unfamiliar and Unexpected Stress"—the title of the NIMH grant proposal—provided the opportunity to complement the organizational database. This randomly-selected victim survey built upon and extended the theoretical insights of Janis (1954), Moore et al. (1963), Williams (1964), Blumer (1966), and others who tried to ascertain the emergent interpretative frameworks that guided people facing the uncertainties of disaster warning messages. Results from this survey were published over the next several years (Drabek 1969b; Drabek and Boggs 1968; Drabek and Stephenson 1971; Drabek 1983a). These writings are testimonials to the dream that had emerged while I was doing the organizational interviews for the DRC.

A quasi-experimental design

In 1967, William H. Key joined the University of Denver faculty as chair of the sociology department. During a lull in student registration while he and I were sitting in the field house, he began describing a series of surveys he and two colleagues—Louis A. Zurcher and James B. Taylor—had completed prior to the 1966 tornado that devastated portions of Topeka, Kansas. The three had been staff members at Topeka's Menninger Foundation—a famed psychiatric treatment and research institution—when the tornado hit. They were involved in preparing a case study of the event with special focus on emergent groups and institutions (Taylor, Zurcher, and Key 1970). My excitement intensified as Bill and I drew out a quasi-experimental research design wherein the earlier survey data could be contrasted with follow-up interviews. Many of those surveyed had been impacted by the tornado, but most had not. Here was a unique opportunity to examine samples of victims and nonvictims and to compare their response profiles both prior to and after the tornado. To my knowledge, this remains the only published study of long-term disaster impacts on family functioning with pre- and postevent data that permitted comparisons of a victim and a control sample (Drabek and Key 1976, 1984). As we pursued this dream, I became much more aware of the complexities involved in both internal (i.e., Were observed differences due only to the tornado?) and external (i.e., To what could we generalize?) validity.

Constructing social maps

By 1976 I had begun to explore new vistas since I never planned to limit my research or writing to disasters. I was chairing the sociology department at the University of Denver which had expanded greatly and had initiated a doctoral program. Work with Gresham Sykes had culminated in a humanistic criminology reader (Sykes and Drabek 1969); J. Eugene Haas and I had finished two textbooks on complex organizations (Haas and Drabek 1973; Drabek and Haas 1974). And the massive database that Key and I assembled needed to be analyzed and written up.

On the last day of July 1976, 139 people died in the Big Thompson River Canyon about 60 miles, as the crow flies, to the northwest of Denver. I was invited by the state department of emergency services to be an observer for a few days so as to participate in an anticipated critique session. This experience—both the failed warning system and the interagency coordination difficulties—stimulated the preparation of a new research proposal. I wanted to document in detail the emergent interagency relationships that would evolve as search and rescue activities were initiated in several future disasters (Drabek et al. 1981; Drabek 1985).

Following much experimentation with datasets created after such events as the 1979 Wichita Falls, Texas, tornado, Hurricane Frederic (1979, Mississippi and Alabama), and the eruption of Mount St. Helens (1980), three things became clear to me. First, the measures and analysis techniques available were woefully inadequate to assess the complex social phenomena we were glimpsing in the field and being told about by those we interviewed. Fortunately, later researchers like David Gillespie and his associates (1993) have pressed this matter much further. Extensions of their work, which built on ours, is clearly a future need. Second, the emergent multiorganizational networks (EMONS) we were trying to assess were the very phenomena that disaster services personnel were trying to manage. Unfortunately, they were rooted in basic theoretical principles of bureaucracy that simply did not give them the perspective required. Too often trained in a military command and control ideology, key emergent qualities of the response often were perceived as problem areas—if they were perceived at all. Dynes (1983, 1994) has elaborated on this theme in several important statements. Third, and finally, even crude social maps of the communication patterns and decision-making structures of such EMONS could be used both to document the behavioral reality that actually occurred after disaster—as opposed to what was supposed to occur—and to help disaster responders better understand and conceptualize the types of systems they were trying to manage. This perspective guided subsequent data collection on the interorganizational linkages that formed regarding earthquake mitigation policy in Missouri and Washington state during the decade of the 1970s (Drabek et al.

structured by a conceptual framework developed by Pennings (1981) who had expanded on Thompson's (1967) analysis of managerial response to environmental uncertainty. In his study of civil defense offices during the 1960s, Anderson (1969) applied Thompson's framework and concluded: "In order to remain viable, organizations must learn to cope with uncertainty. That is, they must establish strategies which enable them to reduce instability and indefiniteness in their internal structures and environments..." (Anderson 1969: 5).

In addition to the local emergency manager, one-hour interviews were conducted with the head or designated representative of seven types of "contact agencies" (law enforcement, fire, elected officials, etc.). These individuals provided much insight into the types of strategies their local emergency manager had used to build and maintain an effective program. Through these interviews, I documented 15 major strategies used by successful emergency managers to maintain the integrity of their programs (Drabek 1990, 1987b). The widespread applicability of these strategies, and their differential use patterns among managers in communities of different sizes, was verified through a telephone interview survey based on a multistage, randomly-selected sample of local emergency managers. Numerous training seminars have been conducted wherein these strategies and the illustrations provided by those interviewed have been used.

Comparative case studies of technology implementation

During the interviews with local emergency managers regarding the strategies used to maintain organizational integrity, I became aware that many were encountering difficulties implementing microcomputer technology. I decided to design a comparative organizational study with which to document this implementation process (Drabek 1991). By interviewing in several agencies—both local—and state-level offices—at two different points in time, spaced about 18 months apart, I was able to obtain information about internal and external organizational changes related to the adoption of this technology. In addition, by selecting some agencies that experienced disasters after they had adopted microcomputers, I documented the types of uses, disappointments, and lessons learned. By circumstance, two of the state agencies and five of the locals that I interviewed initially were impacted by a major disaster. They welcomed me back for follow-up data collection.

While small in number, this comparative case-study design permitted exploration of this unique topic, one heretofore unstudied. In total, five state agencies participated, but two of these were assessed at two different points in time. One—South Carolina—was reviewed after Hurricane Hugo, but not before that event. Twelve local emergency management agencies were studied during the first phase of the project, and I returned to four of these for postdisaster follow-up. Another four that had implemented microcomputers prior to a major disaster were studied also. As Yin (1984) has emphasized, such comparative case studies can yield important insights when heretofore unexplored topics are investigated.

Penetrating the tourist industry

One of the many computer applications that I discovered through the implementation study pertained to disaster evacuation. While much had been learned prior to and after my assessment of the evacuation stimulated by the 1965 flood in Denver, many voids remained. Tourists and other types of transients, for example, had been identified as problems for emergency managers, yet little data had been collected (Sorensen et al. 1987). Through field and telephone interviews, three legs of a stool are being constructed that will provide an important empirical basis for future training and policy review.

First, I interviewed tourist business owners and managers (Drabek 1994b). Initially, 65 were selected from three field sites. The three locations (Pinellas County, Florida; Sevier County, Tennessee; City of Galveston, Texas) were picked because local governments had initiated outreach programs designed to stimulate disaster planning within tourist businesses. Local emergency managers assisted me in securing a diverse sample of tourist businesses that were scattered throughout their respective counties. Through interviews with owners or managers of these businesses, I documented a baseline regarding the history, scope, and type of disaster planning completed. Then six other sites were selected because of extensive flooding (two counties in Washington state) and Hurricane Bob which triggered extensive evacuations along the East Coast as it drifted northward (Carteret and Dare County, North Carolina; York County, Maine; and Cape Cod and Martha's Vineyard, Massachusetts). Again I turned to local emergency managers for liaison assistance, although I selected many of the businesses myself so as to insure diversity (N = 120). In these locations, selected because disaster struck during the project time window, linkages between the tourist industry and emergency managers varied considerably as did the quality of emergency planning.

Second, I explored the customer leg of this stool by identifying tourists (N = 520) and business travelers (N = 83) who were in the areas impacted by five major disasters: (1) Hurricane Bob (1991); (2) Big Bear Lake area earthquakes (California, 1992); (3) Hurricane Andrew (Florida and Louisiana, 1992); (4) Hurricane Iniki (Hawaii, 1992) and (5) the Northridge earthquake (California, 1994) (Drabek 1996a). Small samples of migrant workers (N = 34) and homeless persons (N = 45) were included, too, for comparative purposes. Face-to-face field interviews were conducted with these last two samples, whereas the tourists and business travelers were contacted by telephone. These interviews averaged over 30 minutes each, although many exceeded one hour. Some lodging establishments mailed a letter on my behalf to customers who could indicate a willingness to be interviewed by returning a simple reply card. This strategy was supplemented by locating and using numerous visitor lists from a wide variety of entertainment, cultural, and tourist information facilities. Since neither the challenge nor procedures like those I devised had ever been addressed in the literature, I prepared a detailed accounting (Drabek 1996a: 343-354).

Third, and finally, I have initiated study of the third component of the problem identified years ago—assessment of employee responses. Hurricane Felix took me back to the Outer Banks of North Carolina in 1995. This time, however, the range of organizations for study has been broadened far beyond tourist businesses. Hurricane Fran (1996) will serve as the second event, and I hope to complete the database soon through interviews with employees who evacuated because of other types of threats. Collectively, this database will include nearly 1,500 interviews with managers, customers, and employees. Because the events and communities are varied, comparative analyses will be possible that will address many issues of external validity. Hence, the limits of generalization regarding various forms of evacuation behavior can be specified better than heretofore has been possible. This database already has stimulated some emergency managers (Drabek 1994a) and tourist industry executives (Drabek 1995) to review their evacuation procedures and policies. Given the catastrophic vulnerability that is worsening daily as the tourist industry expands worldwide, the call for action must not go unanswered.

THE STATE OF DISASTER RESEARCH

During the past three decades, many have built on the insights and criticisms offered by those who have reviewed the methodological challenges presented by disasters (e.g., Cisin and Clark 1962; Drabek 1970; Mileti 1987). Within the context of the survey of methods and research goals outlined above, I offer the following observations. While many of these judgments parallel those I formulated over a decade ago (Drabek 1983b), some important changes have occurred.

More is known than has been implemented. This remains true, but the rapid professionalization of emergency management is reducing the gap. This process has accelerated rapidly during the past decade because of expanded and improved training conducted provided by future comparative studies wherein theoretical criteria, not solely randomization, will guide sample selection. Ultimately, the methodological issues of external validity press us to confront fundamental theoretical matters (e.g., What is a disaster?; Quarantelli 1987a, 1995b, 1998). As more diverse perspectives and orientations push the field in different directions, we can anticipate that alternative methods will become more pronounced.

Kreps and his associates (1994) have used hundreds of descriptive disaster interviews completed by DRC staff over a couple of decades to identify basic elements of emergent social organization. Following disaster, important aspects of human responses do reflect domains, tasks, resources, and activities. The four structural codes (i.e., D-R-A-T) identify a taxonomy of emergent social systems. The broad outlines of a formal theory of disaster response has been constructed by this innovative research team who have pressed both new theoretical integrations and appropriate variable measurements into realms heretofore unexplored. This approach to fundamental problems of external validity are exactly the types of recasting and redefinition the field sorely needs.

Increased practitioner-researcher interaction will redefine topical priorities. More so than at any other time, conferences like the annual workshop hosted by the Natural Hazards Research and Applications Information Center at the University of Colorado are stimulating new conversations that tighten the linkages between researchers and practitioners. As research needs are articulated by more professional emergency managers, the research community will be pressed to respond. Old "truths" about disaster myths such as looting, for example, require revisitation and refinement. And just because 14 studies in small Midwest towns hit by tornadoes or floods reported minimal looting, this does not mean that the conclusion can be generalized elsewhere in an uncritical way. New urban centers like Miami, Florida, and Los Angeles, California, may represent very different forms of social constraint. Obviously, as with many of the issues I have raised, this matter is not solely a methodological problem. But it illustrates one edge of this field. Much of what we think we know may not apply to certain threats like terrorism or social systems that have been birthed only recently. As Quarantelli (1993: 32-33) has emphasized, new types of threats are emerging that singly, or in concert with traditional natural disasters, expand the relevance of disaster studies and their complexity, too. And recalling a favorite anecdote from Benjamin Franklin, who was asked about the utility of flying kites in a lightning storm, Quarantelli cautions us about the potential risks of increased researcher-practitioner interaction: "But he [Franklin] posed a question of his own on return: who has saved more lives in the long run, the carpenters who had built better life boats or the astronomers who had plotted the stars, whose knowledge contributed in time to better ship navigation?" (Quarantelli 1993: 16).

All aspects of the disaster life-cycle, and what we think we know about it, must be qualified by the limited and narrow range of societies studied. Starting with an exchange with scholars from Japan (1972), the staff at the Disaster Research Center has stimulated parallel exchanges with researchers in Italy (1986), Central America (1993), and the former Soviet Union (1993). Clearly, literature syntheses like the one prepared by Britton and Clapham (1991) on the Australian experience are essential aids. So, too, are the several journals that have emerged with an international focus (*International Journal of Mass Emergencies and Disasters, Disasters, Disaster Management, Industrial and Environmental Crisis Quarterly, The Australian Journal of Emergency Management*, and the like). While much progress has been made, the overall state of disaster research today reflects the limitations inherent in most studies that have been limited to data collected from citizens of a single nation state. Cross-national databases remain a dream for the future.

FUTURE DIRECTIONS, NEEDS, AND POTENTIALS

As we begin the twenty-first century, disaster researchers will move in many different directions as they apply old methods and can shift from qualitative description to more precise quantitative assessments.

Disaster life-cycle interdependencies

It is a truism to say that disaster recovery may stimulate mitigation activities. But if the disaster life-cycle is divided into even four basic components, the elements of interdependency remain illusive. What dimensions of human conduct require measurement so that this complex of interdependencies can be penetrated? Several long-term historical case studies would appear to be good places to start. Clearly, the conceptualization of the organizational or community system-level processes that are operative remain undefined. Once identified, the conceptualization and measurement tasks could be clarified. Comparative community databases could then be constructed so that the analysis could identify and test alternative theoretical models of this system dynamic. Armed with historical data from two or three hundred communities, a future researcher could produce social models of the life-cycle interdependencies that we now only sense in the vaguest of ways. This research could be conceptualized within the type of "sustainability" perspective recommended by Mileti et al. (1995). And, following their advice (pp. 122-123), it should be placed within a global context also, although that does not preclude applications that are specific to individual nation-states.

Disasters and other social problems

Kreps and I (1996) have advanced an argument that disasters are nonroutine social problems. Without getting into the specifics of that matter, I suspect that some future researchers will apply Stallings' social construction of the earthquake threat to other hazards. Such work would be informative. But the agenda I envision goes beyond the constraints I see within the social constructionist dogma. Disaster events, like the differential distribution of risk within a society, must be placed within the ongoing social life of a community. While I disagree with Erikson (1994) that toxicity must be regarded as the fundamental criterion in a taxonomy of disasters, I do believe that more research on human-caused events will redefine our perspective. One aspect of that redefinition will be to understand better why disasters are relevant to *and different from* other social problems. As these juxtapositions are made, disaster researchers will discover existing voids in their study agenda. As this interface expands, disaster research will both inform and be informed by social problems theory.

Cross-hazards databases

Archival work by Kreps and his associates (1994) illustrates the power of a cross-hazard database. The questions raised by Erikson (1994) regarding the unique qualities of toxic substances in their many forms merit focus. While I believe this criterion, or some social as opposed to physical definition of it, may emerge as a key taxonomic element in our future disaster classification schemes, it is not a panacea. But the broad policy issues related to toxic materials, like assessments of possible psychological impacts on those exposed, require methodological innovations that will produce findings relevant to the needs of both practitioners and theoreticians. Most pressing in the years ahead are methodological innovations that will address this problem of external validity in ways that will advance and inform the theoretical structure of the field. Alternatively, emergency managers who are trying to implement *comprehensive* disaster preparedness and mitigation programs need assistance. Currently, they turn to us for help but are left in a void with research findings that may or may not be at all relevant to their local situations. While I am pleased to find many consulting what some call "the disaster encyclopedia" (Drabek 1986), some of the misapplications reported to me are horrifying. Through rigorous analyses of future cross-hazards databases, we must do a better job.

CONCLUSION

So what can be said about research methods for the future? What advice can one offer after over three decades of conducting research? I will be succinct and limit my answer to three key ideas. First, always follow your dreams. Above all else, start with an interesting question. Second, select or invent the methods required to pursue the question posed. Do not limit your inquiry because the methods are not available or are less rigorous than desired. Third, and most important, always keep in mind the fundamental promise of a sociological imagination. That is, through helping others better understand the social factors that constrain their capacity for freedom, we can have a liberating impact. While difficult, disaster researchers can and, in my opinion, must pursue dual goals as they practice their craft. New theory must be created and old notions tested and revised. But insights for practitioners also must be produced as we join other disciplines in accelerating the professionalization of emergency management. In this way we fulfill the real promise of disaster research—to prevent or ameliorate human suffering.

NOTE

[1] I would like to thank Ruth Ann Drabek for her work on this chapter. Also I thank Bob Stallings who asked me to write it and offered suggestions as to focus, approach, and content. Partial support for the preparation of this chapter was provided by NSF Grant Number CMS-9415959. Any opinions, findings, conclusions, or recommendations expressed in this paper are those of the author and do not necessarily reflect the views of the National Science Foundation.

PART II
CONTINUITIES

The six chapters making up this section reflect continuities in disaster research, methods of long-standing use that will continue to be used in the future. Linda Bourque, Kimberly Shoaf, and Loc Nguyen (Chapter 6) use the results from six postearthquake surveys in California to refute several alleged weaknesses of survey methods in disaster research. Brenda Phillips (Chapter 7) notes recent developments in the use of qualitative methods in the social sciences and shows how these methods will contribute to future studies of disaster. Anthony Yezer (Chapter 8) organizes past research on the economic aspects of disasters under three major topics, describes the models and variables that have been used within each, and shows how standard economic models and their assumptions should be modified in future research. Walter Gillis Peacock (Chapter 9) distinguishes cross-national from comparative research, then identifies both constraints on and prospects for cross-national studies of disasters. Marco Lombardi (Chapter 10) sketches a model of the interactions surrounding media organizations and identifies variables that need to be measured in research conducted from a constructionist perspective. T. Joseph Scanlon (Chapter 11) invites the reader inside his quest for the full story surrounding the 1917 munitions ship explosion in Halifax, Nova Scotia, sharing his experiences in tracking down materials and offering numerous suggestions for those who would engage in future historical studies of disasters.

6

SURVEY RESEARCH[1]

Linda B. Bourque, Kimberley I. Shoaf, and Loc H. Nguyen

Surveys provide a highly viable and excellent source of data about behavior during and after disasters, behavioral and attitudinal responses to disasters, and anticipatory behavior and attitudes about future disasters. Although surveys have increasingly been conducted by disaster researchers (e.g., Mileti and O'Brien 1992; Palm et al. 1990; Bolin and Bolton 1986; Tierney et al. 1996), the traditional approach to disaster research emphasized quick entry into a disaster site where selected informants are interviewed using semi-structured interviews. Disaster researchers' reluctance to recognize the strengths and to advocate the use of well-designed, standardized, population-based surveys reflects both realistic and unrealistic barriers to their use. Some of these barriers are grounded in historical reality, but the availability of new, technologically sophisticated methods for conducting surveys make many of these historical barriers obsolete.

This paper examines the kinds of information that can be obtained from well-designed, standardized, population-based surveys. In so doing, we will demonstrate that some of the issues, which in the past have been considered barriers to the use of surveys, are, in fact, sources of important insights into community responses to disasters and are even among the advantages of surveys. The following aspects of surveys will be examined:

1. The use of standardized surveys to compare community behavior across time, events, and locations.
2. The extent to which surveys represent the population of interest in the aftermath of a disaster.
3. The receptivity of respondents to being interviewed after a disaster.
4. The ability to utilize telephones for interviews after a disaster.
5. The extent to which the data collected in a survey are perishable and subject to memory decay.
6. The use of surveys as quasi-experimental designs for obtaining information on "control groups."
7. The use of surveys as a source of baseline or denominator data for ascertaining what other, more specialized datasets represent.
8. The maintenance of verbal data collected within the context of a survey for later post-coding and analysis.
9. The storage of surveys in archives for use in secondary analyses by other researchers.

In examining the use of surveys as a source of information about disasters, we provide examples from five surveys that we have conducted following the Whittier Narrows, Loma Prieta, and Northridge earthquakes in California; from a survey that Gatz (1996) conducted following the Northridge earthquake; and from the surveys conducted by Ralph Turner and associates in the late 1970s (Turner et al. 1986).

DEFINITION OF SURVEY RESEARCH

Early disaster researchers differentiated between collecting data by questionnaire and by interview (Killian 1956 [Ed.: This is Chapter 3 in this volume]; Cisin and Clark 1962). Questionnaires strictly referred to collecting data by having respondents fill out a structured questionnaire; the method by which the respondent obtained the questionnaire—whether by mail or distribution—

was often unspecified. The use of questionnaires, as defined above, was largely discouraged. Interviews, in contrast, were a recommended way of collecting data, but the interview schedule was recommended to be (and often was) unstructured or semistructured, and the persons interviewed were as frequently selected to be key informants as to be representative of any larger population.

Over the years, the terms "survey research," "sample surveys," and "surveys" have been used interchangeably with the term "questionnaires." This interchangeable use of terms implies that, if a survey is being done, the data are being collected using a questionnaire. In fact, the interchangeable use of these terms is technically incorrect. "Surveys" refer to the units from which data are collected and how those units are selected for study. The assumption is that the population or sample from which data are collected are heterogeneous in their characteristics, behaviors, experiences, and attitudes and that the task of the survey researcher is to adequately and accurately describe that diversity and how these diverse characteristics, behaviors, and attitudes of individuals, families, or groups may or may not be associated. Data in a survey are often collected with a questionnaire, but the questionnaire can be *administered* by mail, supervised self-administration, telephone interview, or face-to-face interview (Bourque and Clark 1992; Bourque and Fielder 1995).

The use of a questionnaire, however, is not necessary for a survey. Data can also be collected from records. We can, for example, take the population of all claims made to the Federal Emergency Management Agency (FEMA) and select a systematic or probabilistic sample of their records and, thereby, "survey" the characteristics of the applicants. In so doing, we never contact an applicant directly for information by mail, by telephone, or in person using a questionnaire.

For the remainder of this chapter, we use the term "survey research" to refer to selecting a sample of respondents—whether individuals, families, businesses, or institutions—in such a way that the sample represents an underlying population. In some instances, all units in the population will be targeted for study; in other instances—probably in most instances—a representative sample will be drawn of the population, generally using probabilistic sampling procedures. For purposes of this discussion, the data collection instrument will be a structured questionnaire, not a record. The questionnaire can be administered either through the mail (e.g., Mileti and O'Brien 1992), by telephone (e.g., Bourque et al. 1993), or by face-to-face interview (e.g., Bolin and Bolton 1986).

CONTEMPORARY SURVEY METHODS

Over the last twenty years, the methods of conducting surveys have been revolutionized. In the 1970s the majority of population surveys were administered using either face-to-face interviews or mail questionnaires, but by 1990 95 percent of U.S. households had telephones (U. S. Census 1990). This close to universal saturation combined with the reduced cost of telephone interviewing, respondents' reluctance to admit interviewers into their homes, reduced English literacy, and the increased availability of random digit dialing (RDD) and computer-assisted telephone interviewing (CATI) systems have resulted in telephone interviewing largely replacing face-to-face interviewing and mail questionnaires as the administrative procedure of choice. While telephones continue to be somewhat differentially distributed across the population with saturation being particularly high in moderate-sized, middle-class households in Metropolitan Statistical Areas (MSAs) outside the South, research conducted to date concludes that studies which utilize RDD procedures result in samples of the universe of households that are as representative as other methods (Groves and Kahn 1979; Aday 1996).

Over the last decade, computer-assisted telephone interviewing (CATI) procedures have increasingly replaced traditional paper-and-pencil methods for conducting telephone surveys. While only 3 of 30 academically-based survey research centers (10 percent) had CATI systems in 1979 (Spaeth 1990), by 1993 21 of 25 such centers (84 percent) used CATI systems for some portion of the studies they conducted (Bourque and Becerra 1993). While more expensive in initial setup, once a questionnaire is entered into a CATI system, a population-based survey can be fielded within a matter of days, sample parameters can be incorporated into the program, and error-free analytical files are available within days of the completion of the data collection. The result is better quality data collected closer in time to the index disaster when respondents are being interviewed about a past disaster.

There is one problem with CATI procedures that must be recognized and has yet to be overcome. Response rates in telephone interviews are determined by and highly correlated with the number and pattern of "callbacks" made (Aday 1996). "Callbacks" refer to the number of times that an interviewer attempts to call a number, first, to determine whether it is a household and thus eligible for inclusion and, second, to complete an interview with the designated respondent within a household. It is well-known that callbacks result in completed interviews more quickly when the *pattern* of attempts is varied across time of day and day of the week. Unfortunately, to date it is easier to vary callbacks productively when interviews are conducted using paper-and-pencil techniques rather than the computer. Setting up queues for callbacks on the computer takes substantial programming and thus increases the relative cost of a callback compared with data collected using paper and pencil.

The Institute for Social Science Research at UCLA was among the first university-based survey research centers to have a CATI system (in 1977), and Turner's 1986 study of earthquake predictions was one of the first studies to be partially conducted on a CATI system. To our knowledge, since that time we are the only research group which has exploited the advantages of using CATI systems to collect data following a disaster. In collaboration with a consortium of researchers from UCLA and the County of Los Angeles, three waves of data were collected using random digit dialing (RDD) and CATI procedures following the Northridge earthquake of January 17, 1994. Earlier we combined traditional paper-and-pencil telephone interviewing data collection techniques with RDD sampling techniques following the Whittier Narrows and Loma Prieta earthquakes (Goltz et al. 1992; Russell et al. 1995; Bourque et al. 1993). Farley and others used a CATI system in interviewing Midwestern residents about the Iben Browning prediction that there would be an earthquake on the New Madrid fault (Farley et al. 1993).

METHODOLOGY

Datasets referenced for examples

The examples used in the remainder of this paper draw upon six datasets: a survey conducted after the Whittier Narrows earthquake of October 1, 1987; a survey conducted after the Loma Prieta earthquake of October 17, 1989; and four surveys conducted after the Northridge earthquake of January 17, 1994. All six surveys were conducted by telephone using a standardized questionnaire. Three of the six surveys (Whittier Narrows, Loma Prieta, and one Northridge survey) were conducted using traditional paper-and-pencil procedures; the remaining three surveys, all following the Northridge earthquake, were conducted on a CATI system. Five of the six surveys were conducted by the Survey Research Center in the Institute for Social Science Research at UCLA, and samples for these same five studies were selected using random digit dialing procedures and were designed to represent households in the California counties from which they were drawn. The sixth survey was conducted by personnel at the Andrus Center for Gerontology at the University of Southern California (USC) and utilized an ongoing three-generational panel sample (Gatz 1996).

Questionnaires

As part of his study of community awareness and responses to earthquake predictions in the 1970s, Turner developed a questionnaire for administration in Los Angeles County should a substantial earthquake occur in the area during the course of the disproportional sampling was intentionally done, the sample must be weighted when estimates of population parameters are desired.

Table 1. Comparison of Response Rates in Earthquake Studies to Response Rates in the Los Angeles County Social Surveys, 1993-1996[a].

	Los Angeles County Social Surveys				Earthquake Studies						
					Whittier Narrows		Loma Prieta		Northridge		
	1993	1994	1995	1996	Hi	Low	5 Cty	SF	SC	W1	W2	W3
Telephone Numbers Generated	3800	3300	2500	3100	639	3790	1100	270	270	2100	500	6400
Total Screened and Ineligible	1436	1459	1167	1423	385	2600	445	152	106	1005	257	3270
Total Useable	2364	1841	1333	1677	254	1190	655	118	164	1095	243	3130
Refusals, Language Barriers, Incapable	1083	597	512	633	48	378	154	20	31	336	59	1101
Status Undetermined	295	387	226	316	15	313	50	15	11	253	88	782
Completed Interviews	986	857	595	706	191	499	451	83	122	506	96	1247
Response Rates (%)	42-48	47-59	45-54	42-52	78-80	42-57	69-74	70-81		74-80	46-60	40-59	40-53

[a] All of the Los Angeles County Social Surveys (LACSS) were conducted on Computer Assisted Telephone Interviewing (CATI). The Northridge survey was conducted on CATI; Whittier Narrows and Loma Prieta were conducted using pencil and paper.

Loma Prieta Sample

Between April 29, 1990, and August 1, 1990, interviews were conducted in both English and Spanish with 656 residents of San Francisco, Alameda, Santa Cruz, Santa Clara, and San Mateo counties. Modified RDD using a prescreened sample of numbers from Survey Sampling was used to obtain a representative sample, with intentional oversampling of predesignated communities in which the Modified Mercalli Intensity (MMI) equaled 8 or 9 (the northwest edge of the San Francisco peninsula, Oakland, and the Boulder Creek-Santa Cruz-Watsonville area). Within contacted residences, all persons over age 18 who resided in the household on the day of the earthquake were enumerated, and one resident was randomly selected for interview using the Kish procedure (Kish 1965).

Within the high-impact areas, interviews were conducted with 83 of the 118 eligible households in the San Francisco-Oakland area and with 122 of the 164 eligible households in the Boulder Creek-Santa Cruz-Watsonville area. In the remainder of the five counties, interviews were conducted with 451 of the 655 eligible households. A minimum of seven callbacks were made when answering machines were obtained and nine callbacks when no answer was obtained. When a respondent within a household had been designated for interview, as many as 12 callbacks occurred to obtain the interview. Response rates were between 70 and 81 percent in San Francisco-Oakland, between 74 and 80 percent in Boulder Creek-Santa Cruz-Watsonville, and between 69 and 74 percent in the rest of the five-county area. In San Francisco-Oakland, interviews were conducted an average of 217 days after the earthquake, and five percent were conducted in Spanish. In Boulder Creek-Santa Cruz-Watsonville, the figures were 223 days and two percent, and, in the rest of the five-county area, they were 226 days and one percent. Because disproportional sampling was intentionally done, the sample is weighted in these analyses for purposes of making population comparisons.

Northridge Samples

After the Northridge earthquake on January 17, 1994, telephone interviews were conducted with three different probability samples of respondents and with three generations of an ongoing, systematic, panel sample of families first studied in 1971.

Wave 1, Probability Sample. Between August 10 and December 6, 1994, interviews were conducted in both English and Spanish with 487 residents of Los Angeles County and 19 residents of 11 zip codes in Ventura County. Only residents of Los Angeles County are included in these analyses. Modified RDD using a prescreened, list-assisted sample of numbers from Genesys was used to obtain a representative sample of Los Angeles and the designated areas of Ventura county. Strata were not created, and no areas were oversampled. Thus, no weights are used with this sample. Within contacted residences, all persons over age 18 who resided in the household on the day of the earthquake were enumerated, and one resident was selected for interview using either the next-birthday method or the Kish procedure within a split ballot experiment.

Interviews were conducted with 506 of the 842 eligible households identified. No-contact cases or those where no one ever answered the phone were called a minimum of 12 times; callback cases, those determined to be households and in which a respondent was designated, were called back up to three times. Assuming alternately that all or none of the 253 numbers of unknown status contained eligible respondents, a response rate of from 46 to 60 percent was obtained. Interviews averaged 48 minutes in length and were conducted an average of 245 days after the earthquake; 18 percent were conducted in Spanish.

Wave 2. Between August 2, 1995, and October 22, 1995, interviews were conducted in both English and Spanish with 96 residents of Los Angeles County. The process by which households and respondents within households were selected was identical to that used in collecting Wave 1 data. Interviews were conducted with 96 of the 155 eligible households identified. Assuming alternately that all or none of the 88 numbers of unknown status contained eligible respondents, a response rate of 40 to 59 percent was obtained. Since the questionnaire used in this interview was reduced to focus primarily on injuries experienced by respondents, interviews averaged 25 minutes in length. Interviews were conducted an average of 577 days after the earthquake, and 17 percent were conducted in Spanish.

Wave 3. Between August 22, 1995, and May 29, 1996, interviews were conducted in both English and Spanish with 1,247 residents of Los Angeles County. The process by which households and respondents within households were selected was identical to that used in collecting Wave 1 data. Interviews were conducted with 1,247 of the 2,029 eligible households identified. Assuming alternately that all or none of the 782 numbers of unknown status contained eligible respondents, a response rate of between 40 and 53 percent was obtained. Since the questionnaire used in this interview was again expanded to include much of the information contained in Wave 1 as well as data on the use of services after the earthquake (which was of particular interest to the Los Angeles County Department of Health Services), interviews averaged 49 minutes in length. Interviews were conducted an average of 712 days after the earthquake, and 23 percent were conducted in Spanish.

Three-Generational Panel Sample. Margaret Gatz and colleagues at the University of Southern California (USC) administered a modified version of the Wave 1 questionnaire to a three-generational longitudinal panel. The objectives were to assess how earthquake preparedness and psychosocial responses to earthquakes differ with age, to ascertain the extent to which earthquake preparedness is diffused through family networks, and to assess the extent to which differences can be attributed to intensity of the earthquake and to experiences in past earthquakes (Gatz 1996).

Families in which at least one member lived within the area affected by the Northridge earthquake with an MMI value of six or greater were identified from the USC Longitudinal Study of Generations (LSOG). A total of 115 families of 127 eligible LSOG families were included in the study. Those lost were due to refusal (n = 1), inability to complete the interview due to physical incapacity of the targeted respondent (n = 3) or to scheduling incompatibilities (n = 3), inability to locate the targeted respondent (n = 2), or death of the targeted respondents (n = 3). Excluding those in which the targeted respondent had died, 93 percent of eligible families participated in the study. Within the 115 participating families, 207 of the 244 potential respondents participated. The final sample contains 38 members of the oldest generation with a mean age of 85.5 years, 99 members of the middle generation with an average age of 66.0 years, and 70 members of the youngest generation with an average age of 42.0 years. Included in these analyses are the 166 persons, regardless of generation, who resided in southern California at the time of the Northridge earthquake.

RESULTS

Earlier we listed a number of issues that have either been raised in the past about the ability to conduct valid surveys in the wake of a disaster or about the advantages and disadvantages of survey data as a source of information about disasters. Here we will address each of these issues utilizing, when possible, examples from the datasets described above.

Comparisons across time, events, and locations

Probably the single greatest advantage of collecting data with standardized questionnaires from population-based samples is that it allows researchers to compare community behavior across time, events, and locations. In an earlier analysis, we compared reports of the preparedness activities that respondents reported completing before the Whittier Narrows and Loma Prieta earthquakes with those reported by Turner et al. (1986) in the late 1970s (see Russell et al. 1995: 756, Table 1). In Table 2 we extend that analysis to look at preparedness activities prior to the Northridge earthquake. (Data on preparedness activities before and after the Northridge earthquake were not collected from the 96 respondents in the Wave 2 sample.)

Prior to the Northridge earthquake, we found that rates of survival activities (having a flashlight, a radio, a first-aid kit, and stored food and water) were dramatically higher in both northern and southern California before the Whittier Narrows and Loma Prieta earthquakes than they were in the late 1970s in the wake of the Palmdale bulge announcement (see Russell et al. 1995: 755-756). In contrast, planning activities—particularly providing family members with instructions for what to do during and after an earthquake—had dropped or stayed approximately the same while hazard mitigation activities (rearranging cupboards, putting latches on cupboard doors, and reinforcing structures) had stayed consistently low.

Table 2. Earthquake Preparedness in California, 1976-1994 (in Percentages).

Actions	Post-Palmdale[a]	Pre-Whittier	Pre-Loma Prieta	Pre-Northridge Wave 1	Pre-Northridge Wave 3	Gatz
Survival						
Have flashlight	11	58	60	48	49	87
Have radio	11	50	48	42	40	78
Have first-aid kit	8	48	42	36	39	77
Stored food	8	38	43	39	39	79
Stored water	8	37	40	42	40	68
Preparedness planning						
Family instruction	48	23	28	11	13	N/A
Earthquake insurance	13	12	17	22	19	46
Neighborhood plan	4	4	4	5	6	N/A
Hazard mitigation						
Rearranged cupboards	10	6	5	8	9	36
Latched cupboards	5	5	6	9	10	18
Reinforced structures	5	4	14	7	7	34
Total N	1432	583[b,c]	550[b,c]	487[b,d]	1247[b,d]	166[e]

[a] From Turner et al. (1986)
[b] Activities done for earthquakes or for both earthquakes and other reasons.
[c] Weighted sample.
[d] Unweighted sample.
[e] Sample of 166 people from Gatz multi-generational panel study (1996).

When we examine similar data prior to the Northridge earthquake from the population-based samples, it is interesting to note that three of the four survival measures (having a flashlight, a radio, and a first-aid kit) had declined since the Whittier Narrows and Loma Prieta earthquakes. Providing families with instruction had continued to decline, but investments in earthquake insurance and mitigation activities had risen slightly in the Los Angeles area. The panel study, however, shows levels of preparedness that are consistently higher than any of the other studies, across all types of preparedness. Possible explanations for this discrepancy will be discussed in a later section.

On the face of it, the downward trends look discouraging for practitioners. If anything, it would appear that investments in the simplest and most publicized kinds of activities actually dropped in the Los Angeles area between 1987 and 1994. There are, however, two potentially confounding pieces of information that practitioners should consider investigating in these datasets before concluding that their activities have been for naught. In these surveys, respondents are first asked whether or not they have done the activity. Second, they are asked whether they did the activity *before* or *after* the index earthquake. And third, respondents are asked whether they did the activity *specifically because of earthquakes, for some other reason, or for both earthquakes and other reasons.*

We suggest that people have increasingly bought flashlights, radios, and first-aid kits not just for use in earthquakes but for use in a variety of situations. We expect that this is particularly true of recent investments in portable radios. Indeed, when we examine whether respondents report having these items, regardless of the reason for which they were purchased, we find that more than 80 percent of the Northridge respondents had working flashlights before the earthquake, 65 percent had working battery-operated radios, and 54 percent had a first-aid kit. Rates of having an item, regardless of reason given for having it, are similarly higher in the Whittier data set, but the differences are not as dramatic: 78 percent have a flashlight; 65 percent have a radio; 62 percent have a first-aid kit; 49 percent store food; and 44 percent store water. Thus, what seems to have changed in the seven-year period between the Whittier Narrows and Northridge earthquakes is respondents' tendency to state that they have the item *because of* earthquakes. By the time of the Northridge earthquake, respondents were more likely to say they had one of these five items for reasons other than earthquakes.

This finding could also be of concern to practitioners. The fact that such a high proportion of Los Angeles County residents have these items but that some proportion fail to associate them with earthquake preparations *may* mean that Los Angeles County residents would not remember that they had them should an earthquake occur and, thus, would not use them. This interpretation is not completely farfetched. Southern California has experienced a significant amount of immigration over the last decade. Immigrants may be less knowledgeable about what to do when an earthquake occurs.

In the Northridge surveys, we asked respondents whether they had immigrated into the United States, whether their parents had immigrated into the United States, and how long the respondent had lived in California. A third of the Northridge respondents (35 percent in Wave 1, 31 percent in Wave 2, 37 percent in Wave 3) stated that they were born outside the United States. Similar or higher proportions reported that both parents were born outside the United States (41 percent in Wave 1, 38 percent in Wave 2, and 43 percent in Wave 3). Immigrants were an average of seven years younger than nonimmigrants, regardless of dataset (37 versus 44 years), and had lived in California only half as long on average (15 versus 33 years). We compared the preparedness activity of immigrants and nonimmigrants and found that immigrants were less likely to store water for earthquakes (47 percent versus 59 percent), less likely to have a first-aid kit (37 percent versus 48 percent), less likely to have done structural reinforcement (7 percent versus 11 percent), less likely to have earthquake insurance (13 percent versus 30 percent), but more likely to have put latches on cupboards (26 percent versus 15 percent) and equally likely to have engaged in the other preparedness activities listed here.

By being able to look at reports of preparedness activities across time and earthquake events, we have increased our ability to identify who may be less aware of what needs to be done to prepare for an earthquake. These analyses suggest that immigrants should be targeted for such programs.

Sample representativeness

One concern that has been raised about doing telephone surveys after a disaster is that the sample from which data are collected are not representative of the population affected by the disaster. Two general objections are raised in this regard. First, it is suggested that telephone surveys will "miss" substantial numbers of persons who do not have telephones or access to telephones *prior* to the index disaster. Second, it is suggested that telephone surveys "miss" those who are dislocated as a result of the index disaster and, therefore, fail to get information on those most affected by the disaster. In analyzing the data obtained from our surveys, we have examined the representativeness of our samples by comparing them to the 1990 U. S. Census (1990). While we cannot definitely answer the two challenges posed, we can state with some confidence what our samples do and do not represent. We suggest that telephone coverage is so pervasive in the U.S. and so quickly reinstated following disasters in the U.S. that the representativeness of any RDD sample, particularly in urban areas, will be as good or better than any other method of data collection, *provided* that the researcher:

- Faithfully draws an up-to-date random digit dialing (RDD) sample;
- Utilizes substantial numbers of callbacks distributed across different time periods and days of the week to maximize all

households in the sample being identified and represented in the sample;
- Accurately and completely lists all adult residents of the household;
- Utilizes a Kish table or other demonstrated method for randomly selecting the designated respondent from within the household for interview; and
- Insures that interviewers indeed interview the selected respondent.

The procedures adopted and the methods used for calculating response rates should follow those recommended by the Council of American Survey Research Organizations (CASRO 1982).

Following the Loma Prieta earthquake, a great deal of attention was paid to, and concern expressed about, the homeless and those who occupied single-room-occupancy (SRO) hotels prior to the earthquake, particularly in the Oakland area. Using the 1990 U. S. Census, we examined the extent to which our sample underrepresented: (1) persons who were homeless or who did not reside in households before the earthquake; (2) people who resided in households without telephones before the earthquake and who remained without telephones after the earthquake; and (3) persons whose telephones were disconnected as a result of the earthquake or whose households were destroyed as a result of the earthquake and who either left the area or remained inaccessible by telephone at the time of the interview (Bourque and Russell 1994: 19-22).

Clearly, persons in category 1 (above) are going to be missed in a telephone survey of any kind. If this is the group of interest for study, a researcher must develop other methods for identifying and contacting them. In terms of persons in category 2, we found that household telephone ownership was high in the Bay Area with 95 percent of households in San Francisco/Oakland having telephones and 98 percent of the households in the rest of the Bay Area having telephones. The unavailability of these households for selection means that we "missed" between 3.82 and 4.23 people in the San Francisco-Oakland area, from 1.59 to 2.44 people in the Santa Cruz area, and between 5.86 and 6.77 people in the rest of the five-county area for a possible maximum miss of thirteen people.

The other group missed, by definition, is those who reside in group quarters, similar to the SROs in Oakland. According to the 1990 U.S. Census, three percent of persons in San Francisco-Oakland, 0.4 percent in Santa Cruz, and 1.5 percent in the rest of the five counties live in group quarters. Had such persons been available for interviewing, we would have interviewed an additional three persons in San Francisco/Oakland, an additional 0.5 person in Santa Cruz, and seven additional persons in the rest of the area. Thus, had all of the above groups been accessible by telephone, we would have interviewed an additional seven people (8.4 percent) in San Francisco/Oakland, three (2.5 percent) in Santa Cruz, and 14 (3.1 percent) in the rest of the area. Clearly, the sample stratum which was most affected in this study by lack of telephones was San Francisco/Oakland.

Finally, we examined whether the rate of disconnected telephones was disproportionately higher in the Bay Area following the Loma Prieta earthquake in order to estimate the extent to which we might have "missed" persons who moved after the earthquake. The percentage of usable telephone numbers (i.e., a number that was neither a business nor disconnected) ranged from 44 percent in San Francisco/Oakland to 61 percent in Santa Cruz. These percentages are comparable to national rates and are actually higher than the rates obtained in Los Angeles County following the Whittier Narrows earthquake, where 40 percent of the telephone numbers in the high-impact area and 31 percent of those generated in the rest of the county were usable. Thus, we did not find an unusually high number of disconnected or out-of-service telephone numbers, and there is no reason to think that mobility either out of the area or within the area was unusually high in the wake of the Loma Prieta earthquake.

The question then becomes: Did this affect our results? Of course this loss affected our results. Certainly a small proportion of the population who were assumed to be highly vulnerable to the impact of the earthquake were missed in this study. On the other hand, we *know* that we missed these people, and we can estimate the extent to which they were missed. Using data available in the actual dataset in combination with 1990 U. S. Census data, we could attempt to interpolate information for this group and to assess the extent to which their experiences might change our findings. On the other hand, the size of the group missed is so small in a total sample of 656 that the actual impact these cases would have on findings—assuming that they all had the most extreme experiences represented in the sample—would probably not substantially change the overall picture that the study provides of the Bay Area following the Loma Prieta earthquake. As noted earlier, if groups of persons without telephones are the major focus of a study, then telephone interviews should not be the method of data collection. If, in contrast, the researcher wishes to get a dependable overall picture of what happened to an entire community during and after a disaster, we suggest telephone interviewing is a very good way to get data.

Responsiveness to interviews following a disaster

Earlier we reported the response rates for our surveys and briefly discussed the importance in number and pattern of callbacks in obtaining completed interviews. We also want to ascertain how receptive people are to being interviewed following a disaster. One way of estimating that is to compare the response rates we obtained in the Whittier Narrows, Loma Prieta, and Northridge surveys with response rates obtained in other telephone surveys conducted by UCLA's Institute for Social Science Research in southern California during the same calendar periods.

Response rates from four annual administrations of the Los Angeles County Social Survey (LACSS) are reported in Table 1 (above). All of the LACSS studies were conducted using CATI. Using methods recommended by the Council of American Survey Research Organizations (CASRO 1982), we calculated response rates by subtracting telephone numbers contacted and screened as ineligible from the total number of telephone numbers generated. To establish the lower boundary of the response rate, we divided the number of completed interviews by the total number of useable numbers. This calculation assumes that all numbers whose status it was not possible to determine are eligible households with an eligible respondent. Included in this category are numbers that are never answered, always busy, and always answered by a machine as well as all combinations of these possibilities. To establish the upper boundary of the response rate, this number is subtracted from the total useable on the assumption that none of these numbers represent a household with an eligible respondent.

We see in Table 1 that the estimated response rates for the LACSS studies from 1993 to 1996 range from a low of 42 percent to a high of 54 percent. At 46 to 60 percent, the response rate for the Northridge study, which was also conducted on CATI, is well within this range. Similarly, the response rates for the majority of the sample in the Whittier Narrows study are within this range *even though* the Whittier Narrows study was conducted using paper-and-pencil techniques rather than CATI and in spite of the fact that paper-and-pencil techniques allow for better queuing of callbacks and thus often yield higher response rates. This occurred because the budget for the Whittier Narrows earthquake was such that fewer callbacks were made outside the high-impact area. In contrast, within the high-impact area of the Whittier Narrows sample and in all strata of the Loma Prieta sample, where many more callbacks were made, the response rates are as much as 20 percent higher than in Northridge and in the low-impact strata of Whittier.

On the basis of these comparisons, we conclude that there is no evidence that persons in households with telephones are any more reluctant to participate in a study after a disaster than they would be at any other time. Clearly, the amount of resources available for locating respondents determines the number of callbacks attempted which, in turn, increases the response rates obtained.

Timeliness of data collection

The inability to collect data in a timely fashion is another argument that has been made against surveys. How true is that in this day and age? In fact, if a researcher had a questionnaire ready to go, a probability sample of housing units already drawn, and the resources to do it, face-to-face interviews with randomly selected residents of households could start immediately after a disaster. The biggest barrier is the high cost involved in moving large numbers of qualified interviewers into an area quickly.

But an additional question is, how imperative is it to enter the area immediately? Many questions of interest cannot be answered if we enter the area too quickly. For example, any information about recovery and rehabilitation, of necessity, cannot be collected until well after the index disaster. It is not even clear that information about use of services, extent of damage, injury, and psychological distress is best collected immediately. For example, in Los Angeles County we are still finding buildings that were damaged by the Northridge earthquake, and there is no definitive information about when, or if, excessive psychological distress—to the level of diagnosable post-traumatic stress disorder—occurs.

Certainly, telephone services do go off or become overloaded in disasters. Table 3 shows the number of respondents who reported losing utilities following the Loma Prieta and Northridge earthquakes. A minimum of 14 percent and a maximum of 26 percent reported that their telephones went off. We did not ask respondents how long their telephones were off, but clearly all of them had telephone service by the time we had the resources to conduct the study.

Table 3. Loss of Utilities, by Earthquake.

	Earthquake		
Quake Caused Utility to Go Off	Loma Prieta[a]	Northridge[b] Wave 1	Wave 3
% Phone Off	21	26	14
% Gas Off	10	4	2
% Water Off	7	8	4
% Electricity Off	68	52	41
Mean Days Post-Quake Interview Conducted	225	245	712
Total N	656	487	1247

[a] Sample is weighted; the weights are 0.96 for the San Francisco/Oakland strata, 0.16 for the Santa Cruz strata, and 1.0 for the rest of the sample.
[b] Wave 2 did not contain these questions.

We suggest that in survey research the largest barrier to quick entry into an area after a disaster is lack of resources. At the time of the Northridge earthquake, for example, the questionnaire used in the Loma Prieta survey was on the computer-assisted telephone interviewing (CATI) facility at the Survey Research Center in UCLA's Institute for Social Science Research. Had the resources been available, we could have modified the questionnaire for use after the Northridge earthquake, drawn a sample, and fielded a survey probably within a month of the earthquake. Instead, the need to obtain funding kept us from entering the field until August 1994, a minimum of 196 days after the earthquake. To the extent that some research questions are better asked later rather than sooner, this delay is not a problem. On the other hand, to the extent that some data are "perishable" and subject either to memory decay or memory enhancement, this could pose a problem for the researcher.

What constitute perishable data?

The three waves of data collected after the Northridge earthquake give us one of the first opportunities to examine the extent to which data really are perishable and to identify which kinds of data are more subject to problems of retrospective memory. Since these data have only just become available to us, we cannot answer this question in any detail at the moment, but we will pick out some questions to examine here. Following the Northridge earthquake, the first wave of data was collected an average of 245 days after the earthquake, the second wave was collected an average of 577 days after the earthquake, and the third wave was collected an average of 712 days after the earthquake.

Data available in Tables 2 and 3 (above) provide the first evidence of how the percentage of events may change as time passes. In Table 2, we see little evidence that with time respondents change their reports of what kinds of preparedness activities they had engaged in before the Northridge earthquake. In contrast, in Table 3 it looks like reports of utility outages decline by as much as half between 245 and 712 days after the earthquake. Whereas 26 percent of the respondents in the first wave reported that their phone was out, only 14 percent in the third wave so reported. Does this suggest that those in Wave 1 over-reported outages or that those in Wave 3 under-reported outages? We cannot definitely answer this question. Further analyses of these datasets might give us information as to whether reporting was different across the population and thus provide some insight into whether certain groups tend to under-report or over-report these kinds of data. On the face of it, we conclude for now that accurate memory of utility outages *does* decay with time.

To examine memory in more detail, we selected a variety of kinds of information that were collected in at least two of the post-Northridge waves of data collection. Included were a selection of factual questions about where respondents were during the earthquake and what happened to them, a series of questions about victimization *in the year prior to the interview*, and two opinion questions. The questions about victimization were included in the surveys because of people's speculation that victimization increases following an earthquake. As reported here, the victimization data are cued to the year prior to the interview, so some victimizations in the Wave 1 dataset occurred prior to the earthquake. All the data are reported in Table 4.

Table 4. Examination of Memory Decay Across Three Waves of Data Collection Following the Northridge Earthquake.

	Data Collection		
Factual Information About Quake and Aftermath	Wave 1 8/10/94- 12/5/1994	Wave 2 8/2/95- 10/22/1995	Wave 3 8/22/95- 5/29/1995
%Who felt the earthquake on Jan. 17, 1994	94	95	92
Of those who felt the quake:			
% Who were indoors at the time of the quake	98	96	96
% Who were in their own home	95	92	87
% Who physically changed locations after the quake was over	63	68	64
% Who report damage within 5 blocks of home	36	43	37
% Who report home damage	39	34	37
Of those with damaged homes			
Mean dollar amount of damage	$14,364	$7,823	$19,553
% Who report home inspected	67	47	66
Of those inspected:			
Who inspected?			
FEMA	20	47	34
City	22	7	17
County	4	7	4
Private Company	15	-	8
Other	17	20	18
Don't Know	22	20	19
% Tag Types			
Red tag	2	-	4
Yellow tag	8	13	9
Green tag	37	31	23
No tag	46	50	48
Don't Know	8	6	16
Of those with damaged homes % who applied for assistance	30	28	35
Of those who applied for assistance %who had difficulty in applying	14	22	18
Of those with damage homes % who evacuated their home for any reason	34	-	48
Injuries			
% who were physically injured	8	7	8
% who were emotionally injured	36	35	32
% who report other members of household were physically or emotionally injured	22	25	22
% who know agencies that worked after the quake	54	-	50

	Data Collection		
Factual Information About Quake and Aftermath	Wave 1 8/10/94- 12/5/1994	Wave 2 8/2/95- 10/22/1995	Wave 3 8/22/95- 5/29/1995
Victimization in the year prior to the interview			
% Robbed	7	5	6
% Say *not* related to quake	100	100	88
% Raped	0.2	1.0	0.6
% say *not* related to quake	100	100	100
% Motor vehicle crash caused injury	3	5	4
% say *not* related to quake	100	100	94
% Had loved one die of accident, homicide, or suicide	6	2	5
% say *not* related to quake	96	100	96
% say life changed for worse in an important way (residence, job, personal relationship)	13	13	10
% say *not* related to quake	84	67	90
% Some other terrifying experience	6	7	8
% say *not* related to quake	80	86	94
Opinions Related to the Earthquake			
In general, on the day of the earthquake would you say that the radio and television programs that you listened to:			
Overreported the sensational aspects of the earthquake	22	-	28
Balanced sensational and helpful ideas	40	-	48
Underreport the sensationalism	5	-	5
Presented jut the facts (not read)	14	-	14
Don't know	3	-	5
Missing	16	-	-
In your opinion, who is most responsible for helping people after our earthquake?			
The government	57	-	55
The people affected	24	-	26
Someone else	14	-	17
Don't know	5	-	2
Total N	487	96	1247

What is striking about the data is the extent to which all the information tends to remain constant across the three waves of data collection. To the extent that reports differ, the differences are almost exclusively in Wave 2 data which, given the small size of that sample (n = 96), may simply reflect statistical variance in a sample that is really too small to stand on its own. In spite of the fact that estimates are most "different" in Wave 2, the differences are not significant across the three datasets, largely because of the different sample sizes.

Looking only at Wave 1 and Wave 3 data, there are some places where increases or decreases in the Wave 3 data probably either reflect actual changes in rates over time or deterioration in memory over time. For example, the percentage of persons who reported having their homes inspected stays constant at 66-67 percent of those who reported their homes damaged, but the identification of who conducted the inspection shifts over time, away from city inspectors to FEMA inspectors, while the percentage of persons who say they "don't know" who inspected their home remains remarkably constant. Another substantial shift is in those with damaged homes who report that they evacuated their homes. Further analyses will allow us to investigate this in more detail, but for now we suggest that these later evacuations may well have occurred when repair work actually started on respondents' homes. In contrast, the slight decline in the percent who remembered agencies that were active after the earthquake may be evidence of memory decline over time, particularly among those for whom the agencies had little salience.

Victimization rates remain constant across the three waves of data collection, and the overwhelming majority of those victimized state that their victimization was not related to the earthquake. Opinions about media coverage during the earthquake and about who is responsible for helping victims of an earthquake similarly remain highly stable over time.

In summary, this quick review of data available across three datasets following the Northridge earthquake suggests that social information about disasters may not be as perishable as we sometimes think and that memories about a disaster remain quite stable for at least some substantial period after a disaster.

Dose-response as a quasi-experimental method for setting up a control

There is no way that a researcher can establish randomized control groups in studying responses to disasters, but the existence of population-based samples does allow systematic examination of whether and how experiences and responses differ across groups within the same community who are differentially exposed to the disaster. In earthquakes, Modified Mercalli Intensities (MMI) provide an approximation of the extent to which an area experienced shaking. Using MMI as an indicator of the extent to which respondents and their homes were "exposed" to the earthquake or the "dose" that they received, we can examine whether reports of damage, injury, and emotional distress differed with MMI. We expect that these three variables do vary with exposure or the "dose" of the earthquake that the respondent experienced. In contrast, there are other things that might not vary with exposure dose. We selected exposures to other kinds of violent events (i.e., robbery, rape, serious car accident) within the year proceeding the interview and opinions about the media and who is responsible for helping victims after an earthquake as examples of variables that might not vary with exposure to an earthquake. These data are reported in Table 5. Note that not every variable is available in every data set.

Table 5. Examining Dose Response by Modified Mercalli Intensity (MMI) and Earthquake.

MMI and Variable	Whittier[a] % (N)[c]	Loma Prieta[b] % (N)	Northridge Wave 1 % (N)	Wave 2 % (N)	Wave 3 % (N)
% of respondents in MMI area reporting damage within 5 blocks of home					
V	N/A	0 (11)	-	-	-
VI	N/A	23 (219)	15 (106)	28 (19)	16 (252)
VII	N/A	41 (202)	26 (260)	42 (56)	29 (698)
VIII	N/A	50 (118)	75 (72)	70 (11)	73 (177)
IX	N/A	-	83 (48)	70 (10)	86 (115)
% of respondents in MMI area reporting damage to home					
V	-	27 (11)	-	-	-
VI	9 (609)	32 (219)	18 (106)	16 (19)	19 (252)
VII	23 (43)	30 (202)	28 (260)	32 (56)	28 (698)
VIII	50 (38)	42 (118)	76 (72)	46 (11)	69 (177)
IX	-	-	83 (48)	70 (10)	81 (115)

Table 5. (continued)

MMI and Variable	Whittier[a] % (N)[c]	Loma Prieta[b] % (N)	Northridge Wave 1 % (N)	Wave 2 % (N)	Wave 3 % (N)
% of those with damaged homes who applied for disaster assistance by MMI					
V	-	0 (11)	-	-	-
VI	7 (609)	1 (219)	11 (106)	0 (19)	15 (252)
VII	11 (43)	15 (202)	22 (260)	22 (56)	26 (698)
VIII	15 (38)	12 (118)	31 (72)	20 (11)	37 (177)
IX	-	-	54 (48)	67 (10)	63 (115)
% of those with damaged homes who report building inspected by MMI area					
V	N/A	N/A	-	-	-
VI	N/A	N/A	42 (106)	0 (19)	31 (252)
VII	N/A	N/A	53 (260)	35 (56)	53 (698)
VIII	N/A	N/A	83 (72)	80 (11)	82 (177)
IX	N/A	N/A	85 (48)	71 (10)	89 (115)
% of respondents in MMI area who reported being physically injured					
V	-	0 (11)	-	-	-
VI	1 (609)	1 (219)	1 (106)	0 (19)	2 (252)
VII	5 (43)	1 (202)	6 (260)	9 (56)	5 (698)
VIII	5 (38)	2 (118)	18 (72)	9 (11)	20 (177)
IX	-	-	23 (48)	10 (10)	24 (115)
% respondents in MMI area who reported being emotionally injured					
V	N/A	N/A	-	-	-
VI	N/A	N/A	25 (106)	37 (19)	26 (252)
VII	N/A	N/A	34 (260)	32 (56)	29 (698)
VIII	N/A	N/A	49 (72)	46 (11)	40 (177)
IX	N/A	N/A	58 (48)	40 (10)	55 (115)
% of respondents in MMI area reporting being robbed in year prior to interview					
V	N/A	0 (11)	-	-	-
VI	N/A	5 (219)	8 (106)	5 (19)	4 (252)
VII	N/A	2 (202)	8 (260)	5 (56)	6 (698)
VIII	N/A	4 (118)	8 (72)	0 (11)	7 (177)
IX	N/A	-	4 (48)	10 (10)	8 (115)

Table 5. (continued)

MMI and Variable	Whittier[a] % (N)[c]	Loma Prieta[b] % (N)	Northridge Wave 1 % (N)	Wave 2 % (N)	Wave 3 % (N)
% of respondents in MMI area reporting being raped in year prior to interview					
V	N/A	0 (11)	-	-	-
VI	N/A	0 (219)	0 (106)	0 (19)	0.4 (252)
VII	N/A	0 (202)	0.4 (260)	2 (56)	1 (698)
VIII	N/A	1 (118)	0 (72)	0 (11)	1 (177)
IX	N/A	-	0 (48)	0 (10)	0 (115)
% of respondents in MMI area reporting being in a car accident that killed or injured someone in year prior to interview					
V	N/A	0 (11)	-	-	-
VI	N/A	6 (219)	3 (106)	0 (19)	2 (252)
VII	N/A	3 (202)	2 (260)	4 (56)	4 (698)
VIII	N/A	2 (118)	3 (72)	27 (11)	3 (177)
IX	N/A	-	6 (48)	0 (10)	9 (115)
% of respondents in MMI area who think the government is responsible for helping victims					
V	N/A	N/A	-	N/A	-
VI	N/A	N/A	55 (106)	N/A	50 (252)
VII	N/A	N/A	57 (260)	N/A	57 (698)
VIII	N/A	N/A	54 (72)	N/A	52 (177)
IX	N/A	N/A	63 (48)	N/A	60 (115)
% of respondents in MMI area who think media coverage of the quake was too sensational					
V	N/A	N/A	-	N/A	-
VI	N/A	N/A	30 (106)	N/A	34 (252)
VII	N/A	N/A	20 (260)	N/A	29 (698)
VIII	N/A	N/A	24 (72)	N/A	17 (177)
IX	N/A	N/A	12 (48)	N/A	18 (115)

[a] Whittier sample is not weighted in this analysis. Persons in the high impact strata were assigned Modified Mercalli Intensities of 7 or 8; all other respondents were assigned an MMI of 6.

b   The Loma Prieta sample is not weighted in this analysis. Because exact zip code maps with Mercalli Intensities were not available at the time of this analysis, the MMIs of some proportion of the sample may be inaccurate.

c   (N) = the number of respondents within that cell of the table. This number is the denominator for purposes of calculating the percentage reported in the cell.

The three types of data that we predicted *would* vary with exposure to the earthquake's intensity and shaking generally do have a dose-response relationship with Modified Mercalli Intensity in all the datasets, although the proportions reporting these events tend to vary across the datasets. (The fact that the percentages reporting damage, applying for assistance, and being injured are lower in the Whittier Narrows and Loma Prieta surveys, particularly in areas with an MMI of eight, may be due to one of two factors. First, as noted in the footnotes to Table 5, the MMI scores assigned to respondents in Whittier Narrows and Loma Prieta need more precision than we were able to achieve for this analysis. Second, there is a possibility that the fact that the Whittier Narrows and Loma Prieta surveys were administered using paper-and-pencil methods while the Northridge surveys were administered on a CATI system has an impact on callbacks and response rates [see our earlier discussion] and may have had some impact on these data.) Respondents in areas where the MMIs were higher are, as expected, more likely to report that there was damage in their neighborhood (within five blocks of their home), that there was damage to their home, that they applied for assistance, and that they were physically injured. In the Northridge datasets, those in areas with higher MMIs are more likely to report emotional injury. In contrast, there is no relationship between proximity to the earthquake and being robbed, raped, or in a serious automobile accident within the year preceding the interview.

The opinions examined, which are available only in two waves of the Northridge surveys, show a very slight dose-response relationship. Persons in areas with an MMI of nine are somewhat more likely to think that government (rather than the victims themselves) are responsible for helping victims after an earthquake, but, interestingly, respondents in areas with an MMI of seven rather than eight are next most likely to think that government is responsible. When respondents were asked about media coverage, persons in the highest impact areas with an MMI of eight or nine were *least* likely to think that media coverage was too sensational.

The availability of data from probability samples where exposure to the disaster varies enables the researcher to estimate the extent to which proximity to a disaster results in different experiences, behaviors, and attitudes. While not as powerful as an experimental design for examining the impact of a disaster on communities, the use of the concept of dose-response provides a viable proxy or surrogate for a controlled experiment and allows for inferences to be made about how the disaster has differentially affected households with, for example, similar household resources.

Population-based samples as denominator data

Population-based samples are useful in determining what a nonprobability sample represents. Gatz and her colleagues (1996) examined how three-generational families were affected by and reacted to the Northridge earthquake. Since the sample was originally drawn from members of the Kaiser Permanente Health Plan in 1971, it clearly is not representative of Los Angeles County residents in 1994. However, since Gatz included questions similar to ours in her questionnaire, we can make comparisons between our three waves of data and her data set. In Table 6, we first see that members of her sample are more likely to be female, to own their own house, and to be on average twenty years older than respondents in our sample. Those residing in the southern California area were slightly more likely to have felt the earthquake and to be alone when the earthquake occurred. They were more likely to report that phones, gas, and water were unavailable after the earthquake and equally likely to report electricity off. Although respondents were *less* likely to state that there was damage in their neighborhoods or that they themselves or family members were physically or emotionally injured by the earthquake, they reported a higher average amount of damage to their homes. Lower reports of injury probably reflect the fact that households of these 166 respondents are smaller and thus contain fewer persons whom the respondent can report were injured. Finally, we see that these 166 persons were *more* likely to live in areas that experienced Modified Mercalli Intensities of eight and less likely to live in areas with MMIs of seven.

Table 6. Comparison of Specialized Population to Probability Sample of Los Angeles County.

	Gatz Sample	Northridge Wave 1	Northridge Wave 2	Northridge Wave 3
% Who Felt the Earthquake	96	94	95	92
% Who were alone at the time of the earthquake	25	22	-	-
% Phone off	33	26	-	14
% Gas off	15	4	-	2
% Water off	18	8	-	4
% Elect off	52	52	-	41
% Who report damage within 5 blocks of home	26	36	43	37
Mean $ damage to home	$22,069	$14,364	$7,823	$19,553
% With self or family members physically or emotionally injured	35	46	41	43
Mean age	64	41	42	41
% Own home	77	47	57	46
% Female	65	53	58	54
Modified Mercalli Intensities				
6	28	21	20	20
7	42	54	58	56
8	22	15	12	14
9	8	8	10	9
TOTAL N =	166	487	96	1247

Judging from these preliminary analyses, it appears that the differences observed between Gatz' families and Los Angeles County residents as a whole are explained by the fact that Gatz' respondents are substantially older and of higher socioeconomic status but were differentially residing in areas of the county that were more affected by the earthquake. Therefore, Gatz' respondents may report higher levels of preparedness activities because of greater economic resources and a tendency to more stability in their lives.

Maintenance of verbal data

In the past, code frames for data collected in response to open-ended questions either had to be created at the time the data were coded and prepared for machine entry, or hard copies of either the questionnaires or the responses to open-ended questions had to be stored until the researcher was ready to analyze them. The availability of computerized data-entry programs now makes this completely unnecessary (Bourque and Clark 1992). Using data-entry programs such as those available from SPSS and other groups as well as the increasing reliance on CATI systems make it possible to store verbal data in machine-readable files at the same time that precoded data are being entered. As a result, it is no longer necessary to create code frames under pressure without sufficient thought, nor is it necessary to store bulky questionnaires in scarce space or in locations where they may be vulnerable to persons not involved in the research project.

Creating archives

The availability of CATI systems and computerized data-entry programs means that clean datasets can quickly become available both to the researcher who initiated the study and to others. Increasingly, archives are being made available for storage where datasets and documentation can be put into the public domain and made available to other researchers for secondary data analysis. Both the Whittier Narrows and Loma Prieta datasets have been archived at the National Information Service for Earthquake Engineering (NISEE) housed in the Earthquake Engineering Research Center Library at University of California, Berkeley. These datasets are also being put into the Social Science Research Archive at UCLA's Institute for Social Science Research.

CONCLUSION

We have examined the kinds of information, useful to disaster researchers, that can be obtained from well-designed, standardized, population-based surveys. We have also examined a number of the perceived barriers that have been used in the past to justify a reluctance to widely adopt the use of survey research in the disaster area and have demonstrated that most of these barriers either no longer exist or are of limited importance in disaster research. Indeed, the supposed barriers may actually be advantageous sources of important information.

The utilization of standardized population-based surveys is especially useful in comparing community behaviors across time, locations, and events. The use of similar instruments across three earthquakes in both northern and southern California has allowed us to examine changes in a number of behaviors relevant to earthquakes. As shown in this paper, we have been able to track the rate of preparedness within California since the 1970s. The use of a standardized instrument also allows us to compare injury rates across different magnitude events.

Concern over the ability to use surveys, especially through telephone interviewing, as well concern about the representativeness of community surveys appear to no longer be realistic barriers. Practically universal coverage of telephones in the U.S. (especially in urban areas) and increasing use of new technologically-sophisticated methodology such as computer-generated list-assisted random-digit dialing (RDD) procedures and computer-assisted telephone interviewing (CATI) have reduced the perceived barriers to accessing the population after a disaster. Additionally, the belief that respondents are reluctant to participate in a survey after a disaster has been shown to be unfounded. These methods, however, need to be implemented appropriately and only if the target is either the general population or one with access to telephones. Certainly a survey utilizing telephone interviewing would not be appropriate to identify the impact of a disaster on the homeless.

The timing of surveys is also an important factor to consider in conducting disaster research. As shown here, while some data may be subject to memory decay, other data appear to be quite stable over time. Still other data, such as estimates of damage and utilization of disaster services, may be unavailable until sufficient time has past after the disaster.

We have also shown that well-conducted population-based surveys may provide some estimates of a "dose-response" and may serve as denominator data for other specialized datasets. The ability to explore how rates of injuries, damage, and other earthquake-induced problems differ by distance from the epicenter or some estimate of the shaking such as MMI is an advantage of population-based surveys. Again, the explosion of technology is increasing the ability to compare survey data to other available data such as peak ground velocity or acceleration. The combination of survey results with such datasets through the use of geographic information systems (GIS) is just now beginning to be explored and will allow for greater accuracy in using survey research in a quasi-experimental design. Likewise, the ability to use survey data as a denominator to compare specialized datasets is increasing. In this chapter we explored one such use where our population-based studies allowed us to identify differences between the general population and a three-generational panel study using a similar questionnaire.

Lastly, we have shown how advances in computer technology have increased both the ease of storage of precoded and verbal data as well as the ability to share data with other researchers. Having verbal data stored in a database at the time of data collection or at the time that precoded data are entered increases both the usefulness of the data as well as their ability to be quickly shared with other researchers.

Overall, we have shown that well-designed, standardized, population-based surveys can provide an accurate look at a population's behaviors and attitudes regarding disasters as well as describe the impact of a disaster on a population. Many of the previously held notions of barriers to the use of survey research have become obsolete in this era of technological advances. Indeed, the future of survey research in conjunction with new technological advances may allow us to answer many of the important questions facing disaster researchers today.

NOTE

1. Data for the Whittier Narrows study were collected and processed with funds from the National Science Foundation (No. 62617), the Natural Hazards Research and Applications Information Center (Purchase Order 494933C1), the Earthquake Engineering Research Institute (EERI M880411), the National Center for Earthquake Engineering Research (Purchase Order R34779), and the Southern California Injury Prevention Research Center under funds from the Centers for Disease Control (No. R49/CCR903622). Data for the Loma Prieta study were collected under funds from the National Science Foundation (No. BCS-9002754). Data for the Northridge earthquake study were collected and processed with funds from the National Science Foundation (Nos. CMS-9416470 and CMS-9411982) and from the Los Angeles County Department Health Services (Purchase Order R41867 and Award No. 953124). The Northridge earthquake panel study was conducted by Margaret Gatz at the University of Southern California under funds from the National Science Foundation (No. CMS-9416271). The authors thank Eve Fielder, Michael Greenwell, Tonya Hays, and Elizabeth Stephenson from the Institute for Social Science Research at the University of California, Los Angeles, for references and resources which they provided for this paper.

7

QUALITATIVE METHODS AND DISASTER RESEARCH

Brenda Phillips

Qualitative research is variously described as field research, ethnography, Chicago School research, or (as if in natural opposition) nonquantitative methods. Typically, qualitative research (QR) involves concurrent data collection (usually interviews, observations, documents, and/or visual records) with data analysis (generally described as the process of searching for patterns or themes in the data). The ways in which qualitative researchers use specific data collection and analysis methods can result in anything from a focused case study to a full-blown, in-the-field-for-months ethnographic inquiry. Most disaster research falls on the interview-based, case study end of the continuum. This chapter argues for continuation of that trend but, significantly, for a wider use of the qualitative research tradition.

Most disaster researchers cite Prince's (1920) Halifax study as the first disaster research. Interestingly, Prince's methodology incorporated qualitative data, a tradition which continues in disaster studies today. Few social science areas of specialization can lay claim to such a rich, established, and long-term qualitative research tradition. This paper proceeds with a brief overview of methodological trends in qualitative research relevant to disaster research, continues by considering misconceptions about qualitative research, links qualitative and disaster research, and concludes with a "wish list" for future qualitative disaster research. The purpose is not to review past studies but to muse about the future. I offer insights from my own experience in learning, teaching, and doing qualitative research and draw from a few others as illustrations of key points.

TRENDS IN QUALITATIVE RESEARCH RELEVANT TO DISASTER RESEARCH

Researchers employing qualitative research have experienced varying degrees of disciplinary support for their methodological choices. Anthropology has historically provided the strongest support, followed by sociology, political science, and psychology. Fluctuations in periodic support have echoed larger research trends. The strong Chicago School fieldwork tradition declined, for example, with the rise of public opinion polling and survey data analysis during the 1940s and 1950s.

Midway through the 1960s, though, Barney Glaser and Anselm Strauss published a widely acclaimed qualitative study, *Awareness of Dying* (1965), and explained their methodology (a qualitative first) in their follow-up work, *The Discovery of Grounded Theory* (1967). Glaser and Strauss's work sparked renewed interest in and support of qualitative methods throughout the 1970s. The first qualitative journals originated in the 1970s, supported by the first significant research guides (Lofland 1971; Schatzman and Strauss 1973; Schwartz and Jacobs 1979; Spradley 1980).

Since the beginning of the 1980s, the number of general qualitative research journals, books, and articles has increased dramatically, as has their quality and impact on a variety of disciplines. New qualitative journals and publications appeared in the 1980s and 1990s including the *Journal of Contemporary Ethnography*, *Qualitative Health Research*, and a Sage Publications series on qualitative methods (now numbering some fifty volumes). Importantly, the *International Journal of Mass Emergencies and*

Disasters published qualitative works from its inception (see observational studies in Carter et al. 1983; interviews in Shamgar-Handelman 1983; interviews in Dombrowsky 1983b). Academics produced advanced qualitative textbooks with interdisciplinary utility (Lincoln and Guba 1985; Bogdan and Biklen 1982). In essence, important qualitative advances occurred in the past several decades that have changed how some qualitative researchers work and which imply challenges for qualitative disaster work.

Probably most significantly, a number of the methodological writings over the past decades have focused on a much-needed area of development: data analysis. While many qualitative researchers used to say (in a quasi-grounded theory way) that we search for "patterns in the data," we now encounter myriad writings on how to analyze data and which data analysis technique to use for specific types of data. Qualitative researchers cannot get away with a loosely-described data analysis write-up any more. Increasingly, qualitative researchers will need to more fully describe their data analysis techniques, even to select a particular strategy such as Spradley's (1980) Developmental Research Sequence or Lofland and Lofland's (1995) structured ways of identifying units, aspects, topics, and types.

Those who say, "We used grounded theory," will be increasingly challenged to identify which grounded theory, due to the significant schism between Strauss (see Strauss and Corbin 1992) and Glaser (1992). To make data analysis even more challenging, see Miles and Huberman's (1994) second edition of *Qualitative Data Analysis* and their detailed, step-by-step strategies for handling and interpreting data. The coming decade will doubtless see us trying and critiquing these various strategies, and refining them even further.

Beyond data analysis, we must reflect on our strategies for how we "write up" qualitative research (Van Maanen 1988; Wolcott 1990, 1994) as well as what our writing choices mean for various respondent and reader constituencies (Richardson 1990, 1994; see also Van Maanen 1995). The commonly-chosen qualitative disaster case study can now be written in a variety of styles (Yin 1984; Stake 1995). When we make choices about how we write up our findings, the format, content, and even tone mean something to the audience for which we write. Because qualitative disaster researchers write for multiple constituencies, we need to thoughtfully and accordingly prepare our findings. Recent writings on qualitative writing compel us to consider the impact of each work we use and the implications of what we intend to say; qualitative work requires sensitivity and an ethical base, from initiation through the final report. It would be even nicer if reviewers for articles would think similarly. I'll never forget the reviewer who recommended that I get rid of "all those interview quotes." I always include interview quotes so that readers can assess my findings and to help practitioners use them in making connections with their own situations. The reviewer further remarked that the quotations should be replaced with statistical tables. I believe that qualitative researchers must educate others in what it is that we do, how we do it, and why we do it.

Disaster research and qualitative research share common disciplinary interests. Today, qualitative research can be found in nursing, dance, sociology, psychology, anthropology, education, political science, family studies, health studies, social work, and communications. Many of these disciplines also contribute to disaster and hazards research. Additionally, qualitative research is at an all-time high of usage and appreciation—and is at a significant point of change. I see more reflection on the process and outcome of qualitative research than ever before. These trends cannot help but impact qualitative disaster research. Before I continue with this thread, however, I would like to address some commonly-held misperceptions that exist and may influence how readers approach this chapter.

COMMON MISPERCEPTIONS ABOUT QUALITATIVE RESEARCH

From teaching graduate-level qualitative research courses, serving on various discipline-based graduate committees, presenting qualitative research, attempting to and actually publishing qualitative research, and experiencing both journal and grant proposal reviews, it is apparent to me that a number of common misperceptions exist about qualitative work. My favorite review experience came a few years ago when one reviewer remarked, "[T]his would be a good project if only it was quantitative.... PI [i.e., the Principal Investigator] should revise and resubmit as a quantitative proposal." Proposal quality aside, I think it is fair to say that the reviewer missed the value, merit, and potential contributions of qualitative research. But this reviewer only reflects a larger concern which flavors the experiences of those working qualitatively. It is not uncommon to hear one's colleagues denigrate qualitative scholarly activity as "not empirical" or "just exploratory." What they are really saying is that the work is not perceived as empirical because it is not quantitative or positivistic—as if these qualities were the hallmarks of good science. The debate is really an argument over paradigms, such as in the debate over positivistic versus naturalistic paradigms (Guba 1981; Lincoln and Guba 1985; Erlandson et al. 1993) currently influencing much of the recent qualitative methodological writings. While the positivistic paradigm carries heavily-institutionalized academic support, the naturalistic paradigm fits more appropriately with qualitative research approaches and, in my opinion, with disaster research (more on this later). Today, qualitative researchers can select from a variety of perspectives and paradigms (such as constructivist, interpretivist, critical theory, feminism, ethnic modeling, and naturalist; see the articles in Denzin and Lincoln 1994).

This unnecessarily dichotomized quantitative-versus-qualitative debate fosters an inappropriate juxtaposition based on the assumption that quantitative research is more objective and that qualitative efforts are more value-laden or subjective. Erlandson et al. (1993: 15), as based on Lincoln and Guba (1985), tell us that objectivity is "largely an illusion." Value-free science is no more possible in positivistic than in naturalistic science. Qualitative researchers, especially those working from naturalistic and feminist paradigms, acknowledge this dilemma and have devised strategies to work through the impact of potential bias.

A related misconception concerns the allegedly poor validity or reliability of qualitative research. However, those who issue such negative salvos lack understanding of methodological advances in QR during the past decade. The terms validity and reliability are not always appropriate in qualitative research, especially for those working from a naturalistic paradigm. The point of this paper is not to give full details, but the terms increasing numbers of qualitative researchers use today include trustworthiness, credibility, authenticity, and dependability. A variety of techniques can be used to enhance each of these characteristics, ranging from prolonged engagement to triangulation to peer debriefers. Qualitative researchers today are also inviting external review of their research process by creating methodological audit trails and employing "outside auditors" (Lincoln and Guba 1985; Schwandt and Halpern 1988; Erlandson et al. 1993).

A final issue within the quantitative/positivistic versus qualitative/naturalistic debate is the concern that qualitative research is not generalizable beyond the context under study. However, Lincoln and Guba suggest " . . . the prevailing view is that science grows through the accumulation of generalizable knowledge. However, . . . total generalization is never possible, even in the physical sciences." Rather than concern themselves with the generalizability issue, many qualitative researchers embed their interpretations within a deep contextual foundation emphasizing the time, place, and circumstances within which a disaster event, response, or process occurs. Contextualizing enables readers to better understand how analysis arises and supports the researcher's theoretical explanation (more later).

An established qualitative tradition exists within disaster research as an appropriate means of conducing scientific inquiry. To maintain and strengthen this scholarly tradition, disaster researchers need to employ qualitative research regularly, to encourage graduate students to do likewise, and to consider paradigmatic shifts and methodological advances made in QR over the past decade. Doing so may mean effectively challenging misperceptions about QR, experimenting with paradigms and collection/analysis techniques. As a way of considering these possibilities, I discuss here why qualitative research is appropriate for disaster research as based on a naturalistic paradigm, note where methodological gaps exist and could be filled in by current qualitative disaster research (QDR), and ponder what the future might hold for qualitative disaster research.

WHY QUALITATIVE RESEARCH IN DISASTER RESEARCH?

Shared history and strong support

A strong reason why qualitative and disaster research fit together well is because they share a common history. For example, much of the increased interest and writings have coincided with the emergence and establishment of various disaster and hazards centers. Concurrently, some disaster and hazards centers have institutionalized the use of qualitative research. It is rare to find a substantive area so supportive of qualitative work.

The long tradition of QDR, as used by single researchers like Prince, became more firmly entrenched when disaster research institutions appeared. Let me give the example with which I am most familiar, the Disaster Research Center (DRC). E. L. Quarantelli and Russell Dynes established DRC in 1963 at The Ohio State University, using fieldwork methods as a major research foundation. DRC's fieldwork tradition extended from the founders back through sociologist Herbert Blumer, Quarantelli's thesis advisor at the University of Chicago. Blumer was part of the Chicago School, a group of social scientists (primarily sociologists) who sent their students into Chicago's neighborhoods to do fieldwork.

The Chicago School legacy of going into the field was passed on to Quarantelli's and Dyne's students, some of whom have become directors of additional hazards institutions. For example, Dennis Wenger, while director of Texas A&M University's Hazards Reduction and Recovery Center, encouraged both quantitative and qualitative work; Kathleen Tierney, codirector of the Disaster

Research Center (now at the University of Delaware) continues to send students into the field with qualitative tools; Henry Fischer and the Social Research Group at Millersville University use interviewing in their studies; David Neal, formerly director of the University of North Texas' Institute of Emergency Administration and Management, uses visual, documentary, observational, and interview data collection methods in his research; and William Anderson, who oversees much of the National Science Foundation's disaster research, provides funding for well-developed qualitative projects.

Other institutions besides those with DRC connections use, fund, or promote the use of qualitative methods. As an illustration, many of the last decade's qualitative disaster projects have been "jump-started" with project funding from the Natural Hazards Research and Applications Information Center (NHRAIC), located at the University of Colorado at Boulder in the U.S. Gilbert White, a geographer, founded NHRAIC in the 1970s; his student, sociologist Dennis Mileti, currently directs NHRAIC in conjunction with Mary Fran Myers, project manager for many of the NHRAIC funded "Quick Response Grants." NHRAIC's work connects researchers to practitioners and serves as a clearinghouse and publisher for disaster studies, and as a conduit for qualitative disaster research to reach both academic and user communities. Because the projects involve quick response into the field, a large number necessarily rely on qualitative research, sometimes subsequently developed into both larger quantitative and/or qualitative studies.

Florida International University opened the International Hurricane Center in the aftermath of the 1992 Hurricane Andrew. Part of the studies that came out of FIU included interviews and observations with low-income victims (Morrow and Enarson 1996). The qualitative tradition of going into the field proved valuable in identifying needs of underserved groups as well as strategies that these groups used to recover from disaster.

Funding and support for projects centralizing or using qualitative research methods have come from the National Opinion

Research Center (via the Army Chemical Center, Department of the Army, in the 1950s), NHRAIC, the National Science Foundation, the Federal Emergency Management Agency (for example, DRC's 1980s emergency-time emergent group study), the National Institute of Mental Health, the Army Corps of Engineers (such as Oak Ridge National Laboratory's Chemical Demilitarization Project), and the National Aeronautics and Space Administration (e.g., the University of North Texas' remote sensing and emergency management project).

General compatibility

The naturalistic paradigm in QR acknowledges the existence of multiple realities, holistic investigation, the mutual influence of researcher and respondents, and the use of thick, rich description to form a context for understanding (Erlandson et al. 1993). Guba (1981, as discussed in Erlandson et al. 1993) emphasizes that naturalistic inquiry focuses on relevance, emergent theory, attention to tacit knowledge, the researcher as the instrument, a flexible research design responsive to the research, and a natural setting over a laboratory setting. These paradigmatic dimensions fit well with disaster research efforts.

Because disasters challenge communities in unexpected ways and have unanticipated consequences, QDR can capture human behavior at its most open, realistic moments. To borrow from Goffman's (1959) dramaturgical perspective, people drop their frontstage behavior, allowing researchers to capture realistic behavior. We get to see backstage behavior, which is best captured by those trained in observational data gathering. Researchers return from the field with data relevant to disasters and potentially useful for building theories of crisis occasion human behavior. Rather than a standardized, preset format, qualitative researchers usually prefer to remain flexible, rendering them more able to capture new ideas and to allow fresh perspectives to emerge from the data being collected. Rich insights can result. For example, much of the recent literature on gender, race/ethnicity, age, social class, and societal developmental levels (much of it qualitative) empirically demonstrates that socially, economically, and culturally diverse communities experience realities at variance from standardized, bureaucratically-streamlined responses (Neal and Phillips 1995; Morrow and Enarson 1996; Miller and Simile 1992).

One personal example I experienced came in the aftermath of Hurricane Andrew. I had spent the day interviewing outreach personnel for a disaster services organization. They had assured me that they were engaging in outreach to the farmworker population and had even been in each of the farmworker housing communities. The following day, I happened upon a local farmworker labor organizer who invited me to visit one of the communities. I abandoned the day's tentative schedule, visited the site, and listened as community leaders asked me how to get in touch with the disaster services organization. It was an entire week after the hurricane hit; the organization's representatives had lied to me. The emergent, flexible methodology (based on a naturalistic paradigm) resulted in papers on organizational barriers to service delivery (see Phillips et al. 1993) and on the ways in which bureaucratized structures inadequately respond to emergent needs (Neal and Phillips 1995).

Furthermore, because qualitative research is grounded in people's actual experiences, the possibility of identifying new, relevant questions becomes more likely. While disaster researchers know that we often see the same stories or "lessons learned" in disaster after disaster, qualitative research bears the possibility of identifying new questions. Qualitative methods and naturalistic paradigms permit the researcher to follow interesting questions and to alter the research design to pursue promising areas of inquiry (Erlandson et al. 1993; Spradley 1980). In so doing, qualitative research can empower and give voice to respondents (particularly disaster managers and victims). And, because so many data often accumulate in qualitative studies, new research questions and theoretical insights can potentially emerge. Although such new possibilities might seem serendipitous, the theoretical sensitivity required to produce innovative notions really results from immersion in the field, deep in the context of people's lives.

The idea of context, an integral feature of naturalistic and feminist inquiry, has supported QDR for decades. Long based in Quarantelli's (1987c) C-Model (see his chapter earlier in this volume; see also Lofland and Lofland 1995; and Strauss and Corbin 1992), context is essential for interpreting qualitative findings and for rendering those findings useful to practitioners. Good qualitative researchers lean on thick, rich descriptions of research settings (Geertz 1973). Reliance on such description provides context— an understanding of the time, place, and circumstances in which a disaster event occurs. Context provides support for emergent theory and for framing researchers' interpretations. Developing a detailed, contextual base for qualitative disaster studies also enables practitioners who rely on the research to make comparisons with their own contexts. As an analogy, I think that most educators recognize that a well-contextualized story beats out a concept and definition any class period. Another useful dimension for providing context is to generate insights unimagined in other settings. An emergency manager in Ohio, for example, might not be terribly concerned about earthquakes but could usefully intuit messages from good description and analysis of earthquake recovery process.

A final reason why qualitative methods and disaster research fit well together is that disaster research has always furnished one of the more unimpeded specialization areas for researchers. In addition to the institutional support and funding for research noted earlier, I believe that many qualitative researchers have found disaster journals more willing to print their work than publications in their own disciplines. Few academics can say that the top journal in their disciplinary specialization (such as a gerontology journal) would accept their qualitative work as willingly as the *International Journal of Mass Emergencies and Disasters*, *Disaster Management*, and *Disasters* (among others) have done. Because of this intertwining between qualitative methods and the substantive area of disasters, qualitative disaster research has been able to make meaningful contributions and even to thrive. I suspect that few disciplines or specialty areas can lay claim to such a multi-generational, multi-disciplinary methodological legacy.

USE OF QUALITATIVE METHODS IN DISASTER RESEARCH: PROBLEMS AND POTENTIAL

Problems/Inadequacies

Disaster researchers using qualitative methods experience several problems or inadequacies, including unexplored levels of analysis, uneven use of data collection techniques, insufficient longitudinal research, and inadequate proposal quality. Several key methodological reviews (e.g., Mileti 1987; Drabek 1986) note that the higher the level of analysis, the fewer studies. This is especially true in qualitative research. More research has been done at the individual or group level, for example, than at other levels. Organizational research, typically using case studies, appears less frequently. Other levels of analysis, such as neighborhoods (for an interesting example, see Guilette 1993) and communities, remain understudied. Qualitative work also offers possibilities for qualitative comparative studies, yet another area that disaster researchers need to develop further. Anthropologists have contributed the most here (Oliver-Smith 1996).

The most commonly used qualitative method in disaster research is undoubtedly interviewing (ranging from fairly focused projects to longitudinal, in-depth studies or even life histories), usually resulting in a case study format. Other data collection methods, such as observation, photography/videography, and documents/records (Bogdan and Biklen 1992; Jackson 1987) appear far less frequently, typically individually or sometimes in combination with interviewing. This latter situation is truly a shame, because triangulated studies result in stronger qualitative research. Combining interviews with additional field methods allows the researcher to enhance credibility and trustworthiness of the data and findings, permits that desirable thick, rich context to develop, and facilitates emergent questions and problems.

A few recent studies have provided some potential for regenerating underused qualitative methods. For example, Scanlon's

(1996) retrospective look at the Halifax explosion uncovered new data, much of it qualitative in nature, including interviews and "unobtrusive measures" (Webb et al. 1981), that is, documents (see especially Scanlon's chapter later in this volume). I hope that we see additional use of documentary evidence, including much of that stored in disaster and hazards research center archives. Existing, unobtrusive measures can be problematic (Musson 1986; Webb et al. 1981), but they can be used creatively as the sole data source or as part of a triangulated strategy. I hope that we see more such QDR in the future.

Visual techniques also remain underused, although the NHRAIC-funded Cohn and Wallace (1992) study of the Exxon Valdez disaster moves disaster research in an interesting direction. A picture is truly worth a thousand words, especially as an educational tool for illustrating one's findings. However, what visual techniques might we apply to disaster studies in coming decades? With the increasingly available camera, video, and computing technology for our use, visual techniques could become a "hot" method (Curry and Clarke 1977; Wagner 1979; Collier and Collier 1986; Hockings 1995). The "native instant" method could even put visual recording devices into the hands of emergency managers, shelter residents, or neighborhood associations (Worth and Adair 1972; Blinn and Harrist 1991). Imagine the data we could collect from the real *emic* perspective.

I also urge qualitative disaster researchers to further explore the possibilities of observational studies. Much of what I have seen in U.S. research has involved observational research done on shelters or in emergency operating centers, usually at the individual or organizational levels of analysis. I would like to encourage us to expand our substantive focus and to employ a range of observational strategies, from complete observation to full participant observation. Why not live in a recovering community for a couple of years, observing at the neighborhood or even community level? Observation also serves qualitative research well as a means of triangulating and strengthening other methods. Nonverbal behavior can tell us as much sometimes as verbal response.

In addition to the above concerns, I see a significant lack of longitudinal research, which qualitative research could supplement. Disaster researchers use the one-time case study far too frequently, when intensive immersion in the field over long periods of time could meaningfully augment our limited-time glimpses of human behavior in disaster. The most understudied disaster phase has always been recovery. Longitudinal qualitative research would fit well methodologically with needed substantive recovery studies, if only academic researchers could spend that much time in the field—and secure the funding to do so!

While qualitative projects have not been the majority of funded disaster studies, it may not be the method so much as the quality of the proposals submitted. To be fair, though, no standards currently exist on how to put together a qualitative proposal. Furthermore, in my experience, most prospective funder's guidelines suggest a quantitative or positivistic framework. Unfortunately, qualitative disaster research does not always fit so readily into existing standardized formats. Until recent years, few works even existed on how to put together a qualitative proposal (Marshall and Rossman 1995). Thus, a fair amount of variation in proposal style and substance has resulted. Part of the confusion comes from the lack of a common format; part of the responsibility rests on the shoulders of qualitative researchers who do not always adequately explain their methodology. Reviewers cannot help but wonder what a qualitative researcher will actually do if funded; likewise, most reviewers remain unaware of methodological advances which should be included in qualitative proposals these days.

Potential

In this section, I concentrate on two trends, both of which are future oriented and will influence qualitative disaster research: the next generation of researchers and computing technologies. Disaster research has recently concerned itself with producing the next generation of researchers (Anderson 1990) in order to continually advance science and encourage fresh perspectives. In my experience, and those of colleagues with whom I spoke recently, the next generation possesses a strong interest in qualitative research. In my own recent graduate qualitative methods class, 22 students from seven different disciplines enrolled. A waiting list develops each semester due to limited space (I can only monitor so many students in the field in a variety of "interesting or exotic" locations), and other departments have offered qualitative courses in recent years to meet student needs (nursing, psychology, education, family studies). Qualitatively-oriented faculty are in demand on our campus to serve on graduate committees. I hear similar stories from other qualitative disaster researchers. In one session at a recent NHRAIC hazards conference, graduate students presented their research—and about one-third had used qualitative methods. Informal conversations with several of these students revealed interest in QDR, but also concern for what professors and future colleagues would think of their methodological choice. Supporting students' choices for either qualitative or quantitative work (or, preferably, both) is essential; promoting high standards for qualitative disaster research is mandatory.

Second, in the 1990s new computer software appeared for managing and potentially analyzing qualitative data (e.g., Ethnograph, Martin, HyperResearch, and QSR Nudist, to name a few; see Weitzman and Miles 1995). These packages tend to be most useful for managing and coding large segments of data such as observational notes and interview transcripts (Richards and Richards 1994; Weitzman and Miles 1995). Eagerly awaited updates might produce packages more useful in analyzing data; for now, however, that task remains largely with the researcher. While my own experiences with such software have been both positive and negative, I find qualitative computer programs generally useful. However, the software on balance is more useful for data management and coding than for actual analysis—the human being is still far superior in telling us what the coded data mean.

Other technological advances can further link computers with disaster and qualitative research. Those interested in using visual data, for example, can employ optical scanners for photographs or slides and can duplicate and store video for potential analysis. (Just think of what a digital camera could do with your computer!) Furthermore, we can now share much of these data electronically, even creating Internet Web sites for use by other researchers and practitioners. (See the several sites annotated in the Appendix.) Whereas disaster research clearinghouses have, in the past, offered quantitative databases for researchers to share, such organizations now might want to offer qualitative databases. Many of us now subscribe to NHRAIC's electronic newsletter, *Disaster Research*; about a dozen qualitative methods discussion lists exist. As electronic journals continue to appear and evolve, disaster researchers will need to consider creating an Internet journal.

One example of a unique QDR/computing technology project is the remote sensing effort undertaken at the University of North Texas where Sam Atkinson and David Neal obtained remote sensing data, created text, and offered "lessons learned" for practitioners and students. (The Web site is http://www.ias.unt.edu:9876.) UNT's remote sensing project based some of its material on qualitative studies, including observational, interview, and visual data, and it planned to gather more as part of this NASA-funded project. We will likely see more of this type of work in the future.

CONCLUSION: A "WISH LIST" FOR DISASTER RESEARCHERS USING QUALITATIVE METHODS

The past decade's rapidly expanding methodological writings serve disaster research well. The future looks very interesting and methodologically challenging. Qualitative and disaster researchers will need to become more competitive and sophisticated in writing proposals, conducting research, and publishing their work. The future looks very promising for qualitative disaster research. Forging ahead on the crest of a methodological, theoretical, and technological explosion, so to speak, many possibilities spill out before us. With this optimism in mind, with "lessons learned" from doing and reading qualitative disaster research in the past, and having spoken with colleagues, I present the following "wish list":

- that qualitative disaster research be used to fill substantive and theoretical gaps in our knowledge base;
- that qualitative disaster researchers become more conversant with ontological and epistemological debates within qualitative research;
- that qualitative disaster research become increasingly interdisciplinary, to drawn upon the methodological, theoretical, and substantive strengths and insights that cross-disciplinary research teams can offer;
- that qualitative disaster researchers submit better-developed proposals reflecting methodological and technological advances, and that theory be considered as an integral part of such proposals (either as an initiating and sensitizing framework or as an end result);
- that reviewers better educate themselves about methodological advances in qualitative research so that they can provide more useful reviews, demand higher quality, and recognize well-developed proposals;
- that qualitative disaster researchers work diligently to incorporate standards for heightening the trustworthiness and credibility of their data and research, such as audit trails, outside auditors, peer debriefers, and triangulation, and that they engage in cognizant and careful data management and analysis;
- that qualitative disaster researchers work to explore and use data-gathering techniques in addition to the often-used interview;
- that qualitative disaster researchers seek out opportunities and ways to conduct studies at higher levels of analysis, especially at the organizational and community levels—and

that these have the potential to foster comparative analysis or to become part of a comparative database;
- that a variety of theoretical models be tried out;
- that qualitative disaster researchers more fully and thoughtfully write up their methods so that we may better evaluate their research and so that their articles can better inform readers of how data were gathered and analyzed;
- that we find ways to share qualitative data with researchers and practitioners via existing clearinghouses and resource centers;
- that qualitative disaster researchers seek out and use technological advances, whether in data management software packages or through sharing research via the Internet with other researchers, practitioners, respondents, the lay public, and students;
- and that qualitative disaster research be exceptionally well-written, in thoughtful, informative, context-rich ways that are meaningful to respondents, practitioners, students, policy-makers, and researchers.

8

THE ECONOMICS OF NATURAL DISASTERS

Anthony M. J. Yezer

The economic problems produced by natural disasters have long occupied the attention of mankind. Along with preparation for war and self-defense, the response to problems of flood, drought, earthquake, and windstorm has been used as a test of the ability of governments and economic institutions to serve a population. In our own time, there is increasing concern with the unwillingness of private insurers to operate in high-risk areas and with the escalating cost of government disaster assistance. In spite of this rationale for studying the economics of natural disasters, a search of on-line sources for books, articles, and other academic publications on the subject provides only a modest number of references.[1] Apparently governments have found little need to examine the economic effects of their policies towards actual or potential natural disasters.

Although the literature is thin, it includes a surprising diversity of theoretical models and empirical results responding to a number of different issues and problems. Unfortunately, these individual approaches to the economics of natural disasters often bear little relation to one another because they proceed from very different assumptions and analyze different aspects of the relation between disasters and the economy. The first task of this review is to organize these approaches under a modest list of headings and to show the relations among them. Ideally, they would all be placed within the context of an overarching general statement of the economics of natural disasters, but this chapter will settle for the more modest goal of identifying distinctive approaches and pointing out differences in their assumptions and purposes.

Probably the most familiar branch of literature dealing with the economics of disasters involves various aspects of economic impact analysis in which the effects of a disaster event on a local or regional economy are modeled. Because it is important to understand the theoretical basis for expecting an economic response to disasters before reviewing empirical models, these economic impact models will be discussed in the last section of this chapter. The next section will discuss attempts to model the theoretical effects of natural disasters within the context of a general equilibrium theory of regional development. An intermediate section considers disaster insurance and mitigation models.

EXPECTATIONS AND INFORMATION EFFECTS IN MODELS OF LOCAL ECONOMIC DEVELOPMENT

Once they have occurred, natural disaster events have direct and indirect effects on a local economy because they damage local capital stock and kill or injure local population. The result of these ex post (after the fact) effects is a temporary reduction in the supply of inputs, some of which are highly specialized to local production. Local output will fall for two reasons. First, a direct effect arises because reduced availability of local capital and labor causes lower output as predicted by any normal production relation. Second, indirect effects arise because the damage to local capital and labor tends to be concentrated in the infrastructure system, particularly transportation and public utilities. Because of the disruption caused by failure of these infrastructure inputs, output falls below what would normally be predicted based on the surviving local capital and labor inputs. Put another way, output will be below what would ordinarily be expected under technically efficient operation of the surviving capital and labor because essential inputs to the production process are missing.

Another indirect ex post effect of natural disasters is a change in the composition of local output. The postdisaster period includes a shift toward production designed to replace the damaged local capital stock and an increase in imported inputs. These changes are often financed by receipt of insurance payments or public transfers, and they may last for two or three years. Finally, the area should return to a long-run equilibrium level of growth in production and input accumulation based on fundamentals characteristic of the economy prior to the disaster. All these direct and indirect effects arise even when the occurrence of the disaster event has no effects on perceptions of the likelihood of future losses. Most of the literature on the economics of natural disasters focuses on analysis and measurement of direct effects along with some subset of likely indirect effects depending on the purpose of the study.

The theoretical role of expectations and informational effects

There is substantial evidence that natural disaster events have informational effects whether they occur or not and that the economic consequences of these ex ante (before the fact) effects may be larger than either the direct or indirect effects. Informational effects arise ex ante because disaster events are inherently local in character and expectations for future events, both their frequency and severity, are very uncertain. As noted by Kunreuther and Kleffner (1992), the range of probability estimates for serious earthquakes offered by experts is often sufficiently broad to argue for major mitigation expenditures (at the high probability end) and no mitigation (at the low end). The occurrence or nonoccurrence of disaster events over a period of time allows for substantial updating of these expectations. Surveys of households before and after disaster events show a substantial change in attitudes toward purchase of insurance and expenditures for mitigation. Brown (1972) provides an early discussion of the likely informational effects of disaster events and models a process in which prior expectations of disaster probabilities are subject to Baysian updating based on recent disaster experience. Note that, in such models, nonoccurrence is just as important as occurrence in its effect on expectations. Recent experience is added to the individual's information set, and expectations are updated periodically.

The potential importance of disaster expectations was demonstrated by Ellison et al. (1984) through use of a small-region econometric model of the Charleston, South Carolina, area that included investment equations in which expectations played a role. They were then able to simulate the effects of an earthquake under alternative ex ante informational scenarios: a fully-anticipated event, an unanticipated earthquake, and an earthquake announcement with no subsequent earthquake. There are a number of features of this modeling effort that are noteworthy and will be discussed later, but the essential informational feature is the comparison of an unanticipated event with a true prediction and a false prediction. Because the model was structured so that migration of labor and capital (investment) depended on expectations, credible earthquake predictions that are not believed have a substantial negative effect on output in the affected area. These negative effects would probably have been larger if the model had incorporated an option to invest in mitigation efforts. Ellison et al. (1984) reach conclusions that should be sobering for those who predict disasters. While they find that the present discounted value of local production when a local earthquake is correctly predicted is greater than the present discounted value of production when there is an unanticipated earthquake scenario, this gain from correct production of an earthquake occurrence is far less than the fall in present discounted value of production when an earthquake event is incorrectly forecast.[2] The losses from an incorrect prediction exceed the gains from a correct prediction.

If false announcements can have substantial ex ante information effects, then it follows that disaster events themselves may have information effects because they influence future expectations. This view is advanced in Brown (1972) and tested empirically in Yezer and Rubin (1987). Papers by MacDonald et al. (1987), Shilling et al. (1989), and Brookshire et al. (1985) relate the asset prices of housing in a given area to proximity to natural disaster hazards, flood planes, or earthquake faults. The standard finding is that houses farther from the hazard sell for higher prices that appear to reflect differences in insurance costs.[3] Bernknopf et al. (1990) demonstrate that the announcement of a probable future disaster in Mammoth, California, had significant negative effects on both real property investment and residential property values but not on recreational visitation. These results suggest that there is a market response, at least on the part of investors, to new information on disaster probabilities in a specific area. One limitation of these studies is that they involve comparisons of a single change in disaster probabilities. While this is sufficient to show the importance of information effects of natural disasters, it gives little insight into the dynamic component of expectations formation. It would be useful to know more about the process that individuals use to formulate their disaster expectations.

Empirical evidence that disasters have significant informational effects

In addition to the Bernknopf et al. (1990) demonstration of specific announcement effects in Mammoth, California, Yezer and Rubin (1987) tested a model in which the divergence between actual disasters and disaster expectations was related to changes in a local economy. First, they developed measures of the expected frequency of natural disasters for cities based on their previous disaster history. Comparison of this expected frequency, based on historical disaster rates, with subsequent disaster experience over a four-year period allowed them to compute the unanticipated component of recent disaster rates, based on the divergence between actual and expected (historical) disaster rates.[4] Second, they identified the rate of change in the price of housing services as one appropriate indicator of the local economic effects of disaster expectations. Third, they estimated a model relating the number of disaster events, the number of unanticipated disaster events, and other factors to the change in local housing prices. The test results were consistent with expectations that the number of disaster events had no significant relation to house price changes, but the unanticipated component of recent disaster experience was negatively related to the rate of house price appreciation. Specifically, occurrence of one unanticipated disaster event was estimated to lower house prices by two percent, or a negative effect of several hundreds of millions of dollars in a city with 250,000 households.[5]

Theoretical effects of disaster expectations on development

It is useful to compare models that include the information effects of unanticipated disasters with the results of models that ignore information effects and implicitly ignore local differences in disaster expectations. The evidence from case studies of disaster events is mixed. Many have shown significant long-term effects, both positive and negative. The record also contains observations of little or no effect (see Friesema et al. 1979). Rossi et al. (1982) report results indicating that natural disaster concerns are not particularly important among public officials, many of whom might be charged with dealing with their consequences. A major econometric study of a large national cross-section of disaster events occurring during the 1960-1970 period, which was conducted by Wright et al. (1979), found no long-term effects on population or housing trends. While this study might be criticized for using only population and housing units as indicators, the Yezer and Rubin (1987) results suggest that no significant effects would be observed even if housing or land price changes were being analyzed. Failure to find significant economic effects of the rate of occurrence of disaster events merely indicates that, on average, disaster events occur where they are expected. It does not imply that unanticipated increases in the frequency or severity of disaster events could not have significant economic consequences or attract the attention of public officials.

Models of natural disasters with information effects have significant implications for patterns of residential land use. These implications have been developed by Frame (1998), applying the standard (monocentric) urban model to a city with differentiated levels of natural disaster expectations. This application of the urban land market models has much in common with earlier modeling efforts by Scawthorn et al. (1982). Frame (1998) models a closed urban area in which a fixed number of households, earning a given market wage at a central location, choose housing quantity and location to maximize utility from consumption of housing and a composite good whose price does not vary.[6] Within the urban area there are different location-specific expectations of the probability of damage from natural disasters.

Because households are free to move throughout the city, the price of housing services must fall in less attractive areas—i.e., areas with larger commuting cost to center city jobs or higher expectations of disasters. This is not a particularly strong assumption, given that it has been demonstrated in empirical studies by Shilling et al. (1989) and Brookshire et al. (1985). All of these studies have demonstrated the presence of price discounts for increases in expected risk, but empirical evidence on other market reactions, such as lower rates of investment or lower population densities, have not been the object of substantial testing.

Based on this assumption of household mobility, Frame (1998) is able to establish the following theoretical propositions regarding the effects of increasing the expected frequency of hazard events. Producers of housing respond to the lower price per unit of housing services in areas with greater hazard risk by bidding less for undeveloped land in such areas and producing housing services with a lower ratio of capital to land—i.e., the ratio of housing services per unit of land is lower in areas with greater hazard risk. In areas characterized by a mixture of high commuting cost and/or high hazard risk, land prices may be depressed to the point where no housing is built. The effect of the lower supply price of housing on quantity of housing demanded in areas with greater hazard risk is ambiguous because lower price encourages consumption and decreases the significance of any damage that is done, but greater hazard risk also makes housing consumption less certain. Because the effect of greater hazard risk on quantity of housing demanded by households is ambiguous, it is also not possible to prove that increasing risk lowers the density of population. However, this effect seems likely and certainly is true when increasing risk lowers developed land rents below the level needed to promote any residential development—i.e., when land is withheld from residential use, population density falls to zero.

The model developed by Frame (1998), like the earlier efforts of Scawthorn et al. (1982), generates downward shifts of housing prices and development densities in parts of the urban area where disaster probability is high. This result fits intuition well, with natural hazards operating in a manner similar to any natural barrier that raises costs of producing housing. It is likely that analogous results could be obtained for noxious plants producing outputs on such sites, although an additional consideration would involve any supplemental wage compensation required to attract workers to a hazardous site.

Frame (1998) also provides a number of results on the effects of disaster insurance. Provision of unsubsidized insurance, at a price reflecting the actuarial value of future losses, allows households to diversify away geographic risk. Because the premium payment increases with expected loss, provision of unsubsidized insurance does not change the negative relation between hazard probability and the price of housing and the density of housing production per unit of land. If insurance is subsidized, the effect of hazard risk on housing prices and on the density of housing supplied is attenuated compared to the unsubsidized or no insurance cases. The positive effect of insurance subsidies on housing prices was noted empirically by Shilling et al. (1989). While these predicted effects of providing actuarially fair or subsidized insurance may generally agree with intuition, a final result provided by Frame (1998) is likely surprising. The effect of insurance, even subsidized insurance, on the density or welfare of population in high risk areas is ambiguous.[7] It would be interesting to extend these results to open-city models in which one city is subject to greater natural hazard risk and must compete for capital and labor against another city.

The Frame results are compatible with models of information updating in which recent rates of disaster activity or announcements of new disaster forecasts are used to update disaster expectations. This implies that either the recent rate of disaster events or announcements concerning such events can have information effects to the extent that expectations of future disaster events in that area are changed. The result will be a change in the prices of housing and land and in desired density of development. These predictions appear to be testable.

MODELS OF THE INSURANCE MARKET, MITIGATION, AND POSTDISASTER AID

The rising cost of insurance claims, both private and government-sponsored, in the last decade has stimulated substantial interest in various aspects of the provision of insurance against losses from natural disasters. Among the most important issues appear to be: (1) reluctance of individuals to purchase insurance and invest in mitigation efforts; (2) the interaction among insurance, mitigation expenditures, and postdisaster aid and their effect on economic development; and (3) the design of financial instruments appropriate for capitalizing private insurance against major disaster events. Examples of the literature on these topics will be considered in turn. All three have been the object of significant recent research efforts motivated in no small part by the rising costs to both private insurance companies and government of natural disaster events in the United States during the 1990s.

Behavior and perceptions of property owners in high-risk areas

Significant efforts have been made to assess whether individuals in high-risk areas perceive those risks and act upon them in their insurance and mitigation decisions. The National Flood Insurance Program (NFIP) has served as a "poster child" for nonparticipation. Kunreuther (1978) noted that, in the first four years on the program (1968-1972), only 3,000 of 21,000 eligible communities with substantial flooding history participated in the program, and fewer than 300,000 homeowners voluntarily purchased a policy. In spite of the fact that NFIP was subsidized, participation was only accelerated by threatening communities with denial of federally assisted or guaranteed postdisaster aid, specifically construction or mortgage loans, to property owners in nonparticipating communities that had been identified as special flood hazard areas.[8] Palm et al. (1990) have documented the failure of homeowners and even mortgage lenders to seek earthquake insurance in high-risk areas.

Taken at face value, this stylized fact regarding failure to insure has, in the view of some, called into question the use of standard models of the economics of insurance in the special case of natural disasters. Kunreuther and Kleffner (1992) and Kunreuther (1996) have argued that homeowners do not behave as if they were maximizing expected utility in their decisions to purchase insurance or engage in private mitigation efforts.[9] They hold that behavior most closely resembles the contingent weighting model in which individuals place different weights on the probability of a hazard event and on the contingent loss should it occur. If individuals place low weights on the probability of infrequent natural hazard events, they will act as if the expected utility of insurance and mitigation is lower than that perceived by experts. The failure of the public to use seat belt restraints has been rationalized in this way. Kunreuther and Kleffner (1992) consider simple benefit-cost estimates of a loss mitigation measure, bracing pre-1940 homes in California so that they would not slide off their foundations. They find that low rates of mitigation appear to suggest low probabilities of hazard events—or low weighting of those probabilities. Overall, it appears that property owners in areas subject to natural hazard risk underspend on insurance and mitigation efforts compared to standards that are justified by sound benefit-cost analysis.[10] Even where building codes require cost-effective mitigation efforts, Kunreuther (1996) reports serious cases of underinvestment and failure to comply, resulting in significant additional damage, as in the case of Hurricane Andrew.

In contrast to the evidence that property owners in high-risk areas ignore those risks, research also indicates that their behavior changes with additional information. While property owners are observed to be "insensitive" to levels of risk, they do respond to "changes" in risk. Most dramatic is the finding by Palm et al. (1990) that, prior to the Loma Prieta earthquake, there was no relation between proximity to the San Andreas fault and the rate of purchase of earthquake insurance, but that after the earthquake the purchase of insurance increased and insurance rates fell with distance from the fault. Survey evidence indicates further that attitudes toward the importance of earthquake insurance shifted substantially after the Loma Prieta earthquake. There is also evidence that market prices reflect learning behavior. Brookshire et al. (1985) find that publication of information on the distance from properties to established fault lines causes house prices to fall within the earthquake-prone zone. Shilling et al. (1989) found that NFIP subsidies to pre-FIRM structures were capitalized into house values. As noted in some detail above, Yezer and Rubin (1987) find that housing prices respond to unanticipated disaster events (i.e., to differences between the frequency of floods, windstorms, etc.), during a given period and the previous history of such events. Thus it appears that the standard insurance model, in which expected loss and the variance in expected loss motivate behavior, works in a dynamic context. Changes in disaster events have the predicted effect on property owners perceptions and behavior as well as on market prices.

Research on the economic behaviors and responses of property owners in high hazard risk areas are at least partly consistent with standard expected utility models employed to explain expenditures for insurance and mitigation. Surely, more research is needed to resolve these issues, and they are important to public policy toward insurance, mitigation, and postdisaster aid. One possible explanation for the failure of property owners in high-risk areas to make insurance and mitigation decisions that appear economically justified is that they are selected into high-hazard areas based on their perceptions of risks. If markets provide price discounts for properties in areas with greater hazard risk, those properties will be differentially attractive to individuals who fail to perceive the risks. Alternatively, if there is a risk distribution, those whose expectations of actual risk is at the low end of expert opinion will likely occupy properties in areas rated as highest risk by the experts. With low probability events where expert opinion on frequency has high variance, particularly events such as earthquakes, it would not be surprising if property owners in the highest risk areas believed that risks were at the low end of the distribution of expert opinion. Subsequent occurrence of an event could have a dramatic effect on these expectations, as survey evidence and behavior have shown.

Interaction among insurance, development, mitigation, and postdisaster aid

From the point of view of the property owner, insurance, mitigation, and postdisaster aid are all substitutes, providing either funds to cover losses or lowering losses contingent on a natural disaster event occurring. As noted by a number of authors, establishment of public subsidies for insurance or mitigation or the provision of postdisaster relief have the effect of shifting the burden of loss, both loss expected ex ante and loss experienced ex post, from the property holder to others.[11] Unsubsidized insurance shifts the burden of loss ex post to those entities who have purchased the contingent liability but not ex post as actuarially sound premiums reflect expectations of loss. These issues are part of current public policy debates regarding the establishment or extension of government insurance programs and requirements for mitigation in the form of zoning.[12] Another response has been the Coastal Barriers Resources Act of 1982 which prohibits federal government programs, including the NFIP, from activity in areas designated as part of the Costal Barriers Resources System (CBRS). New development in such areas is totally dependent on private funding for infrastructure, mitigation, and insurance. This has led to the curious situation in which, along a given beach, successive areas are full participants in NFIP and beach nourishment programs while neighboring areas rely entirely on private-sector development and insurance.

The expectation that government subsidized insurance and mitigation followed by postdisaster aid will lead to overdevelopment and under-mitigation in high-risk areas is largely based on partial equilibrium models in which the expected cost of hazard events born by property owners is reduced by the subsidy. There is some empirical evidence in Schilling et al. (1989) that subsidized insurance is capitalized into land values. This is consistent with Frame's (1998) general equilibrium result that insurance subsidies raise land prices, house prices, and the physical density of development, while the effect of insurance subsidies on the density of population in high-risk areas is ambiguous, except that some areas previously considered too risky for development are developed.

Cordes and Yezer (1998) test directly for the empirical effects of subsidized insurance through the NFIP and mitigation in the form of beach enhancement by the U.S. Army Corps of Engineers. They estimate a model of beachfront residential development, specifically the rate at which new building permits are issued, for a panel of beach communities that allows for the observation of communities with and without insurance and enhancement programs and also observation of given communities before and after programs were implemented. The results suggest that provision of subsidized insurance under the NFIP encouraged higher levels of real estate development in beachfront communities. However, subsidized mitigation in the form of Corps of Engineers beach enhancement had no such effect on development. These results may appear surprising because only a modest fraction of property owners in beachfront communities chose to maintain insurance in force. Nevertheless, insurance availability appears to have encouraged new construction, while beach enhancement failed to do so.

Overall, a number of papers have produced theoretical results suggesting that provision of subsidized insurance or mitigation efforts tends to increase the density of real property development in high-hazard areas. The implications of insurance for both development and mitigation in such a general equilibrium setting have been neglected along with the implications of postdisaster aid. Finally, there is little empirical evidence on the size of these real estate development effects. The effects of insurance availability and/or public mitigation efforts on the decision to develop high hazard areas are clearly important and should be the object of additional testing.

The interaction between development and mitigation efforts and the provision of postdisaster aid is well recognized as an important policy concern but has not been the object of substantial formal modeling. A number of authors, including Rubin (1982), have noted that systematic withholding of disaster relief from individuals who have failed to mitigate risks is not politically feasible. Most disaster relief schemes clearly include a measure of equity and efficiency for reasons discussed in Dacy and Kunreuther (1969). These implicit equity-efficiency tradeoffs have been examined explicitly by Butler and Doessel (1981), and a scheme for partial loss compensation, with the replacement rate varying inversely with income, developed as a remedy. The full set of tradeoffs among ex post disaster relief expenditures and ex ante decisions regarding development and mitigation expenditures was developed by Lewis and Nickerson (1989). In this model, risk averse households living in areas varying in hazard risk can invest either in risky structures or in risk-preventing mitigation. While "expected" findings of excess spending on risky structures and underinvestment in mitigation in the presence of disaster relief are obtained from this model, the results are complex. Mitigation expenditures designed to prevent small losses are undertaken because those losses do not result in aid, but there is underspending on efforts to avoid substantial losses where relief is expected. The authors conclude with a list of potential research extensions that are still waiting to be made.

Design of financial instruments for capitalizing private insurance

In addition to the concern expressed by Kunreuther (1978, 1996) and others that property owners fail to demand actuarially fair insurance against hazard losses, there is also concern with the supply side of the hazard insurance market. Insurance markets normally provide policies in which the price of coverage reflects differential risks of loss plus a margin for normal returns and administrative costs. Companies maintain sufficient capital to cover expected losses and, in periods where losses are high, are able to raise additional capital at moderate cost. Private insurance should price risk appropriately and provide incentives for deterring investment in high-risk areas and for mitigation efforts, in the form of reduced premiums. A number of sources, including the Congressional Budget Office (1992) and Kunreuther and Roth (1998), have noted that private insurance markets have failed to perform this function for property owners in areas with high hazard risks. Nine smaller insurance companies failed in the wake of Hurricane Andrew while others tried to reduce coverage or asked for increased premiums. The public-sector response was to insist on continuing coverage at current rates. Clearly, neither the initial structure of the market nor the regulatory response fit the normal insurance market model.[13]

In response to the problems faced by private insurance companies providing coverage in areas where expected hazard events are particularly severe, a recent literature has considered innovative financial instruments and changes in tax law that could provide substantial capital for companies supplying disaster insurance. Problems identified in the current system as noted by Cummins and Geman (1995), Lewis and Murdock (1996), and Jaffee and Russel (1997) arise from the extraordinary payouts of claims in excess of the flow of premium revenues. These payouts produce periods of negative cash flow and profits which reduce the accumulated capital reserves of insurance companies when major disaster events occur. Four possible solutions to this problem have been considered. Obviously, premium rates must be sufficient to cover expected losses plus normal profits and administrative costs.[14] Second, private reinsurance markets could be further developed to better diversify risk across the nation and even the world. Third, tax law changes could delay recognition of profit on insurance against natural disaster events by substituting a multi-year accounting period for the current annual period used to compute taxable income. Actuarially fair insurance will generate apparently large accounting profits over short periods followed by large losses when disasters occur. Unfortunately, it may be difficult for insurance firms to recapitalize after particularly large losses even if they have tax advantages associated with a large-loss carryforward. While corporate profits taxes may appear to diversify some of the risk of large claims to the government, the very uneven nature of disaster losses appears to reduce the value of the diversification benefits.[15]

The fourth solution to the capitalization problem has captured recent attention and involves the design of contingent liabilities whose value depends on the amount of disaster claims losses in any given period. Purchasers of such liabilities earn high returns in periods where claims are negligible and low or even negative returns when major disaster events produce high claims rates. This line of inquiry has been spurred by specific attempts in the private sector to develop new instruments including "Act of God" bonds and catastrophe futures and options.[16] Jaffee and Russell (1997) note that accounting requirements, tax provisions, takeover threats against firms with large cash surplus, and regulatory constraints all limit insurance company capital held for catastrophes. Nevertheless they argue that primary problems of disaster insurance reside in capital markets and that some form of government-sponsored reinsurance may be required. In order not to crowd out private-sector innovations, the government reinsurance program would only apply to disaster events in which aggregate losses exceeded some substantial threshold, say $20 billion (U.S.). The ability of insurance markets to serve areas subject to large-scale disaster losses is an active topic of research involving both insurance and capital market issues. Actual innovations are producing new opportunities for analysis even as this review was being written.

EMPIRICAL MODELS OF ECONOMIC IMPACTS OF DISASTER EVENTS

These models trace the likely economic effects of a disaster event. In contrast to the theoretical models of effects on housing markets and residential real estate development, these models emphasize effects on industrial output, although effects of rebuilding of the residential capital stock are usually included. The effects of disasters are generated by simulating a baseline economic scenario and then comparing this with one or more disaster event scenarios. The appropriate comparison is economic activity with and without the disaster event. It is important to model effects so that all costs can be measured without double-counting. For example, a model should estimate both the cost of damaged property and subsequent output loss due to inoperable equipment. However, it is not appropriate to add both the value of damaged equipment and costs of replacing that damaged equipment. A model should also be capable of estimating costs over time, because a cost experienced today is worth more than a cost years after the disaster event. Finally, it is important to distinguish positive models of the most likely effects of a disaster and normative models that attempt to reallocate resources in the postdisaster period in order to minimize losses.

There are two distinct modeling approaches, regional econometric models and interindustry (input-output) models. In either case, some special adaptations of the standard regional model are necessary if it is to be used for estimating effects of disasters. Specifically, disasters lower subsequent output levels by disrupting the availability of inputs locally and the ability to transport inputs from alternative locations. Most regional econometric and input-output models are designed to estimate the effects of fluctuations in national demand for regional products on regional output, assuming that input supplies are very elastic. Therefore, modeling effects of supply disruptions as well as efforts to rebuild damaged local capital stock requires some special adaptation of the usual model.

The two primary examples of regional econometric models adapted to study regional disaster impacts are Ellison et al. (1984) and Guimareas et al. (1993). Regional econometric models consist of a series of equations characterizing output, usually disaggregated by a few broad sectors, employment, consumption, labor supply, wages, and prices with additional specialized modules, often including local government and housing. In order to capture the effects of natural disasters on local economic activity, the econometric model should include equations reflecting the process of local investment and capital stock accumulation and then relate local output levels to both capital and labor inputs. Special attention to residential real estate production and accumulation is also needed if the recovery process is to be modeled. The model should allow both labor migration and local capital investment decisions to be modified by expectations in response to disaster events. It is also important to include a role for transfers from private insurance and public disaster aid. The regional econometric model is usually estimated using annual data often available for only 30 or 40 years. With so few observations, the complexity of individual equations is limited. Ellison et al. (1984) argue that it is important to disaggregate the region into subareas when disaster events are localized within an urban area. Data are available to disaggregate Metropolitan Statistical Areas by county. Overall, substantial modification of traditional regional econometric models is necessary if these models are to be used in disaster impact evaluation. A guide to these needed modifications along with practical suggestions for their achievement in the face of data limitations has been provided by West and Lenze (1994).

Regional interindustry models are traditionally demand driven within which changes in exogenous national demand for regional exports drive the local economy. However, these models can been substantially modified to include supply-side feedback effects, and their superior industrial disaggregation allows them to trace specific interindustry effects of supply disruptions. Earlier examples of the adaptation of input-output approaches include Cochrane (1974) and Rose and Allison (1989). A regional input-output modeling effort requires that the extent of local interdependence among industries be identified. The extent to which local industries rely on inputs of locally-produced intermediate product is made apparent in the modeling process. Accordingly, the output effects of an interruption of these sources of local supply can be determined. Of course, it is difficult to determine the effect of a disaster on the availability of inputs imported from outside the area to replace local supplies. But the input-output model can be used to estimate the effects of local supply interruptions under alternative assumptions on the availability of imports.

It is also possible to adapt interindustry models to an optimizing mode and to draw implications for the effect of alternative postdisaster responses on local output. The optimizing model works to achieve an allocation of infrastructure inputs to minimize the output effects of disruption. This literature has reached very sophisticated levels with models such as Rose et al. (1997) in which programming models are used to allocate scarce electricity disrupted by a disaster event. This model is remarkable in the extent of geographic disaggregation which allows for the simulation of effects of electricity outages at the substation level. The model operates in normative mode to identify patterns of electricity restoration that minimize the loss of local value added in the postdisaster period.[17]

Although there is a tendency to think of using economic impact models to estimate the likely losses from disaster events, their greatest value may be in simulating the effects of alternative approaches to postdisaster recovery.

CONCLUSIONS REGARDING THE LITERATURE ON THE ECONOMICS OF NATURAL DISASTERS

There are two obvious conclusions from this review. First, the economic effects of natural disasters is a remarkably under-researched area. The total number of citations to research in economics of disasters is very small, and the number of potential topics is quite large, leaving a very thin rate of coverage. Second, various approaches to the economics of disasters have remained segregated. Theoretical models including information effects have had little effect on economic impact modeling. There are few empirical tests of information effects. The relation among insurance, mitigation, and postdisaster relief has been examined, but the exploration is far from complete. One branch of literature has concentrated on the response of households and residential housing markets while the economic impact literature has dealt largely with firms. Hopefully this review will inspire additional efforts to fill in gaps between the literatures and, more important, to produce models that integrate the economic effects of disasters on both households and firms.

NOTES

1 For example, at the end of the 1990s the on-line index to the *Journal of Economic Literature* listed only thirty-eight citations under the topic "natural disaster," seven under "earthquake," and eighteen under "floods."

2 The loss from an incorrect prediction is computed as the difference between present value of production under a no-earthquake baseline scenario and present value of production when an incorrect earthquake forecast is announced to the public.

3 A number of studies have found that events with low probability but high losses are reflected in housing prices. For example, Baker (1986) has shown that hazardous waste leakages reduce property values near the leakage site.

4 The expectations hypothesis regarding market responses to disasters implies that, if the frequency of disasters in each city during the 1980s were identical to prior expectations, then the observed disaster rate in each city would have no effect on economic activity. Unanticipated disasters are equal to zero in this case. If actual disaster experience were significantly higher (lower) than expectations, the expectations hypothesis suggests that disaster expectations would rise and the consequent negative (positive)

effects on employment, housing, and land rents discussed above would be observed. For example, the occurrence of three floods during the 1960s in an area expected to have one (three) [five] floods per decade should have a negative (neutral) [positive] effect on expectations of flood danger and a corresponding positive (neutral) [negative] effect on the local economy. In an area expected to flood three times per decade, the danger of flooding has already been discounted at that frequency and is reflected in both land values and levels of employment and population. As unanticipated disasters rise from -2 to 0 to +2, the local economy experiences increasing negative effects.

5   The mean asset price of housing in their sample was $64,000 in 1979. For a city with 250,000 housing units, the two percent fall in asset price implies a decline of over $300,000,000 in the aggregate asset prices of housing units due to the effects of a single unanticipated disaster event in a city where disaster expectations had been zero. A loss of 2 percent in housing asset price implies a fall of approximately 8 percent in the underlying land values. If this fall were extended to all values of developed land throughout the city, the economic effects would be far larger than the $300,000,000. Unfortunately, extending estimated information effects from residential to commercial and industrial real estate is difficult due to data limitations and potential modeling issues.

6   Hazard events lower utility of households through the damage that is done to housing and the consequent reduction in consumption. The loss function is sufficiently general to allow for a direct financial cost reflecting personal danger to the household. By assuming a closed city, the model allows wages and population to vary independently. An open-city model would hold utility constant by allowing intercity migration to equalize welfare.

7   Insurance subsidies increase the welfare of current landowners by raising housing prices, but they do not increase population density or the welfare of those living in the high-risk area after housing prices rise. The reason for this counterintuitive result is that, when insurance subsidies raise housing prices and the quantity of housing produced per unit land, they also raise the demand for housing per household, which makes effects on density of households per unit of land ambiguous.

8   While the subsidy component of the NFIP for structures constructed since Flood Insurance Mapping (or January 1975) is likely small, older structures receive a significant subsidy and yet have significant nonparticipation rates. This occurs in spite of requirements that lenders require flood insurance as a condition for giving a mortgage because owners have allowed policies to lapse after a modest period without hazard losses. Overall, only about one-fourth of homeowners in flood-hazard areas have purchased NFIP insurance.

9   Kunreuther (1978) first argued that property owners fail to purchase insurance and spend for mitigation some time ago.

10 See Palm et al. (1990) for survey evidence on the failure of property owners to adopt any mitigation actions in earthquake-prone areas of California.

11 Note particularly the extensive discussion in Kunreuther (1996).

12 See, for example, Congressional Budget Office (1992) for a discussion that treats insurance, mitigation, and disaster relief as substitutes and the possibility of moral hazard problems associated with public expenditures for any one of the three. Kunreuther and Kleffner (1992) make the case for mandatory mitigation.

13 See Kunreuther and Roth (1998) for a discussion of hazard insurance regulation and subsidies, including the California Earthquake Authority and the Florida Joint Underwriting Authority.

14 Regulators in areas experiencing high loss rates may be under pressure from constituents to keep premiums low and maintain coverage rates when companies attempt to retreat from the market, as in Florida after Hurricane Andrew.

15 This argument, that corporate taxation of insurance company profits reduces the attractiveness of high-risk business such as disaster insurance, contrasts with the normal theoretical result that profits taxes increase risk-taking because the variance in after-tax profits is less than that of before-tax profits.

The literature argues that large disaster events create huge losses and that the consequent threat of bankruptcy may deter recapitalization of companies even if they have large tax losses.

16 The pricing of catastrophe futures and options is taken up in Cummins and Geman (1995).

17 The modeling effort includes numerous clever tricks that allow the authors to overcome limitations in the availability of data at the local level. These should be useful in a variety of research applications.

9

CROSS-NATIONAL AND COMPARATIVE DISASTER RESEARCH

Walter Gillis Peacock

Pick up almost any work on comparative research, and it will invariably begin with an attempt to distinguish and clarify it from other forms of social science research. The distinction, however, is far from clear (Lane 1990; Przeworski and Teune 1970; Grimshaw 1973). Many researchers suggest that the very essence of all social science research is comparison (Swanson 1971; Lieberson 1985). Regardless of our unit of analysis or method, as researchers we are explicitly or implicitly comparing our observations either against each other or against some theoretical or ideal type. This position suggests that there is little reason to consider "comparative" research a special form of social research and is tacitly confirmed by the lack of sections devoted to comparative research methods in most social-science research texts.

Other researchers suggest that comparative research is inherently different, although the nature of its distinctiveness is identified in a variety of ways. For some, the essence of comparative research is cross-national comparison, regardless of method or unit of analysis (Kohn 1989a; Ragin 1987; Øyen 1990a). Others suggest that comparative research demands macro-level comparisons, although some controversy remains concerning the level of scale and even the abstraction of these units. More often than not the units are defined as a society, nation, country, world-region, or even the world system, and these units are defined as real (Ragin 1987). However, given the slippery nature of these concepts, still others consider intra-national comparisons between regions, communities, or even among cultural, racial, or ethnic groupings as legitimate.

This latter position suggests that comparative research need not be cross-national. Despite the fact that much of the research I have engaged in is cross-national, living and working in Miami certainly makes me question rigid definitions demanding that research be cross-national for one to properly apply the label of comparative. The heterogeneity found in Miami and increasingly elsewhere in the U.S. and many other nations demands that we remain on our theoretical and methodological toes, particularly when we are interested in comparing among a variety of populations, neighborhoods, or jurisdictions. Many issues encountered when conducting research in Guatemala, Mexico, Peru, Italy, and Cuba, for example, will also be encountered when conducting research in Miami. Indeed, I have been guilty of failing to consider certain processes that were important in other countries, only to discover after it was too late that similar processes occur here as well. In this light, it is certainly possible to consider intra-national research to be comparative, particularly when focusing on ethnic, racial, class, or gender issues.

It is beyond the scope of this chapter to solve the problems and issues related to the controversies about the nature of comparative research and why it might be different from other forms of research. For my purposes, the essence of comparative research is the explicit comparison of similar units that are at least initially perceived to vary in terms of their structure, organization, culture, or context, and these factors are generally considered important determinants of variations in the phenomenon under study. This characterization is probably inherent in most cross-national research situations where intra-national variation is ignored, and the focus becomes international differences. And yet, comparative research can certainly be undertaken within a nation or society when comparing variations in structure, context, or culture among more micro-level units. While I will, for the most part, focus on more conventional comparative work involving cross-national comparisons, much should also be relevant for intra-national comparative research.

DOES THIS DIFFERENCE MAKE A DIFFERENCE?

I do not necessarily consider cross-national and comparative research qualitatively different from other forms of social research. The same issues confronted in dealing with everything from the logistics of survey research to the most esoteric issues of estimation procedures and latent structure analysis will potentially confront one engaged in both forms of research. However, particular and peculiar issues can confront comparative researchers that might not be encountered in the normal course of social investigations. Indeed, it is often the theoretical and methodological issues safely ignored in the course of normal research that can become threats to validity in comparative situations. For example, issues such as the simple interpretation, and sometimes translation, of every single word used in a questionnaire as well as our ideas about what constitutes a household, what is the structure and function of local government, and how cost evaluations are determined must in the comparative situation be clearly identified and defined. Every step in the research process requires considerable forethought. In other words, it is not that the problems and issues related to comparative research are so different but rather that they are more pronounced and compounded, allowing little to be taken for granted (Øyen 1990a).

Comparative and cross-national research can offer additional challenges. Often these research projects require multi-national and multi-disciplinary teams (Øyen 1990b; Hantrais and Mangen 1996b). These situations demand and require the sharing and modification of ideas and perspectives. Our cognitive maps and models will be challenged, demanding that we modify, develop, and incorporate new ways of thinking about the world and how research is conducted. Initially these can be frustrating experiences, particularly when one is anxious to get on with the research. Despite the frustration and time that is often entailed in dealing with these difficulties, they can also have long-run intellectual payoffs, generating new perspectives, ideas, concepts, and techniques that would not have otherwise have emerged.

The intellectual and scientific gains achievable through cross-national and comparative research, however, cannot occur if we seek only to impose our ideas and methods in the new situation. Social science research conducted in the early post-World War II era often suffered from this problem, and mistakes continue to be made (Calderon and Piscitelli 1990). Of course, certain methodological principals and techniques are not open to modification without severely compromising the research; however, if we simply impose our theories, concepts, and methods without critically evaluating their applicability in different research contexts, very little will be learned and even less achieved. Kreps and Drabek (1996: 144) recently observed that, "It is an interesting exercise to speculate how our prevailing images of disaster responses, and the conceptual tools used to assess them, would differ if the preponderance of events studied had occurred in Poland, Mexico, Iran or Kenya." I would agree and, while lamenting the lost opportunities, suggest that this speculation can be furthered by future work with colleagues from these and other areas.

My goal is to offer at least one person's perspective on the future of cross-national comparative research related to disaster. I will not attempt to summarize the history of cross-national comparative research in general nor in disaster research in particular. These things have been done elsewhere (e.g., Armer and Grimshaw 1973; Drabek 1986; Dynes 1988; Oliver-Smith 1996; Kohn 1989b; Hantrias and Mangen 1996b). What I will offer are some reasons why cross-national comparative research is likely to increase and touch on some of the methodological problems and issues likely to be encountered. Lastly, I will suggest some potentially fruitful areas that should be explored.

INCREASING OPPORTUNITIES FOR CROSS-NATIONAL COMPARATIVE RESEARCH

The future of cross-national research appears to be much brighter than it may have been in the not-too-recent past. This stems from a variety of changes. First and foremost, there now exist much greater appreciation and realization of general globalization of all processes and problems (Dynes 1988; Øyen 1990a; Teune 1990). Of course, many social and physical scientists have long recognized that issues of poverty, underdevelopment, resource depletion, climatic changes, and the like cannot be understood unless they are put into a global context. Similarly, to understand the production of and increases in vulnerability and risk demands an understanding of how they fit into larger processes of social, political, and economic development and underdevelopment, population growth, etc. Even the most conservative politician recognizes the internationalization of finance, labor, production, insurance, and, even more generally, of the information revolution. Furthermore, the United Nations' International Decade for Natural Disaster Reduction (IDNDR) and other disaster reduction initiatives by the Organization of American States (OAS), the Office of Foreign Disaster Assistance-Agency for International Development (OFDA-AID), the European Community Health Organization (ECHO), and the Pan-American Health Organization (PAHO), to name a few, have fostered increased awareness of the international connectivity of disasters and the necessity of seeing disaster reduction as a global need. These initiatives, as well as the changing appreciation of the global nature of many problems, have increased funding opportunities for international disaster-related research.

Equally important is the growing population of disaster researchers world wide. Consistent with the increasing realization of the importance of disaster mitigation, planning, preparedness, and emergency management activities among all nations, we have seen increased interest in developing training and research capacities in institutions of higher education and training. The International Sociological Association's Research Committee on Disasters has been supportive of these efforts, providing opportunities for researchers to share their knowledge and experiences, and other international associations have emerged. For example, LA RED (Red de Estudios Socales en Prevencion de Desastres en America Latina) is a network of researchers in universities, public organizations, and nongovernmental organizations dedicated to disaster reduction and prevention in Latin America. The growth of these associations not only supports researchers and provides mechanisms for distributing their results, but also stimulates the formation of multi-national research teams, enhancing comparative research opportunities.

Yet another factor enhancing the possibilities of comparative research is the growing wealth of national and cross-national datasets and emerging resources related to disasters. The last several decades have seen increasing emphasis by both national and international entities such as the World Bank, the International Monetary Fund, and the United Nations on the systematic collection of national indicators related to economic and social development. These datasets form the backbone for additional efforts at developing indicators of political development, economic inequality, resource depletion and utilization, trade and production flows, and purchasing power (e.g., Gurr 1990; Jaggers and Gurr 1996; Taylor and Jodice 1983; World Resources Institute 1990). Recently we have begun to see efforts at developing cross national datasets on disasters and disaster impact by OFDA-AID, the International Federation of Red Cross and Red Crescent Societies (1995), and LA RED. The participants at the Hemispheric Congress on Disasters and Sustainable Development (1996) held in Miami called for increased efforts to create disaster inventories and datasets such as LA RED's DESINVENTAR, an inventory of disasters in Latin America. The growth and proliferation of national public opinion polling throughout the world, and the inclusion of questions related to environmental issues and risk, may also contribute to the growing body of data. And lastly, increased use of Geographic Information Systems (GIS) (Dash 1997; see also the chapter by Dash later in this volume) may also facilitate the generation of large cross-national data bases. While the utilization of these national and cross-national data resources will not be without problems, the net effect will be the creation of much greater opportunities for cross-national comparative research (Lane 1990).

METHODOLOGICAL APPROACHES TO CROSS-NATIONAL COMPARATIVE RESEARCH

It would be bordering on methodological arrogance to suggest that certain forms of comparative research, be they characterized as qualitative, quantitative, case study, cross-national, time-series or longitudinal, or cross-sectional surveys, take precedence over others. In fact, it can be difficult to cleanly characterize the many forms comparative research can take because that characterization depends upon the level of analysis at a given point in time. Mileti (1987) correctly points out that a large national study, regardless of how quantitative its analysis may be, becomes a case study with an N = 1 when our interest is comparative. And yet, labeling work a case study should not be a point of denigration. Ragin (1987) has noted that case studies have had longer-lasting and more profound effects on the social sciences than have many of the most elaborate cross-national studies based on a relatively large number of cases. The point is that all procedures and methods can potentially offer findings that can further our understanding of disaster-related phenomena.

For many, the future of social science research depends upon the development of more quantitative approaches, particularly as they relate to the testing of theories and ideas of social structures and processes (e.g., Blalock 1969). Further development of national and cross-national datasets and increased funding of longitudinal and cross-sectional surveys should facilitate quantitative comparative research, although they will not be without problems, particularly those related to data comparability. Most comparative research of this type will depend upon either cross-sectional or longitudinal data collected in one or more counties. Whether data from a single nation is analyzed and its findings are compared with research elsewhere or data from multiple counties are combined in some fashion for direct empirical comparison (cf., Bates and Peacock 1993), the central issue facing the researcher is the legitimacy and appropriateness of the comparison and subsequent generalizations. Do the data accurately reflect the nation or macro-units of interest? Are the comparisons between survey sites and/or nations legitimate? What generalizations can be appropriately generated by the findings? The answers will depend upon the nature of the specific questions and the types of answers sought. However, researchers must provide the consumers of the research with sufficient information to make these evaluations.

As disaster researchers attempt to utilize cross-national datasets to carry out multi-national comparisons, many methodological problems must be addresses. Some of these include problems of aggregation, intra-national heterogeneity or variability, and specification. The level of aggregation required to compare nations as well as subnational units such as states, counties, or communities can obscure variability and heterogeneity, resulting in very coarse measures at best. This problem was evident when attempts were made to examine the relationships between disasters and change (e.g., Wright et al. 1979; Friesema et al. 1979). We must ask ourselves if, in the pursuit of data, our theoretical questions have been so compromised by operationalization processes carried out at such a high level of aggregation that our tests bear little in common with the initial questions and concepts. Additional problems arise from obscuring intra-national variability or the variability among certain nations in favor of limited between-nation variability. This can result in erroneous empirical findings. Scheuch (1990), for example, suggests that when nations are inappropriately collapsed into categories, often labeled using theoretical-sounding concepts such as "periphery," "less developed," or "centrally planned," attention focuses on the limited variation among categories rather than the considerable within-nation differences, and erroneous findings are generated. Finally, being able to appropriately specify our ideas in modeling can be difficult in cross-national research. When attempting to specify a statistical model, for example, one tends to quickly run short of degrees of freedom the more elaborate the model, again compromising the testing of our theory.

Much can be learned by examining problems encountered by earlier cross-national comparisons, particularly in sociology, political science, and economics (Andorka 1990; Berting 1979; Kohn 1989b; Hantrais and Mangen 1996b; Hoover 1989). While some of these problems cannot be resolved, perhaps their severity can be lessened if, in the development of our datasets, attempts are made to lower the level of aggregation. This may well mean that, for the short run, comparative research must be undertake intra-nationally, but in some areas such as in the European Economic Community cross-national comparisons at lower levels of aggregation may be possible (Cheshire, Furtdo, and Magrini 1996). Ragin (1987) has offered a technique that attempts to wed both more qualitative aspects of case study work and the systematic testing procedures associated with more quantitative methods. His Boolean method of qualitative comparison attempts to deal not only with the problems of aggregation and the loss of rich data but also with the problem of diminishing degrees of freedom.

While opportunities for cross-national research may increase, the staple of comparative research will remain the case study, utilizing both qualitative and descriptive methods as well as survey approaches (Bertaux 1990; Ungerson 1996; Turner 1990). Sound and rigorous historical, anthropological, ethnographic, or otherwise more qualitative case study research is fundamentally important for all social research because of the insights, depth, and richness of perspectives it can provide (see Erickson 1976; Oliver-Smith 1992). Having worked in many developed and underdeveloped areas, I know the importance of this type of research—without it, much of what we attempted to do would have been impossible. This in-depth research can provide an inexhaustible supply of insights and important secondary data from which to challenge current perspectives and offer new lines of theoretical and research development. Both Prince's (1920) Halifax study and Wallace's (1956b) study of the Worcester tornado are exemplars because they stimulated, perhaps even altered, the direction of subsequent lines of research. Currently more qualitative work related to gender, family, organizational network change, and policy-related issues are forcing a reevaluation of the nature of disaster research and beginning to set new agendas.

Interestingly, a qualitative case study approach may play an additional critical role in parts of the world where many cultural and national variations are diminishing. It has been suggested that comparative research should shift its interest from focusing on the emerging commonalities and trends to preserving the cultural and national uniqueness that are quickly being overcome and destroyed (Sztompka 1988; Øyen 1990a). I think it is also possible that such research might document and record the successes and failures of disaster-related policies and organizational "solutions," whether directed at mitigation or recovery efforts, before they are forgotten, neglected, or cast aside and rejected. It may well be that many workable solutions were and are being rejected, not because they failed, but because they are perceived to be old, less modern, old fashioned, or simply part of the old regime.

THE PROBLEM OF EQUIVALENCE

The major difficulty one encounters in cross-national and comparative research revolves around the issue of equivalence or comparability. This issue permeates all levels of comparative research from the conceptualization of the question, through the selection of countries or sample sites, to the measurement of concepts. It is here too that we find one of the fundamental difficulties of cross-national comparative research. The essence of comparative research is to compare some unit of analysis on the basis of certain attributes. If, however, our measures are dissimilar across objects or through time, selectively biased, or prone to varying amounts of error or, even more fundamentally, if the observational units are not comparable in the first place, then we are left with nothing, or worse—erroneous findings.

Issues of equivalence are most easily seen with respect to national-level indicators and statistics. Regardless of the topic under consideration, the cross-national researcher must be concerned about the measures being employed, not only what they mean, but how they were collected and compiled, as well as timing issues. For example, some researcher may be interested in exploring social disruptions related to disaster and seek to use crime statistics of various sorts, divorce rates, or suicide data. However, these indicators do not necessarily mean the same thing across countries, and they may be compiled by very different types of agencies using unique procedures. The definition of what constitutes a criminal or antisocial act much less what constitutes spouse abuse, abandonment, divorce, or even something like voting, employment, or unemployment may not be equivalent among nations. Equivalence, both in terms of the basic nature or definition of the measure and its timing, should not be assumed.

Even where there appears to be a good deal of agreement and a long-standing tradition of using certain types of indicators, a more careful analysis can yield some surprising findings, suggesting the use of alternative measures where possible. A simple example is Gross Domestic Product (GDP) which is readily used, at least in its per capita form, as a indicator of national development, output, or even well-being. While there are many problems with using this measure (see Block 1990), a more subtle one concerns its conversion into a common metric, U.S. dollars. More often than not, GDP-based measures use some form of exchange rate, such as a sliding average, for converting into dollars. However, many factors, such as political decisions or relative strength of a currency in world currency markets and not simply the relative purchasing power of a currency, can impact exchange rates (Balassa 1964; Kravis, Heston, and Summers 1982). As a consequence, using GDP per capita may not accurately reflect the output levels of a nation much less its level of development, particularly since certain types of nations will have their currency systematically undervalued using exchange rates. Alternative procedures, such as purchasing-power parity conversions, better reflect differences among nations at a given point in time as well as changes through time (Summers and Heston 1988; Peacock, Hoover, and Killian 1989). Indeed, attempts to compare nations using measures that are dependent upon monetary conversions, such as costs of living, should not automatically adopt exchange rate conversions when purchasing-power parity or similar conversions are possible (Bates and Peacock 1993). Cross-national measures of disaster impact are likely to require such conversions.

The definition of disaster has long been a point of contention and discussion for disaster researchers. The conceptual confusion is mirrored in empirical research where one can find every form and type of event varying across all levels of magnitude discussed under the general label "disaster." In cross-national and comparative research, issues of type and magnitude must also be addressed. For researchers employing a case-study format, the issue is less problematic, and they are relatively free to describe the nature of the event and its impact given the data at hand. The problem comes when attempting to compile the results from multiple cases studies in some understandable fashion (cf., Drabek 1986; Blaikie et al. 1994).

In cross-national or other forms of comparative work where differences among a set of cases are being empirically compared, a common typology and measure of disaster impact must be clearly and unambiguously defined. This will undoubtedly be problematic, although there have been some attempts to discuss the issues (e.g., Wright et al. 1979; Friesema et al. 1979; Hoover and Bates 1985; Bates and Peacock 1987). Some possible measures include mortality and injury, structures destroyed, and residences destroyed as well as some measure of economic loss. In each case, it will be important to standardize, such as converting to a common unit or metric, and modifying or converting to a relative impact figure, thereby controlling for variations in economic, infrastructural, or population base. In other words, losses in the billion dollar range for a relatively small economy would have a much greater impact than in a larger community or nation. It will be important to assess impact relative to the ability of a nation, or unit being analyzed, to sustain or overcome the impact.

Problems of equivalency are equally important when considering the "things" we are comparing (Teune 1990). When we are discussing households or families across nations or even within the same community, are they the same? And what of even more complex social units like governments, legislatures, city governments or councils, code enforcement boards, relief agencies, relief programs, nongovernmental agencies, emergency management operations, communities, and even nations themselves? Clearly there can be considerable variation in structure, organization, and function among units similarly labeled. Part of the purpose of comparative research is to ascertain these differences and similarities and thereby better understand their consequences for the particular issue under study. However, in the process one often assumes equivalence with respect to certain phenomena or attributes of phenomena, thereby bracketing them off from study or analysis. Comparative researchers must constantly reconsider the question of equivalence. Interestingly, it is often in the context of working with a multi-national and multi-disciplinary team of researchers that assumed equivalencies are challenged. Through such challenges of our assumptions of equivalence, intellectual blinders can be removed, opening new, perhaps fruitful, avenues for comparative analysis.

Given the difficulties in using secondary data, it may be assumed that survey methods offer techniques for ensuring comparability, at least with respect to most measurement issues, since they are under the researcher's control. However, even here there are a number of problems that must be addressed. Primary among them is the lack of a convenient "tool box" of measures that can be employed when operationalizing theoretical concepts. Most often researchers reach into their own tool box, reproducing faulty measures they have employed before. This difficulty is most readily seen when attempting to measure abstract concepts related to risk, attitudes, and value assessments. Indeed, many attitudinal and value orientations may be very difficult to translate across national, cultural, or ethnic lines. Similar problems arise when dealing with concepts such as recovery, restoration, preparation, and the like. Bates and his colleagues have devoted considerable effort to developing a cross-nation measure of disaster impact and recovery for households (Bates and Peacock 1993, 1992, 1989; Peacock et al. 1987). This work draws upon the work of others (e.g., Belcher 1972; Bolin and Bolton 1978) to develop a single measure that should prove useful in cross-national comparative work. The field might well benefit from other attempts to develop measures for theoretically significant concepts in the future. Perhaps this, too, will be a spin-off of increased cross-national and comparative research.

THE FUTURE OF CROSS-NATIONAL AND COMPARATIVE RESEARCH

There are, of course, a multitude of research topics that can and should be pursued in cross-national and comparative settings. Any attempt to delineate the most important topics is bound to fail—too much will be omitted. Disaster research is too broad a subject area and disaster researchers come from too many different disciplines to possibly list all of the areas open to comparative and cross-national research.

While it is difficult to identify particular areas of research, I consider the next few decades to be potentially exciting and promising. Along with the growing numbers of disaster researchers internationally, heightened awareness of disasters as global problems, and the growing stores of data becoming available, there are also intriguing theoretical perspectives beginning to take form and spawn research agendas. Central among these is the linking of disaster and development perspectives which began in the late 1970s and is now emerging into broader ecological perspectives addressing issues related to sustainable development, social, political, and economic inequality and urbanization. A central theme of these perspectives is that disaster researchers must seek to understand the broader social, political, and economic factors producing differential vulnerabilities within social systems and differential abilities to subsequently respond to impact (see Blaikie et al. 1994; Bates and Pelanda 1994; Kreps and Drabek 1996; Peacock et al. 1997). As a consequence, while our common focus may be on disasters and disaster-related phenomena, as researchers we must draw upon the latest empirical and theoretical work from our individual disciplines to research these factors. Hence, issues that are the focus of our individual disciplines should also be pursued and explored in the disaster context. These issues might include many of the following found in variety of disciplines: markets and market imperfections; race, gender, and political economy factors; political participation, stability, and democracy; policy formation, implementation, and enforcement; interest groups and their formation; public planning and decision-making; conflict and its mitigation; interorganizational networks and coordination; financing and assessments of risks; and the structure of ecological networks and systems, to name a few.

To produce quality cross-cultural and comparative disaster research demands good researchers from a variety of disciplines including anthropology, sociology, political science, geography, economics, planning, business, environmental sciences, and other related disciplines. Cross-national and comparative research, much like new methodological techniques, can offer the opportunity to explore new issues or even reconsider old issues in new ways. However, undertaking such research will not ensure quality findings nor save poorly thought out research designs or theory. The decisions made concerning measurement, sample selection, and other methodological issues will be contingent in large part upon theory and methods within our individual disciplines, at least initially. However, the production of quality cross-national and comparative research will not only impact disaster research and theory, but will also have impacts upon our individual disciplines as well.

One additional comment: Øyen (1990a: 2-3) suggests that politicians are calling for more comparative research because of a perceived need to increase their understanding and policy options.

I have difficultly seeing many politicians, particularly ideologically-driven politicians, drawing lessons from social science research in general, much less research based upon other nations. If Øyen's assessment is true, however, cross-national and intra-national comparative research has much to offer. Comparative research into why governments adopt certain policies and positions, the long- and short-term consequences of policies, how special interests shape policy related to building code and land use regulation, enforcement, and modification, who wins and who loses, and the conditions under which policies optimally perform will prove useful. However, to ensure such research informs political debate and influences policy, as researchers, we too must become more involved. That, however, is the subject of another paper.

NOTE

1   The author would like to thank Betty Hearn Morrow, Nicole Dash, and Hugh Gladwin for their helpful comments.

10

MEDIA STUDIES

Marco Lombardi

The arrival on the scene of the study of mass emergencies and risk analysis represented an important step forward in the world of communication, not only because of its theoretical aspects, but also because of its ability to influence policy formulation. This takes place in a context which includes at least three groups of social actors, namely the public, experts, and political decision-makers. They make up a network of interdependent and reciprocal relationships which combine to form a particular social structure of the phenomenon of risk. In such a complex and changeable context, risk communication has contributed to the "shaping and assessment of the relationship created between the risk itself, risk analysis, social response and the socio-economic effects arising from political decisions made" (Kasperson et al. 1988: 180).

Obviously, neither the risk itself nor a specific interpretation of that risk can be at the heart of research. The former is a likely source of controversy because it is seen from a variety of standpoints. The latter would, by definition, favor the views of only one of the parties involved.

The focal point, then, of risk communication research and practice is the exchange of information between all the parties concerned. Risk communication experts' approaches can immediately be divided into two categories: one whose approach is rather narrow and studies message transmission in a single direction from experts or administrators to the public-at-large; and one which is more far-reaching and looks at the reciprocal feedback between the public and risk managers (Krimsky and Plough 1988).

Two phenomena have emerged from the more serious cases of environmental accidents and damage. These have led analysts to consider risk management as a process of global communication, that is, a social process. The first of these is *the dynamics of risk amplification/attenuation*, involving every passage of information from face-to-face to the mass media. The second is the continual *interaction between scientific and experiential reasoning*, creating a perpetual conflict. Both have important political implications.

The communicative approach offers a conceptual outline which makes it possible to cope with both these phenomena equally. Most importantly, it has helped to clarify the ethical implications of risk communication in society and has considered conflict as a means of understanding and solving social formations.

The conceptual outline of risk communication is based on the assumption that *social groups alter the contents of risk by amplifying its details*. In this way, the phenomenon or event which gave rise to the risk has uncontrolled, long-distance effects which modify the structure of relationships. Subjects entrusted with power, for example, "lose face" when confronted with the uncertainty of the moment. Such a situation is clearly a breeding ground for controversy or social conflict. Let us have a closer look at the key notions behind the communicative approach.

NOTIONS OF RISK COMMUNICATION

Amplification

The concept of social amplification of risk derives from criticism of the "perceptionist" view. According to Kasperson et al. (1988), risk events not only strike at the system of perception but also interact with the social structure of a given group, influencing its members' experience of the event. In their turn, the behaviors which emerge have secondary effects (economic and social consequences, stigma, etc.). They then go on to reach the most abstract of institutional levels (administrations, producers of culture, mass media), bringing about yet another order of effects. This process is called "amplification" if the effects of the events are maximized, "attenuation" if they are minimized. The theory of social amplification of risk, as laid down by the before-mentioned authors, maintains that the characteristics of a given system of information, along with the characteristics of public response, are essential elements in the determination of the nature and extent of the risk itself. In short, in the concept of social amplification, risk is transformed into a multi-faceted phenomenon which surpasses both the mere fact of the incident and manipulation of the natural environment. This concept comes to include a wide range of factors: visible and reported events; the direct and indirect impact of a risk event (including far-off and future ones); the various communicative agencies or stations ranging from the media to the "grapevine"; the various moments of communication (transmission, codification, reception, decodification, reaction), each of which hides either filtering or amplification of signals; and information mechanisms of amplification (dramatization, volume intensification, disputes) and of response to the information flow.

In conclusion, risk is not only the result of a fact but also the communicative interaction between the fact, a given social structure, and a given culture or system of values. Risk communication is not limited to the activities of the media but also includes all the implicit and explicit messages between the social groups involved. In this way, problems of negotiating environmental policies, their implementation, and public involvement in environmental initiatives all become the focus of communicative research and practice.

Credibility and responsibility

Loss of credibility and trust in the eyes of the public during technological emergencies is symptomatic of communication interference but, in turn, derives from the fact that risk has its roots in sociology. Moreover, if risk perception depended on individual psychological components, it would be hard to understand how certain opposing reactions regarding high-risk plants are so widespread.

Even psychometric and perception research has shown that there is a sociocultural organization of risk perception, common to all members of a community. This is found within a social context which defines the individual's experience of nature and technology. That which Wynne (1987) defined as "the dialectic of credibility" occurs in the same way. Following accidental trauma, the community is alienated from its source of credibility. The man in the street does not want to feel responsible for accidents, never having been informed of the likelihood of their occurrence. The experts disagree with each other over the interpretation of data. The authorities are incapable of taking responsibility for the necessary action, thus losing legitimacy. At this point, the building of social responsibilities takes place. Divided and polarized, having to rationalize something, the individual members of the community come up with a credible explanation and develop an explanatory process. This is the origin of symbolic images, epitomizing both the collapse of the myth of technological infallibility and the total lack of social responsibility.

What emerges from this is the absence of negotiation and, therefore, of social communication. When this is not replaced by the authorities, the community is forced to transfer the responsibility for the event from real agents to imaginary ones, to the detriment of its own internal cohesion. Cognitive-affective involvement in risk events does not, therefore, vary individually but is a social resource. It is necessary to learn from the attributive mechanism expressed by ordinary people. It is necessary to act, not so much to ascertain risk, as to maintain the structure of the group's social relationships which, in an interactionist model of human nature, is worthwhile in itself.

In a comparative study of five cases of environmental risk, Krimsky and Plough (1988) noted two sources of social rationalism:

technical and scientific expertise; and the social and cultural experience of the man in the street. Each is necessary, but neither is able to stand alone. They communicate with each other using five linguistic elements: intention, information content, audience direction, information source, and information flow. Each of these has a potential for receiver distortion or manipulation.

Conflict

The rise of social conflict seems currently to be an integral part of risk situations. As we have seen, the communicative approach to risk analysis is a result of the structuring of questions concerning the environment and risk, depending on the type of dispute and controversy. However the initial definition of the situation comes about, it always ends in social conflict. This rule seems to be true for any source of controversy, from the nuclear question to the collapse of a dam wall. Even when there has been no overt politicization by those involved, the technological manipulation of natural components or the responsibility for an accident brings the political-institutional element into play. This is because technology is not neutral as regards economic interests, legislative power, and social inequalities.

Edwards and von Winterfeldt (1984) classified typical risk-associated conflicts into three opposing types: work environment, risk environment, and economic development environment. Giving each case of conflict two dimensions, content and process, the authors came up with six types of environmental/technological controversy: data, degree of probability, definition of events, cost/benefit estimate, assessment of the cost/benefit distribution, and social values acquired as a result of the assessment. The explosion of conflict as a valid form of social communication, studied and described by risk analysts, has yet to produce new forms of social regulation on a political level. Suffice it to say that the conflict is often contained within local boundaries, working on the specific awareness of risk in a particular area, and is considered at an end once the event or the source of risk has been brought under control.

However, the fact remains that, both on the level of political action and on the level of human values, there are insufficient resources to deal with the controversy stirred up by fresh environmental damage.

According to Stern (1991), no communication about risk can be free of political implications or party prejudice. The transfer of information from experts to nonexperts and vice versa necessarily involves the following: a need for information; a need for secrecy; and more or less explicit feelings of uncertainty and values which are, to a greater or lesser extent, self-centered. Conflict is therefore unavoidable and is implicated even in the most cautious and exacting messages. From the experts' point of view, conflict is already aroused in the choice of line of research and evaluation of results. Instead of trying to unify different points of view, it would make more sense to bring them out into the open. The basic resource is the *scepticism of the man in the street*. At first, this is tested to find out which analogies are most often used by a given culture to calculate the extent of the risk, which language is most easily understood, to what degree people are aware of the fallibility of science, etc. In short, study should not focus on the code content of the messages (information theory does not help to resolve conflict) but rather on the social structure of the group involved.

MASS EMERGENCIES, THE PUBLIC, AND THE MEDIA

The first studies of psycho-social variables of perception of environmental risk maintained that ecological information and sensitization depend more on individually formed values, which have a social reference, than on the characteristics of a person's system of perception. In other words, they depend on the ability to derive positive meaning from environmental information. This is closely linked to the typical dynamism of magical and primitive thinking (exorcism of fear, contradiction, indifference, "Me-world," etc.) and so is not sustained by scientific information (which does not touch on the emotive aspect of fear) but by that which is "enthralling." In other words, it is linked to information which is anchored to value references already recognized by the group and by individuals. If this is not the case, the information serves either to amplify the anxiety caused by the catastrophe (which amounts to passive acceptance of the technical data) or to exclude a sense of individual responsibility for collective problems (removal or refusal of the data).

Risk perception research, too, seems to support these hypotheses. The analysis of the psychometric matrix (Slovic 1987) has continually revealed the peripheral role of technical-scientific data in the estimation of risk acceptability. This phenomenon explains the well-known gap between statistical evidence and the perceived interpretation of risk (Valentini 1992). Also in this case the role of information is judged to be "strategic" for the way in which it makes it easier to understand low-profile phenomena which are subject to a high level of cognitive distortion. For this to happen, the communication inherent in risk needs to be developed starting from the dominant conceptions and information needs of the audience (Cannell and Otway 1988). De Marchi's (1991) study deals with a population's information needs when faced with pollution risks. This pilot study showed that people tend to contextualize risk, to transform an abstract concept (environmental safety) into a concrete experience. Even when provided with abundant and extremely detailed engineering or administrative communications, people tend to retreat into their own private world of meanings and symbols. Therefore, regardless of the quantity and quality of information available, people feel confronted by a lack of knowledge and only give credit to information coming from sufficiently familiar or high-profile sources from their own culture.

The results of empirical studies, therefore, suggest a very complicated picture of the sender-receiver relationship in communication, especially when this contains elements of probability as in the case of risk. A purely semantic approach as referred to by Eco and Fabbri (1978), based only on the understanding of the message, was shown to be insufficient to explain the interference of the cultural context, particularly when it came to the "credibility game" played by all the information sources in a given area. On the other hand, more stucturalist approaches, which try to correlate the reception of environmental messages with the structural characteristics of the population (with special reference to age, academic qualifications, and occupation), could not be used to form generalized models. This is mainly because they do not take into consideration individual preferences underlying ethical values in attitudes towards the environment, etc.

STUDY OF THE MASS MEDIA AND RISK

Recent Italian research into local and international cases has identified and described the information models adopted by the various media when faced with disasters, emergencies, or situations of ecological risk. The models were obtained by reconstructing the behavior of individual operators at the time of a disaster such as the Valtellina landslide (Di Bella 1987), the Chernobyl cloud (Lombardi 1988a, 1993), pollution of the Seveso (Mascherpa 1990), the case of Acna-Valle Bormida (Colombo 1995), and the toxic cloud caused by Sandoz over Reno (Ghiglione 1992). These analyses concentrate on a description of the information style which presents a generalized trend towards the catastrophic. This is also explained by the type of phenomenon described at the time of the above-mentioned events. They are complex subjects which require a certain amount of basic technical information in fields such as geology, chemistry, political economics, medicine, etc., and they have *low visibility*.

We must remember that these characteristics are common to all the basic themes and subthemes of the environmental world. Furthermore, the various media behave in different ways. If television emphasises the amplification role of the news, leaving little space for comment, the press is open to charges of excessive ideology or falsification of the event, as happened during the Seveso case. On the other hand, TV can offer a more controlled and up-to-the-minute service because it makes use of the visual element more than the press does. However, this, too, could be distinguished by the strong rhetorical power of images while nevertheless passing on overly-simplified, emphatic, and overly-symbolic information. The various phases of the spreading of news which condition the attention curve of the audience are outlined in the above-mentioned research.

For example, in the case of Chernobyl, it can be maintained that " . . . the circulation of news is in the public interest" (Lombardi 1993: 54). This underlines the adaptable strategy of the media when confronted with crisis, a strategy which leads it to reflect only one part of the social reality (that of the minority affected by the disaster). The resulting impression is of neutrality and detachment from the event. Such behavior can also be called "change of perspective strategy" because of the way in which the media are able to react to public interest in a given piece of news by focusing more efficiently on it. Moreover, this research has highlighted the choices of coverage and documentation and the principal information effects achieved by referring to these indicators: content (with particular reference to often-sensationalist headlining, poor correlation between image and text, use of melodramatic language, and redundant information); and communicative strategy (use of official and unofficial sources, facts obtained directly or through press agencies, coverage of the subject in all its ramifications, and time taken to pass on the news).

A fairly disorganized picture emerges from this, showing a system of information which chases after news and dwells more on tragic and dramatic details than on the reasons behind choices made or protests voiced. Both TV and the press are presented as spontaneous informers, driven only by *questions of saleability and newsworthiness*. They appear to carry out only a limited information function without having any preventive role. The emphasis on turning environmental events into a spectacle has been attributed by some to the lack of technical experience of the operators, who were taken by surprise by the wave of technological accidents that swept over Europe in the 1980s.

The situation in the U.S. seems to be more organized and

articulate than that in Europe, perhaps due to the fact that the media, and in particular the press, have long played an active role as consensus receivers. Hansen (1991) proposes a classification of research concerns, arranged as follows:

(1) the study of *agenda-setting*: measuring media interest in the theme of risk and emergencies. In this way, we can obtain an idea of the almost random series of alternating attention "peaks" and "troughs" in the media. This research is based on the general hypothesis that there is a correlation, albeit slight, between communication emphasis and the social prominence of a subject. In our field, this is seen as valid for general matters and for secondary news. However, there is still an area of public opinion, equivalent to specific information, which is not directly influenced by media choices;

(2) the study of *problem diffusion*, based on the correspondence between information flow and the shaping of public opinion. Some of these studies have established a connection between public awareness and information provided by the media, but only on a short-term basis. However, long-term analysis has shown that public opinion remains substantially unchanged regardless of the activity of the media. This could be explained by the fact that risk communication passed on by the media is frequently found to be weak and lacking in substance. It is centered more on the political effect of the news than on scientific-technological facts and bringing adaptability rather than creativity to the fore (Wilkins and Patterson 1987);

(3) *audience study*, that is, the direct influence of the media on public opinion. The media's social influence is relevant as it represents the only source of information for the public, which is passively watching the development of the event. (This is particularly true if the event is invisible or geographically distant.) The receptive behavior of the audience should also be correlated with newsmaking (i.e.,

the process that occurs between the fact itself and its becoming "news"). A few years ago, Mazur (1981) posed the question of the *media's responsibility* as an amplifier of events and shaper of fluctuating opinions, going back as far as the scientific debates of the 1960s. He upheld the need for regulation of the influence of the media by means of more rational choices of information coverage;

(4) the study of the *internal organization of the media*, whereby newsmaking depends on the strategic choices made by each individual information organization: the presence of environmental reporters, use and availability of sources, and solid relationships with people active in the so-called environmental forum (scientists, pressure groups, government, local authorities, and technical agencies). Anderson (1991) observed after a case study that there was growing media professionalism, especially in the search for credibility and in target selection. Scientific disclosure and sensationalist reporting were becoming more highly differentiated, depending on the anticipated audience;

(5) the study of the *cultural and ideological values* of the informers. These have identified stereotypes embedded in journalists' use of certain metaphors, reference to value concepts (What do we mean by health?; by risk?; etc.), and affiliations with an ideological background (individualism versus universalism, innovation versus conservation, hierarchical versus the liberal ideal, or fatalism versus rationalism). A recent study of the consistency between the information coverage of a subject by newspapers and its ideological trend shows that the media are responsible to a great extent for public support or shelving of interest in environmental quality. Therefore, the media must guarantee consistency and correctness when talking about themes, problems, and solutions and not merely perform some kind of balancing act between diverse professional groups (academics, technicians, journalists, etc.) In the four newspapers which were

examined, the consistency between political-environmental orientation and coverage was satisfactory. However, the authors maintain that a qualitative analysis of the contents would have better assessed the level of information correctness.

To summarize, some interpretative models of the relationship between the mass media, public opinion, and news of environmental risk and emergency can be outlined. The most popular is the constructionist one which sees the media's discourse as a package (Gamson and Modigliani 1989) or rather as a picture or interpretative key determined by three factors: cultural interest (which gives value to the news, contrary information, and secondary news); activities in support of the facts (experts, movements, governments, etc.); and communication practices (symbolism, the catastrophic, etc.). However, we must beware of thinking that this makes the public a passive receiver. On the contrary, it influences the packaging in various ways (e.g., by what the journalist believes to be the orientation of the audience).

Another interpretive model is that based on the actual "text," a model put forward by Corner et al. (1990). This "textualization" shows the process of the selective transformation of a media message expressed in harmonious rhetoric into a particular public opinion. In the case of environmental emergencies, the authors maintain that the public is influenced by three factors: exposure to audio-visual messages (which provide persuasive evidence of certain phenomena rather than others); the listener's personal evaluation dynamics (e.g., wariness of prejudices expressed by the media); and finally, the ability to distinguish between possible and probable facts and therefore to distance oneself from the assertions made in the messages themselves.

SOME WORK TOOLS

The preceding pages represent a brief outline of methodological guidelines for the analysis of communication and information passed on by the mass media. However, the range and complexity of means of communication make it impossible to define empirical analytical tools that would be valid for all types of media, or even for one single type at different times. Therefore, we need to find other tools which, while particularly useful for research carried out on the press, would nevertheless provide general criteria valid for other types of media. This methodological proposal has as its objective the qualitative analysis of the media's contribution, but also uses instruments of quantitative analysis which "measure" the qualitative aspects worthy of further investigation.

Field investigation can be organized into four phases: problem-setting, coverage, narrative style, and communicative role. Each of these has its own diverse objectives and variables. These alone cannot guarantee reaching the objective, but they do supply quantitative information which is useful to the researcher's interpretative work.

Problem-setting

The analysis is of the controversy and proposed solutions to the problem. This means understanding the definition of the problem (problem-setting) which leads to the subsequent definition of the role played by the media. Variables to be taken into consideration include: number of records (numbering from 1 to N, the articles collected); name of the newspaper which published the news; type of article (headline/photo, editorial/cover story with a by-line, agency/press release/unsigned article, letter to the editor, advertisement); and section of the newspaper in which the article was published (feature, local news, regional news, letters to the editor, cultural or scientific section).

Coverage analysis

By surveying the distribution of published articles in terms of frequency (both in general and according to specific matters dealt with), it is possible to qualitatively evaluate the subject (i.e., themes covered). The same can be done at a quantitative level (e.g., space taken up within a particular time period). Variables to be taken into consideration include: day of the week of publication; date of publication; page number where the article is published; number of columns taken up by the article; number of photos accompanying the article; position on the page (e.g., top, middle, or bottom); height of the piece in millimeters; width of the piece in millimeters; area occupied by the piece in square centimeters (new variable created in the data processing center); and area taken up by the piece as a proportion of the total area of the page (new variable created in the data processing center).

Narrative style

A semantic analysis of the article allows us to identify the recurrence of certain linguistic constructions which, for example, refer to a "conflictual," "alarmist," "rhetorical," "technical," etc., viewpoint. In addition, the identification of certain focal keywords, of dramatic elements, euphemisms, and particular constructions (e.g., "Buts" and "Ifs") help identify the narrative style. Variables to be taken into consideration include: type of news communicated (story, commentary, in-depth technical report); emphasis of headline, analysis of how frequently certain key words occur (e.g., disaster, emergency, deaths); analysis of the opposing conjunction "BUT" and its meaning in relation to the headline and the extent to which it is eye-catching: absent, present, and "pessimistic" (e.g., 10 deaths, BUT there could be many more); or present and "optimistic" (e.g., 10 deaths, BUT this is unconfirmed/could be an over-estimation); and analysis of the conditional "IF" and its significance in relation to the headline and the extent to which it is eye-catching (absent or present).

The role assumed by each means of communication in the controversy

The identification of the origin of the news (in other words, the source) is central to the definition of the media's point of view.

For example, reference to "official" or "unofficial" sources and "experts" or "politicians" certainly contributes to the identification of the communicator's "reference paradigm" or "ideology." Also, the communicative model chosen (strategic, informative, etc.) becomes a useful reference for the researcher engaged in this fourth phase of his/her work. It is not closely connected to the identification of specific variables, but it is closely linked to the comprehensive evaluation of previously collected empirical data. The variable to be taken into consideration here is the identification of the source of the news.

CONCLUSION

In conclusion, many researchers and scholars of mass emergencies and risk analysis today agree on focusing their research activities on communication. Communication is seen as a social process, something that is fundamental to the understanding of both crisis management and to the various activities which precede and follow crises themselves. On the other hand, information, as a product of communication, is merchandise that has great importance in many of our relationships, both on a micro- and macro-level. This brief account aims to stimulate the debate which is already active in the scientific community and also to provide food for thought as to the working tools that might be used in research which is constantly face-to-face with empirical reality.

11

REWRITING A LIVING LEGEND:

Researching the 1917 Halifax explosion[1]

T. Joseph Scanlon

> The historian represents the organized memory of mankind, and that memory, as written history, is enormously malleable. It changes, often quite drastically, from one generation of historians to another—and not merely because more detailed research introduces new facts and documents into the record. It changes also because of the changes in the points of interest and current framework into which the record is built . . . . The historian cannot avoid making a selection of facts, although he may attempt to disclaim it by keeping his interpretations slim and circumspect.
>
> C. Wright Mills, *The Sociological Imagination*, p. 144

Historical research has limitations. Records are lost or destroyed. Some sources are dead. Others are alive but their memories dim. Persons have taken records with them and kept them in private hands. However, there are also advantages. Some records that were private or secret have become public; some persons will produce records or talk about past events, though they would not have been cooperative at the time. In addition, some statistics and comparative data will exist only because time has passed. Sophisticated methods of analysis may reveal things that were not evident years ago. In Halifax, for example, oceanographers have located the depression left by the explosion, pinpointing the precise location of the ship that blew up.

The excitement comes when previously untapped information allows new insights into what seemed to be a well-told story. For example, the aunt of a chemistry professor at Carleton University was the daughter of the Halifax harbor master. She typed her father's account in 1917 and kept it. It revealed that he had attempted to stop the munitions ship from entering the harbor. (She was interviewed after her nephew—who knew of the author's interest in the explosion—mentioned she was a survivor, though he did not know about her father's role.) Historical reconstruction of a disaster has another advantage. Disasters are so dramatic that many vividly remember what happened even three-quarters of a century earlier. They also remember if someone told them of an experience and are happy to pass that person's name along. In Nova Scotia, the explosion of 1917 is a living legend: everyone loves to talk about it.

PREVIOUS RESEARCH

In 1977, in a note on methodology in Quarantelli's *Disasters: Theory and Research*, Verta Taylor wrote, "Documentary and archival material has nowhere been collected and analyzed in the innovative fashion it could be in disaster research" (Taylor 1978: 276). This is less true now. Working two years after the incident, Scanlon (1979) used interviews as well as published and unpublished material to reconstruct what happened in the first 24 hours after Cyclone Tracy hit Darwin, Australia. He and Angela Prawzick took a similar approach in reconstructing first response seven years after the 1978 San Diego air crash (Scanlon and Prawzick 1985). In Darwin, he found enough data to contradict the version in a book by an Australian general (Stretton 1976). In San Diego, there was enough material to allow a comparison with the 1985 Gander air crash (Emergency Communications Research Unit 1987; Scanlon and Sylves 1990). Scanlon also used unpublished data from Halifax in his chapter for the book written in honor of Henry Quarantelli (Scanlon 1994) in which he tests current theories against the response in Halifax.

For historical research, there are any number of starting points—newspapers, books, academic or other articles, theses. There are official papers, reports, minutes, logs, and letters. While much of this material will be found in libraries or archives, it may be far from obvious which libraries and archives are worth visiting. After that, finding material becomes even more challenging. There are private papers ranging from notes written in a scribbler to diaries to typed memos. Tracking these down means poking around in basements or vaults. It also means using unconventional techniques, making one's interests known, and following trails from family, friends, professional colleagues, or even strangers to written sources, then trying to fit the material into a pattern. The same process is used to find survivors or those who remember what survivors told them. One Halifax survivor was found in Bangkok.

While this search may seem haphazard, it flows from Carleton University's research on rumors, and is systematic. It is similar to that described by C. Wright Mills in *The Sociological Imagination* (Mills 1959: 196). Starting in 1970, Carleton students responded to unexpected events, selected a sample, found out how those in the sample learned of the event, then followed a person-to-person information trail to the media or eyewitnesses (Scanlon 1971, 1974a, 1976, 1977a). Once, they traced the flow of news as far as 10 stages and found eight eyewitnesses to the murder of a police officer (Scanlon 1974b; Wotherspoon and Scanlon 1974). On another occasion, they reconstructed the flow of information after a mudslide led to evacuation of a logging town, a town without local radio, television, or its own newspaper (Scanlon, Jefferson, and Sproat 1976).

To follow the flow of information through a community—and trace links that sometimes connected strangers—the Emergency

Communications Research Unit (ECRU) developed techniques such as the seven-day interval—going to the same place precisely seven, 14 or 21 days after an incident. (People are creatures of habit and do the same thing at the same time every week.) This technique allowed the researchers to find and interview someone described only as "a stranger in a doughnut store" or "a woman at a table in a bar." Successes like this taught the value of detail: one fact may be enough to locate an interpersonal contact. It also helped those involved learn how connections can be used to acquire information. Results from this research were reported in the *Canadian Journal of Sociology* and elsewhere (Scanlon 1972; Erickson et al. 1978; Richardson et al. 1979; Scanlon 1985b). However, the methodology was never described.

Although ECRU did mainly quick-response research, follow-up trips proved productive. Four months after a destructive tornado, it was realized that a handwritten record of who lived in a mobile home park on the afternoon of the tornado was still on a wall board in the park office. On occasion, ECRU worked a considerable time after an incident. In 1981, an ECRU field team used snowball sampling techniques to learn about hospital fires which happened up to two years earlier (Scanlon and Hiscott 1982). The same techniques were used for other research. In 1977, Scanlon and a colleague developed a history of the Sikhs in Vancouver for the United Nations (Scanlon 1977b). In 1983, Gillian Rutherford, a fourth-year journalism student, looked for those who were in an Ottawa apartment building when a Soviet diplomat, Igor Gouzenko, defected in 1945. Starting with assessment rolls, city and government directories, she found all the individuals involved, even though 40 years had passed, an achievement investigative journalist John Sawatsky called "amazing" (Scanlon 1985a; Sawatsky 1984: x).

Some ECRU techniques are directly applicable to historical research. When ECRU did quick-response research, it made its presence and its interest known to the affected community, telling people in person and using the mass media. This has also proved productive in researching the 1917 Halifax explosion. ECRU also tried to obtain records before they were erased, something those studying Three Mile Island found most important (Rubin et al. 1976). This is less productive after 75 years, but Halifax records have been found in private hands—once just as a person was going to dispose of them. In following up incidents, ECRU found it useful to search for search professional journals: for "Day One in Darwin," Ross Laver found 30 documentary references in such journals (Scanlon 1979). Similar sources were useful for research on Halifax partly because persons write things for their colleagues that they do not disclose in other less specialized publications. Finally, ECRU developed a format for asking questions of participants in such events. The most important was, "How did you first become aware something was happening?" followed by the soft, open-end, "Anything else?" People remember their first contact with a dramatic event, and chronology is the best way to reconstruct what happened. Even after three-quarters of a century, the same approach proved effective. (The question appears similar to the ones asked by the Disaster Research Center at the University of Delaware but is quite different. DRC researchers ask persons where, when, and from whom they heard [Wenger 1989: 246]. ECRU experience and John Sawatsky's work on effective interviewing suggests "How?" and "What?" are far more productive than "Who?", "When?", and "Where?")

THE INCIDENT

The Halifax explosion occurred at 9:05 a.m., December 6, 1917, after a French ship, *Mont Blanc*, and a Norwegian ship, *IMO*, collided in the city's inner harbor. *Mont Blanc* was carrying aviation gasoline and explosives, and the collision set it on fire. Eighteen minutes after the collision, she blew up with one-seventh the power of the first atomic bomb. There were 1,963 dead and 9,000 injured. The North End of Halifax was in ruins or on fire, and there was also substantial damage in Halifax's sister city, Dartmouth, across the harbor. In the harbor itself, ships were battered, both by the explosion and by a tidal wave it created. The tug *Hilford* was lifted six meters and dropped in wreckage on the docks. Communications and transportation were disrupted. The fire department lost its senior personnel including its chief. Initial search and rescue was done by family, neighbors, and passers-by, then by hastily organized groups. Victims poured into physician's homes where physicians tried to help despite their own injuries. The hospitals were swamped; patients were on floors, under tables, and, despite the chilly weather, even outside on the steps. In the railway yards, 374 freight and passenger cars and five engines were tossed about. The tracks were covered with debris. Railway telegraph lines were down. The situation was not helped by a major snow storm which struck at noon the next day.

Despite these problems, there was an organized community response. Recognizing the seriousness of the situation, the acting mayor, city clerk, and chief of police asked for and got military assistance. In addition, at the acting mayor's request, a railway official drove to the nearest place where communications existed and sent an appeal for help to nearby communities. Then, less than two and a half hours after the explosion, at a meeting at City Hall, committees were appointed to deal with search and rescue, food and transport, registration, and the creation of a morgue. Within six hours, the first relief arrived by sea. Soon after that, trains started arriving from Nova Scotia communities such as Truro, Kentville, and New Glasgow and from Moncton, New Brunswick, with physicians, nurses, medical supplies, firefighters, and fire equipment. Within days, there were trains carrying building materials and skilled workmen. Soon, Halifax struggled with congestion from over-response. It appealed for persons to wait to be asked before sending supplies. Inspectors boarded trains at nearby Truro and told those not on official business to postpone their visits. Although early disaster scholars did not know about this (e.g., Fritz and Mathewson 1957), it marks the first thoroughly documented incident of convergence (Scanlon 1992a).

Disaster scholars know of the explosion and the response to it because of Samuel Henry Prince. In 1917, he had a B.A. and M.A. in psychology from the University of Toronto and was curate of St.

Paul's Anglican Church in Halifax. In 1919, he left St. Paul's to preach at St. Stephen's Protestant Episcopal Church in New York City and study for a Ph.D. in sociology at Columbia. His dissertation, *Catastrophe and Social Change* (Prince 1920), a study of the response to the explosion, is recognized as the first scholarly study of disaster (Drabek 1986; Dynes, De Marchi, and Pelanda 1987; Anderson 1978). The author's interest in Prince got him interested in the explosion (Scanlon 1988, 1992b, 1994). What follows describes where that interest led.

NEWSPAPERS

One starting point both for getting an overview of what happened and for creating a list for follow-up later was newspapers. While newspapers are notorious for making errors of detail, including errors about disasters (Barton 1962: 258; Dynes 1970: 26; Scanlon 1972; Scanlon et al. 1978), they provide a broad picture of an incident and, because they are dated, help establish a sequence of events. They also provide names of organizations involved. For example, the Halifax papers mention the YMCA, St. Mary's College, and the Knights of Columbus. Each of these organizations can be contacted to see if they have kept records. The value of newspapers does not stop as time passes. There are anniversary editions 1, 2, 5, 10, 25, 50, even 75 years after an incident. These will include interviews with and articles by persons who have not previously told their stories.

While local newspapers are important, newspapers away from the scene are also useful. The newspaper in Truro, the next major center to Halifax, printed the name of every survivor brought to it. The paper in Saint John, New Brunswick, printed interviews with survivors who passed through by train. The papers in Providence interviewed physicians from Rhode Island after their return; their comments to reporters differ from their reports to the Canadian Army Medical Corps. The papers in Boston covered the Massachusetts response to Halifax in detail; all sent reporters by train to Halifax. The papers in Toronto reported business losses.

(Many companies had head offices in Toronto.) The *Toronto Daily Star* also carried an account by Stanley K. Smith about how he covered the explosion. Smaller newspapers printed personal correspondence. One in Charlottetown, Prince Edward Island, has a letter describing how a bank teller moved survivors along the waterfront by boat. One in Lewiston, Maine, published a 12-page letter from a wireless operator on the Canadian depot ship *Niobe*. (It was tracked down by the sailor's wife who sent a copy to the Maritime Museum of the Atlantic.) There was information in European papers, especially those in Norway, home of *IMO*, and in France, home of *Mont Blanc*. The *Sandefjord Blad* in *IMO*'s home town had an obituary of *IMO*'s captain, Haakon From.

Newspapers also give a sense of the times. The explosion occurred 10 days before Canada's most divisive election, which divided the country on ethnic lines, attitudes that help explain why the English Canadians blamed a French ship for causing the explosion. (Many French Canadians were opposed to Canadian involvement in the war in Europe.) Reading the election ads makes history come alive. The explosion also occurred just as Lenin and Trotsky seized power in Russia. One Paris daily had its whole front page devoted to photos of Lenin and Trotsky haranguing crowds in Russia. This context is important—as the draft text of the author's book on the disaster indicates:

> If it had not been for the war, neither *IMO* nor *Mont Blanc* would have been anywhere near Halifax. *IMO* would have been whale hunting in the Antarctic and *Mont Blanc* trading along the coast of South America. If it had not been for the war, *Mont Blanc*, even if by some chance she had come to Halifax, would not have been carrying munitions. If it had not been for the war, *IMO* would have left in the evening after it got coal instead of being forced to wait until dawn. If it had not been for the war and the activity that created in the harbour, the pilots would not have been overworked.

Canadian newspapers, including most weeklies, are in the Public

Archives of Nova Scotia in Halifax or in the National Archives in Ottawa. They are on microfilm and can be borrowed through interlibrary loan. Norwegian ones are in the national library in Oslo, French papers in the national library in Paris. Language skills are an asset, but anyone familiar with the basic details can puzzle out the essence of what was reported. Most librarians are helpful to foreigners: it took half an hour to get a pass in Paris, less to get access in Oslo. The only paper not located was the weekly in nearby Kentville, which sent physicians and nurses to Halifax by train the day of the explosion. The crucial issue has disappeared.

Anniversary stories are not confined to the community affected or to print media, nor are anniversaries only media events. In 1936, the *Evening Standard* in London carried an article by the first officer on *Niobe*, Lt.-Cdr. Allan Baddeley. He sent a pinnace to take *Mont Blanc* in tow, unaware she carried munitions (Baddeley 1936). He mentions that when the explosion happened he was standing on *Niobe*'s bridge with Captain Walter Hose, later appointed to the official inquiry. In December 1995, CBC radio interviewed Halifax's current fire chief. The next day someone called the program's "talkback" line to say that the fire engine used in 1917 was not the first of its kind. Anniversaries may also be marked by a memorial service or a special presentation. These attract those involved, their families, and others who are knowledgeable and interested in the event. Hundreds of survivors showed up for a recent reception for survivors in Halifax.

OTHER PUBLICATIONS

Dramatic events inspire fiction, some of it autobiographical. Of the four novels set around the explosion, two reflect their authors' experience. McKelvey Bell, who wrote *A Romance of the Halifax Disaster* (Bell 1918), was chief Canadian Army Medical Corps officer; his hospital scenes are real. Hugh MacLennan, who wrote *Barometer Rising* (MacLennan 1941), was a schoolboy who saw for himself the effects of the explosion. His book is a Canadian classic and has made the explosion a legend and has—because of its historical inaccuracies—left many persons with inaccurate impressions of what happened. Anyone writing about the explosion needs to be aware of the misleading public image created by MacLennan. MacLennan's novel is not his only contribution to explosion literature. In 1938, he wrote a brief account of his personal experience in the *Lower Canada College Magazine*. He wandered through the streets in the south end of the city immediately after the explosion and—to his surprise—was alone:

No one has ever explained why the streets in the south end of Halifax were empty for so many minutes after the great explosion but they certainly made up for their quietness by what happened during the remainder of the day. (MacLennan 1938)

The explanation seems simple enough. Most persons in the south end were cleaning up broken glass and other minor damage to their homes, totally unaware that the North End of the city had been devastated. The other two novels are by Jim Lotz (1981) and Robert McNeil (1992). Lotz's book is a fanciful story involving Leon Trotsky based on the fact Trotsky was in Halifax prior to the explosion. McNeil's includes some anecdotes based on family memories.

There are three major nonfiction books. Archibald MacMeachan's official history was not published until 1978, but it is based on first-hand accounts though few are from persons who lived in the North End of the city (Metson 1978). Many typescripts of those accounts are at the Public Archives of Nova Scotia in Halifax. In *The Town That Died,* Michael J. Bird pays more attention to the collision and the immediate response (Bird 1962). Although part of his account seems based more on imagination than research, other parts of his information came from British and American naval records, available at the Public Records Office in London and the National Archives in Washington, and from interviews with some of the sailors involved. Janet Kitz's thoroughly researched *Shattered City* takes a longer-term view using the records of the Halifax Relief Commission (established in 1918) letters from many of the survivors, and interviews. She also had access to the material collected from the bodies of the victims (Kitz 1989).

There are many other relevant books. In his autobiography, Thomas Raddall describes how Chebucto School became a morgue (Raddall 1977). In his, journalist Kelly Morton describes the scene at his home (Morton 1986). Prince's biographer, Leonard Hatfield, reports that Prince used his Hupmobile as an ambulance (Hatfield 1990). Dean Jobb deals with the criminal aspects in *Crime Wave: Con Men, Rogues and Scoundrels* (Jobb 1991). Mary Ann Monnon used records and personal contacts for *Miracles and Mysteries* (Monnon 1972). Her book illustrates the endurance of the explosion as a legend; it reached a ninth printing in 1992. There are unpublished manuscripts such as Dwight Johnstone's "The Tragedy of Halifax: The Greatest American Disaster of the War" (Johnstone *circa* 1919). There is an instant book, Stanley K. Smith's *Hearts Throbs of the Halifax Horror* (Smith 1918).

Many of these authors also have links to the explosion. Thomas Raddall was a student at Chebucto school in 1917 when soldiers asked him to help turn that school into a morgue. Kelly Morton's father was a physician, and Morton saw first-hand what happened at his home. Mary Anne Monnon's father was one of the survivors. Janet Kitz uses her mother's experience as well as letters collected from survivors. Stanley K. Smith came to Halifax on the first train from Saint John, New Brunswick. Archibald MacMeachan kept a diary that includes information not in his official history. (His diary is in the Dalhousie University archives along with other diaries kept at the time.) Dwight Johnstone's father was chair of the Dartmouth Relief Commission, his sister a secretary with the Halifax Relief Commission, his brother a physician. His father's status helped Johnstone learn and report how New Glasgow assisted in the response, an account that led the author to that community for a most rewarding visit. His connections as a journalist allowed him to get from the *Toronto Globe* all the initial news bulletins on the day of the explosion, bulletins which reveal that it was two and one-half hours before there was any indication how serious the situation was.

Two pieces by social welfare specialists appeared in *The Survey* within weeks of the explosion (Carstens 1917; Deacon 1917). Both deal with the relief effort, a subject not well covered in disaster research (Taylor 1978: 272). There are three articles on the response in medical journals, two written at the time (Foster 1917-1918; Tooke 1918), and the other more recently (Connelly 1987). Two are personal accounts from physicians. One includes statistics on the types of injuries. There is a fairly comprehensive overview in a high quality Maritime publication, *Atlantic Advocate* (Jefferson 1958). There is an article in the *Canadian Pharmaceutical Journal* (Aldrich 1985) which profiles the pharmacist who took charge of the Victoria General Hospital. The article revealed that Charles Puttner was an unnoticed victim of the explosion; he collapsed from exhaustion three days after the explosion and never fully recovered. An article in *Maritime Merchant* (Henry 1905) profiles the deputy mayor, Henry S. Colwell, the man who later would put the community's emergency response together. (The mayor was away campaigning as the Liberal Roman Catholic Unionist candidate in the federal election.)

More information came out in 1992 when St. Mary's University held a conference on the seventy-fifth anniversary of the explosion. The results were published as *Ground Zero* (Ruffman and Howell 1994). This book includes everything from an analysis of the legal aspects by a Halifax lawyer to a review of the importance of Prince's dissertation, from an analysis of how professional women were treated to a delightful account of the experiences of one elderly woman. It has the first look at the explosion by scientists since a Dalhousie University physicist, Howard Bronson, reported his findings in 1918 (Bronson 1918). There is even an analysis of the tidal wave created by the explosion using computer modeling to test contemporary accounts. The footnotes contain scores of references to material on the explosion including articles on its effects on city planning (Adams 1918), housing (Ross 1919), architecture (Clarke 1976), nursing (Bligh 1920; Ross 1923), education (Fergusson 1976), the tsunami (Greenberg et al. 1993), the impact of the explosion on Dartmouth (Martin 1957), and the role of the Dartmouth ferries (Payzant and Payzant 1979). These were only a start.

The material cited above appeared in such diverse publications as *Public Health Nurse, Construction: A Journal for the Architectural Engineering and Contracting Interests in Canada, Marine Geodesy, Canadian Nurse,* and *Journal of Education.* One reference describes two versions of a folk song on the explosion, recorded in 1938 (Creighton 1961). In "Post Cards on the 1917 Explosion," Bernard Kline quotes one card which says the cards are not a complete record. "The authorities will not permit cameras near the harbor where the explosion occurred" (Kline 1994). Records in the archives show this is correct; soldiers seized cameras and made persons swear film would be taken to the censors once it was processed (Brown 1917).

ARCHIVES AND LIBRARIES

Some archives and specialized libraries are very useful, others less so; all have information. However, there are hierarchies of archives. Beginners to historical research should start where there is an interest in their topic. Major archives are more useful to persons who know what they are looking for. For the explosion, the place to start was the Public Archives of Nova Scotia (PANS). As well as the records of the Halifax Relief Commission and other miscellaneous documents, it has the remaining documents collected by Archibald MacMeachan and the draft manuscript on the medical response by a Dalhousie University professor named Fraser Harris. It has reel after reel of microfilm of letters and other material including the surviving records from the morgue. It has the typed minutes of the Halifax Relief Commission, starting 11:25 a.m. December 6, two hours and 20 minutes after the explosion. These reveal disagreements, such as the debate about when the emergency period ended and when physicians could get paid. Most important, PANS's staff is well informed about and interested in the explosion. Many attended the 1992 conference at St. Mary's to see their material being used. They are more than ready to assist scholars starting out on explosion research.

The National Archives of Canada has military records from 1917 including those that were secret. These include daily records compiled by the Canadian Army Medical Corps. It has records that show how the Canadian Army tried to account for or recover its missing supplies, including bandages, stretchers, and blankets. It has the papers of the Governor General, the link between Ottawa and London, and those of Prime Minister Robert Borden, who abandoned the election campaign to go to Halifax. It has the appraisal of the chief medical officer, McKelvey Bell, as to how emergency medicine should be organized for a catastrophe. It has material which shows that an inquiry was less than ambitious in chasing the real cause of the explosion. A letter from the inquiry's counsel notes that *Mont Blanc* was a French ship loaded in the United States and sent to Halifax by the British; it would not be appropriate to have the inquiry examine the role of Canada's allies while the country was at war (Henry 1917). Another letter opens up questions of conflict of interest. Among the normal clients of the lawyer for the inquiry were the agents who represented both ships involved in the collision, *IMO* and *Mont Blanc*.

There are many smaller but useful archives. The one at Dalhousie University has two diaries kept at the time of the explosion. One records a woman's trips to the morgue to search for her sister. The other tells how a physician performed emergency surgery on the couch of Professor MacMeachan, author of the official history. The Beaton Institute in Sidney, Nova Scotia, has a letter from an American in the Canadian Army. He describes, among other things, conditions at the morgue. Augusta, Maine, has that state's official report, telling that the governor acted after consulting with Massachusetts. The Colchester Historical Society Museum in Truro has a box containing unsorted records of the relief efforts in Truro, including correspondence with Halifax.

St. Francis Xavier University in Antigonish has a Roman Catholic publication telling how 125 families were wiped out in St. Joseph's parish. Pat Townsend at the Acadia University Archives in Wolfville found material not only in her own section of the university library but also books and theses relevant to the explosion. The minutes of the board showed, for example, that the president had gone to Halifax to assist with the emotional recovery and that the School for the Deaf and Dumb moved to Acadia. A thesis provides a profile of George Graham, head of the Dominion Atlantic Railroad, the first person to send a message calling for help. The staff at smaller archives have contacts as well as documents. The staff at the Old Kings Courthouse Museum in Kentville, not far from Halifax, not only produced railway timetables but tracked down a man who remembered his mother's experiences.

The National Archives in Washington has a huge file on the first American ship to respond, the *von Steuben*, revealing that it was a German raider converted into an American troop ship. The Guild Hall in London has records of shipping movements from 1917. They made it possible to determine the name port of registry and destination of ships in Halifax on 6 December and to discover the name of an American tramp steamer that entered harbor ahead of the *Mont Blanc*, a name not revealed at the inquiry. The Public Records Office at Kew Gardens outside London has the dispatches Rear Admiral Bertram Chambers sent to Admiralty the day of the explosion. Once secret, they are now public. One of them reveals that it was decided never to load another ship the same way as *Mont Blanc*. The staff at the Boston Public Library found the obituary of the man who headed the Massachusetts response to Halifax. He was owner of a bank, a prominent Republican, and a philanthropist, which explains why he fitted in with the conservative establishment in Canada. The library at the Massachusetts state capital was a gold mine. It has the records of the Halifax-Massachusetts Relief Committee and shows how discreetly and efficiently Massachusetts eventually helped the survivors. The archives at Memorial University have the original hand-written crew lists for *Calonne* and *Curaca*, two British ships severely damaged in the explosion.

Archives have maps, blueprints, and photos. For example, there are two photo collections in Halifax, one in Dartmouth, another in Toronto, and a fifth in Boston's state capital library. One shows the *Hovland* in dry dock. (It is possible to read her name.) At the National Archives in Ottawa, there are maps of the harbor, one with numbers of all the piers as they were prior to the explosion. (Since many eyewitness accounts refer to pier numbers, this was crucial.) In Halifax, a helpful archivist suggested another resource—fire maps showing street addresses for 1914. With those, it was possible to see that casualty treatment centers were in a semicircle around the impact area. The Public Archives of Nova Scotia also has the city directories for 1917 (a year when the street numbers were changed) which, when used with morgue and search records, both at the same Archives, allows the impact area to be precisely defined. At the Bedford Institute, there are maps of the harbor bottom. One shows the depression left by *Mont Blanc* exploding, eliminating any doubt about her precise location. No such data were available in 1917.

The whistle which was sounded to alert the citizens of Truro is in the firefighter's museum in Yarmouth. The *Lulan*, the steam pumper sent from New Glasgow, is restored and on display across from New Glasgow City Hall. The chief of police in New Glasgow provided a tour of the West Side School, which was used as an emergency hospital. When a document was found describing how the school became a hospital, it was easy to follow. Even though things change over time, tours are always of value. In Sandefjord, the author spent hours walking through the local cemeteries. In Norway, tombstones carry not just the name and date of birth and death of the deceased but the occupation. Two matched precisely names on crew lists found in Oslo. It was possible to work out how old they were in 1917. Despite spending several hours in the cemetery, the author missed the grave of the *IMO*'s captain, Haakon From. Knowing that From had been killed in Canada, he was not expecting that; only later did he learn that From's body was shipped home to Sandefjord and buried in the family plot. Friends in Sandefjord found the grave and showed it to the author during a subsequent visit. They also came up with a photo of Jurgens M. Osmond, the senior partner of the firm that owned the *IMO*. (In

Norwegian "I" is used instead of "J"—*IMO* is Johan Martin Osmundsen's initials.) Osmond, incidentally, is described as "Danish American," which was a reminder that Scandinavian countries did not always have the same boundaries and that names can change over time. (The author also discovered this when he asked someone in Oslo where "Christiana" was located; he was unaware that was the previous name for Oslo.)

Sometimes, polite persistence is necessary to find information. In Washington, some ship's logs were located easily, but the one from *Old Colony* was not. *Old Colony* later became a British ship, but on 6 December she was in Halifax harbor flying the Stars and Stripes. With that information, the log was found. The librarian at Palais de la Reine, Chateau de Vincennes, in Paris where naval records are kept, said there were no crew records for *Mont Blanc*. During the war, all French merchant ships were taken into the navy. A second search turned up the hand-written crew list from Bordeaux. Persistence is easier when you know something. That is why it is best to start with a local archive. Many archivists become more helpful when you explain what you have done and what you have found, satisfying them you are a serious scholar.

OFFICIAL RECORDS

Halifax is the political, educational, and business center of Nova Scotia, so the impact of the explosion was province wide—and the province responded. That response is recorded in town minutes. Most are still in vaults or basements. Burrowing among old minute books is a filthy but rewarding task. In 1917, many clerks pasted correspondence in these books. The author located several original letters sent by Halifax asking for support for its plea for federal assistance.

Fire departments record their calls, and many departments still have entries from 1917. It seemed strange when three nearby communities—Kentville, Wolfville, and Windsor—had no record of a response to Halifax though there were records of other calls that month. When it became clear that this was not an omission, it also became clear that these communities on the Dominion Atlantic Railroad sent physicians and nurses to Halifax but not firefighters. This was in sharp contrast to communities along the Intercolonial Railroad such as Truro, New Glasgow, Amherst, and Moncton, which sent physicians and nurses *and* firefighters and fire equipment. This led to a major discovery: that the response from a community depended on the railroad it was on, for railway communications determined the response. Once that was known, a thesis in the Acadia library proved extremely valuable. It showed that the response the day of the explosion fitted the railways' usual style. The Dominion Atlantic was a subsidiary of Canadian Pacific, not inclined to deal with local politicians or the agencies they manage, including fire departments. The Intercolonial was a government railroad integrated with the communities it served. Its agents immediately contacted local government officials and sought and got their help.

In searching for information about the explosion in small towns, it was possible to develop a simple but effective routine. First, visit the town clerk to see if there are surviving minutes from 1917. While at the town hall, ask about homes for seniors and names of elderly citizens. Next visit the town library; most towns have some form of clipping service and copies of local histories. At the library, check to see if the town has a historical society or archives. Next visit the fire department. Fire departments are social organizations and are well informed about and proud of their history. In New Glasgow, the chief's son is a historian; he had the town records from 1917 that he found at a yard sale. After that, visit the homes for seniors. Seniors are delighted to talk to visitors, especially when it is obvious that the visitor finds what they have to say valuable. Another useful checkpoint is the local historical society or any resident historian.

It is extremely important to tell everyone what you are doing, and that means *everyone*—the desk clerk, the hotel maid, the swimming pool attendant, the parking attendant, storekeepers, service station attendants. Incidents like the Halifax explosion are the stuff of legends, and everyone is interested. By telling people about your research, you allow word of your interest to spread. In March 1996, for example, the author met a university classmate at a golf show. He mentioned he was working on a book on the explosion. The classmate then told him—in considerable detail—how in 1968 in a bar in the disreputable part of Tours, France, he had talked to the chef from the *Mont Blanc*, the ship that exploded. He even kept a note about that conversation in his personal diary. In June 1966, a man attending a lecture came up and revealed that some of Samuel Henry Prince's papers are filed in Saint John, New Brunswick, under someone else's name. The man lives in a home on what was once the Prince farm and played the organ at Prince's funeral. Publishing findings can also be important. When the author published an article about the role of the railroads, that led to an editorial in a Halifax paper, and that led to a letter to the editor. When the author called the letter writer—his name was in the Halifax telephone directory—he discovered the man knew how the first radio message from Halifax was sent and who sent it. (Until then—well after the first version of this chapter was published—the author was unable to explain how Havana, Cuba, had been able to send out information on the explosion. It now became clear that Havana naval radio had head that first radio message which was not sent in code.)

In Halifax and Dartmouth, it seems everyone has a connection. When Gillian Osborne and the author were mapping in the North End, they chatted with a police officer. He immediately came up with the names and addresses of relatives who had been in the explosion. A note posted at the YMCA produced a phone call from a survivor the following day and—incredible as this may seem—16-mm film of the explosion. In Truro, the nearest major center to Halifax, the town clerk chatted with the author for about an hour (he wanted to hear what the author had found to date), then personally went into the vault and found the minutes from 1917. A visit to homes for seniors turned up a man who tried to get on board the special train leaving for Halifax. Another call found the daughter of the local newspaperman who was the first outside reporter to cover the explosion; he drove from Truro and filed the initial Canadian Press story. Seniors were located by a local minister whose son played basketball at Carleton, the author's university, before moving on to graduate work at the London School of Economics.

Such visits do not require advance notice. A business card and an explanation plus a willingness to share information is enough. Quite often, if persons cannot help they will suggest others who can. A librarian at the Dalhousie School of Pharmacy in Halifax suggested calling a retired professor. He remembered the article on the pharmacist who ran the hospital. In Halifax, the police department has records of who was arrested before and after the explosion. They show that most prostitutes and bootleggers survived. Victoria General Hospital has some records from 1917. The police records were made available after an informal request, probably because the author is an instructor at the Canadian Police College. The medical records were provided after a formal application for their use. The Archives at Canadian Forces Base Esquimalt on Canada's West Coast has the yearbook from the Royal Naval College with notes about the cadet injuries. (The college moved west after the explosion.) It is open to the public. There are histories of Camp Hill and Victoria General Hospital. There is a history of Stellarton, whose firefighters went to Halifax. Even semi-confidential records are available to legitimate scholars, with the unstated assumption that discretion will be used. One record turned up an account which stated that a soldier had stolen money and jewelry from the corpses. It seemed inappropriate to use his name.

Many books on other topics will touch on a major event. A history of dentistry has information about dental reconstruction for victims. A history of the Canadian Press news agency discusses CP coverage of the explosion. A book on blacks in Canada says a new black school was damaged by fire. Books on fighting ships identify the warships involved. Some of these were located by browsing at book sales and visiting stores that sell used books. One visit turned up the autobiography of Bertram Chambers, rear admiral in Halifax in 1917 (Chambers 1927). Unfortunately, he stopped his story before the explosion. (The idea of going to old book sales came from the author's neighbor who runs a used bookstore in Ottawa.)

Statistics can be revealing. A colleague, Ross Eaman, suggested looking up U.S. Census and immigration figures. Tied to the fact that many Americans came to Canada to fight in the war, this helps explain why New England responded so quickly. The following is from the draft manuscript of the author's book on the explosion:

In the decade prior to 1917, more than 700,000 Canadians emigrated to the United States, many to New England. U.S. census data shows that in 1920, 1.8 million Americans had at least one Canadian parent. A woman in Ayer, Massachusetts lost 13 relatives in the explosion: her mother, father, three sisters and eight of their children. A woman in Providence, Rhode Island lost her sister-in-law and two nieces and nephews. To Americans, the explosion was a catastrophe close to home.

MEMORIES

Those who were at least five years old when the explosion occurred are now 87. Those who were teenagers are 94 or more. The visits to homes for seniors turned up one or two persons with a vivid memory in each town. The author's daughter came along (she helped make older women feel more comfortable). Many recalled their own experience. Others recalled what someone told them. One lady recalled that her mother had been with her grandfather watering a horse, and the horse reared up when the ground shook. They were 360 kilometers from Halifax. Other memories included:

- A former Truro firefighter who still remembers the trainload of injured arriving from Halifax the day of the explosion, especially one woman carrying a dead baby in her arms. His memory is so vivid because his house was next to # 9 siding, where the train stopped;

- The town clerk in Bridgewater had no personal memories of the explosion, but her grandmother, now in her late nineties, worked in Halifax as a nursemaid. She called her grandmother on her speakerphone and asked about her memories. She said later some of things she had never heard before;
- A man in Amherst whose family were the main luggage and coffin manufacturers, a firm that shipped 3,000 coffins to Halifax after the explosion. He recalled his mother telling him they packed some coffins with sandwiches to make sure the men who went with them would have something to eat;
- A man now living in California told in vivid detail how the windows of his school, the Halifax County Academy, blew in during morning prayers. He recalled how, once he left school, his brother, who ran a store, sent him to warn the family there might be a second explosion;
- A man living in Ottawa not only recalled how news spread along the party telephone line in his home town of Lovat, Pictou County, Nova Scotia, but being a physicist, he was even able to describe how and when that telephone line was constructed and how it worked.

One man talked about his own memories then, half an hour later, phoned the Colchester Museum to say that his sister would have clearer memories. She did. The explosion was a life experience for those alive at the time, the way adults today feel about the Kennedy assassination. First memories are especially vivid. So are memories of things that were recorded at the time, even though the records have been lost. These memories may seem trivial at first, but they start to tie together. After a number of stories were collected from persons who were rescued by someone, it appeared as if women did most initial rescue work. That made sense. In 1917, women were at home and men were either at work or at war. This is how that anecdotal information was woven together in the draft manuscript:

The Halifax explosion killed something like a thousand people, critically injured several hundred more, left five to six hundred trapped in burning wreckage and left 9,000 survivors with everything from severe lacerations and burns, to broken bones, cuts and bruises. All through the North end, these survivors tried to assist each other. Usually, the first to help were women like Annie Chapman. They were taking care of their children while their husbands were at work or at war. Next came the neighbours, both women and men. Later, as helpers poured into the North end, the rescuers were mainly men. This included men who came from work to search for their families, visitors who came from the downtown hotels, soldiers and sailors.

The woman mentioned, Annie Chapman, was a young mother who pulled herself out of the rubble of her house then rescued her infant son. She wrote out a detailed account in a school scribbler several weeks after the explosion. Her son found it when she died. It was one of the many private accounts turned up during visits to small communities.

The Halifax newspapers listed the names of those who were in hospital. The Halifax Relief Commission listed those who were identified at the morgue. Because the explosion affected so many persons, one way to find relatives of survivors is to note any unusual names in these records and call persons with those names in the current phone book. There were a lot of Boutiliers in the explosion. There are two columns of Boutiliers in the current phone book. Since everyone is interested in the explosion, all such calls get a positive response. There were three persons named Beiswanger killed in the explosion. The current phone book lists no one by that name, but there are Beeswangers. A call to them produced no direct links to the Bieswanger's, but it led to a family history and a family friend who was in the explosion and is still alive. Every person called was friendly and helpful and interested in hearing about the research.

FAMILY, FRIENDS, AND COLLEAGUES

Information was also discovered by following a trail, starting with a person, ending with a record. This was done with help from family, friends, and colleagues. Doing this requires conviction that any lead is worth following—and dogged determination. One search began with the author's great niece, who works in Oslo. (The author went there because *IMO* was a Norwegian ship.) One of her staff is the daughter of a man whose best friend is director of the Maritime Museum. His files included a letter from a relative of *IMO*'s owner. He suggested the whaling museum in Sandefjord. That led to Tonnessen and Johnsen's book, *The History of Modern Whaling*, which revealed that Captain Haakon From, *IMO*, and *IMO*'s home port, Sandefjord, were part of Norway's whaling tradition, making them among the world's most experienced sailors (Tonnessen and Johnsen 1982). In Sandefjord, the author visited the local newspaper, asking it to run an article and hoping to attract attention (Omvik 1994). It led to a call from the family of one of *IMO*'s crew then, months later, to letters from the two daughters of *IMO*'s first mate. They were visiting Sandefjord and heard about the research. One now lives in Norway, the other in Staten Island, New York. Later, after the *Mont Blanc*'s records were discovered in Paris, it became clear the sailors on both ships came from small, coastal communities.

At the Sandefjord newspaper, someone said a man who repaired typewriters had lost his arm in the explosion and that his grandson was a policeman. The grandson was located with the help of the daughter of a woman from Dartmouth, Nova Scotia, who lives in Sandefjord. She married a Norwegian sailor during the Second World War and is a friend of Gillian Osborne's family. (Osborne is a former student and colleague, coauthor of a chapter in *Ground Zero* [Ruffman and Howell 1994].) The woman's help proved crucial; the grandson's name is not the same as his grandfather's. However, he spoke English and produced a pay book from *IMO*, a photo of the ship, and a letter from the Canadian government expressing regret at the loss of an arm, with a check for $100. He also had a clipping of an interview with another survivor published in the *Sandefjord Blad* and patiently translated it, and he had an oil painting of *IMO* done by his grandfather. (The only copy of that painting now hangs in the author's living room.) Another document in private hands is the diary of a cadet at the Royal Naval College. An amateur military historian in Esquimalt has it on Canada's West Coast. The author's son, who is a naval officer, met him when the base library fell under his direction. Other documents in the library showed that many cadets who were in the explosion became senior officers in the Royal Canadian Navy.

A chat with a wireless operator and wartime naval officer at a Carleton University basketball game led to the Naval War College at Newport, Rhode Island. Its librarian, Maggie Rauch, provided information on U.S. ships involved and suggested other sources in Washington for their logs. Might there be similar material available in England? The author asked his insurance broker, and he faxed a colleague who deals with Lloyd's of London. He suggested going to the Guild Hall where there are Lloyd's records of daily ship movements in and out of Halifax for 1917. The librarian said there would be more detailed material at the Public Records Office, Kew Gardens, and at St. John's Memorial University in Newfoundland. One document was from a naval officer who surveyed the damage by traveling by boat along both sides of Halifax harbor about an hour after the explosion. A chat at another basketball game revealed that the wife of Carleton's athletic director grew up in Nova Scotia and that her 96-year-old aunt, who lives in Truro, was working in Halifax the day of the explosion. Her boss called her to take dictation so she was away from her desk by the window when it shattered. She and a friend did registration and inquiry at the YWCA the day after the explosion.

Samuel Henry Prince's papers were discovered after sociologist Kurt Lang mentioned at a meeting in Washington that he had met Prince in Halifax and that Prince had been teaching at King's College in Halifax. Gillian Osborne learned that Prince had died but that his papers were with his first graduate student, Rt. Rev. Leonard Hatfield in Port Greville, Nova Scotia. Hatfield was working on his own book (Hatfield 1990). But he sat in his study and read from the Prince papers while the author and Osborne scribbled notes. (Some of those papers are now in the archives at King's College.) Osborne also found that Prince's most quoted source, the unpublished manuscript by Dwight Johnstone, is now in the Public Archives of Nova Scotia. (It had been in private hands for fifty years.) While in the Archives, the author asked if there was anything on medical aspects of the explosion. That led to Fraser Harris's manuscript, a valuable source for the medical response. (Harris was supposed to write a medical history of the response.) Impressed with the wealth of material available, the author phoned the National Archives where a former classmate suggested other documents.

A former dean at Carleton University, an historian, neighbor, and friend, dropped off several references. One was an unpublished manuscript by a survivor. He and his family were trapped in their home when the doors jammed but were rescued by neighbors. The head psychologist with the London (Ontario) Board of Education is a King's College graduate who knew Samuel Henry Prince. (The author met the psychologist at a seminar in London.) He referred the author to a prominent government official in Ottawa who also knew Prince. That man's mother, Carrie Best, is in New Glasgow. She went with her father five days after the explosion when he picked up victims at the train station and took them to the West Side School. She is one of Canada's most distinguished black activists (twice awarded the Order of Canada), and she suggested contacts in the black community in Halifax. Legend has it that only one black person was killed in the explosion, but the author has found four names of blacks were who were killed and a morgue record that states that among the dead were 10 "Africans." The legend is wrong. The editor of the Dalhousie University alumni magazine (she is a Carleton journalism graduate) offered to put a note in her publication. It generated a score of letters and e-mail messages, even a query from some film makers in Hollywood. One note came from a man who studied with Howard Bronson, the physicist who did the original scientific work on the explosion.

Not every trail was productive. Someone said that the explosion was folklore and that this was mentioned by a professor from Maine during a summer course in the Maritimes. That lead to a visit to the University of Maine at Orono and a search of files at its folklore center. It is possible a professor from Maine discussed the explosion as folklore, but there are no records of such comments. In France, it was possible to discover the exchange of telegrams between the French government and its officials in Washington. It was not possible to locate a similar exchange between Norway and the Norwegian consulate in Canada though both the foreign office and the State Archives in Oslo were helpful. A call to the St. Mary's University archives revealed that the priests who ran the college in 1917 took their papers with them. Yet that call proved productive in two ways: the author learned that there is an association of archivists in Nova Scotia, and his attempt to track down those priests at the Jesuit archives in Toronto led to a priest who knew a priest who had lost an eye in the explosion. A visit to the archives at Mount St. Vincent not only led to historical records but also the discovery that the archivist was a nun with vivid memories of the ships involved. Her father had written the history of Dartmouth.

For every apparently unproductive trail, there were unexpected discoveries. A man whom the author met while playing golf in Cavendish, the storied home of *Anne of Green Gables*, took the author to a former captain of the ferry between Nova Scotia and Prince Edward Island. When the author said that the prime minister, Sir Robert Borden, traveled from Charlottetown to Pictou on a boat called the Avon*more*, he said the boat was the Aron*more* and that it saw service as a deep sea rescue tug during World War II. The *Aronmore* is mentioned in *The Grey Seas Under*, a story of tugs and ocean rescue (Mowat 1956). *Aronmore* was a federal buoy boat, and its logs at Transport Canada in Ottawa should show when the Prime Minister left Charlottetown and arrived at Pictou. A chat with a man at a seminar for the Steel Company of Canada led to his mother who was the daughter of John Cranwell, the Salvation Army man in Halifax. She had a clipping of *War Cry*, the Salvation Army publication, a publication found in another useful source, the Salvation Army's Heritage Centre in Toronto. There it was possible to identify the key Salvationists who went to Halifax, when they got there, and what they did when they got there. It turned out they did the second careful survey of the impact area. The first survey was directed by Clare MacIntosh, Lady Superintendent of St. John Ambulance. Her report is mis-filed in Public Archives of Nova Scotia; It was found because a graduate student in history happened to be looking at that file and was sitting across from the author. (Another researcher turned out to be the granddaughter of one of the blacks killed in the explosion.)

The police officer who found his mother's scribbled account of what happened called the author after seeing a note about his research in a newspaper. The man mentioned that his father, serving overseas, tried desperately to learn if his wife and child were safe. His story confirmed what archival documents showed: that after the explosion the bad news was flowing from Canada to Europe rather than the other way. Corporal Frank Rickets who was with the Canadian Army overseas was wired as follows: "Regret have to advise you wife injured out of danger boy killed recent explosion" (General Office Commanding, Military District No. 6, 1917). That telegram is in the National Archives of Canada. Incidentally, while people lead to records, records also lead to people. The article on the man who was acting mayor revealed he had founded Colwell Brothers, Inc., a men's clothing store in Halifax. The store still exists, and one phone call produced his grandson who has a family history and recalled his grandfather. A letter to the editor of the *Halifax Chronicle-Herald* produced 10 letters from relatives of survivors in two weeks. One woman's father and grandfather went to Halifax to search unsuccessfully for her uncle's body. He had been best man at her father's wedding on 3 December, three days before the explosion.

It is valuable to return to a place or a source to check something or just say "Thank you." People remember things once they know you are interested. The archivist at Acadia University produced another document when the author dropped in to say "Thank you." When the author called another source's widow to check on her husband's background for this chapter, she chided him for not calling an Ottawa woman, Evelyn Dakin, a survivor. Mrs. Dakin, then Evelyn Welch, recalled that when she and her family were told to flee because of the threat of a second explosion, her grandfather, George Lovett, refused to go. A veteran of the expeditionary force that suppressed the Northwest Rebellion, he got out his blunderbuss. Then he sat outside on the front porch steps holding the gun. His daughter stayed with him. Convinced Halifax was under German attack, he vowed, "I am going to stay here and I am going to get the buggers." He was 80.

INFERENCE

The anecdote about George Lovett and his daughter is more than an amusing story. It suggests that when there were warnings of a second explosion, some declined to leave. Several other accounts were found of persons who worked on despite those warnings and of those who went in and out, carrying injured and/or supplies. There are lots of stories of flight behavior, but it clearly was not universal. It is this step—fitting information into patterns and using inferences to increase its meaning—which takes research away from mere description.

One story about the explosion is that a railway telegrapher, Vincent Coleman, started to flee, then returned to warn incoming trains—a decision that cost him his life. (The story is featured as a vignette on Canadian television.) The message was recorded by operators down the line. What is not known is how Coleman knew *Mont Blanc*'s cargo. That was secret. It seemed possible that Coleman handled a telegram reporting either the departure of *Mont Blanc* from New York or its arrival outside Halifax. Both mentioned the ship's cargo. The problem is that Coleman was a railway telegrapher, not an employee of Western Union or Canadian Pacific. The next logical step was to ask: Who knew *Mont Blanc*'s cargo and, of those, who had contact with Coleman? That made the answer easy: Lt.-Cdr. Murray, the convoy liaison officer was on *Hilford* when it passed *Mont Blanc* right after the explosion. The pilot shouted to him that *Mont Blanc* was going to explode. *Hilford* then went to Pier # 9, and Murray went into the station to use the phone. That was where Coleman worked. Perhaps Coleman overheard Murray talking. Perhaps Murray told him. The crew of *Hilford* knew, so it was no longer much of a secret.

Information about how the railway telegraph works was acquired from former telegraphers, one of whom has a telegraph key in his basement and uses it to keep his touch. While earlier researchers report that the railway used the Morse code, these men said the system used was American Landline. (It uses spaces as well as dots and dashes for various letters.) A copy of that system was found in the Kentville archives; its significance was recognized because of those conversations. After learning how the system worked, the author concluded messages from Halifax must have been relayed to Intercolonial headquarters at Moncton by the operator at Truro. In a railway publication discovered later, the man who sent those messages says that is exactly what happened.

Once links are established, inference may suggest their significance. Many persons arrived in Halifax in private railway cars. These included the heads of the Intercolonial and Dominion Atlantic railways, the Prime Minister, the head of the Massachusetts relief group, the head of Canada's largest store (Sir John Eaton), and the head of an American railroad. These cars were parked close together: the head of the Massachusetts response saw Prime Minister Robert Borden out for his morning stroll. One can infer that these persons would talk to each other. That may help explain why Massachusetts got continuing concessions from Canada, such as no duty on goods shipped across the border. Similarly, Thomas Raddall describes the first bodies received at the morgue as blacks from Africville. The "black rain" which followed the explosion turned many things black. Survivors took weeks to get the black out of their skins. It is reasonable to infer that, while Raddall saw bodies that were black, it is more likely that the explosion blackened them than that they came from the ghetto known as Africville.

Inferences can also be made from a railway pass issued to a Methodist minister who arrived in Truro the day after the explosion;

it allowed him to continue to Halifax. The existence of the pass confirms that passes were issued and that some form of checking was going on. Since the man was both a minister and a trained member of St. John Ambulance, it helps indicate what sort of credentials would allow someone to continue. Another kind of pass was issued to Dwight Johnstone, Samuel Henry Prince's friend. It allowed Johnstone access to the impact area in Halifax itself. That suggests there were levels of passes—one would allow you to enter the city, but another was needed for access to certain locations. The existence of passes—especially one issued the day after the explosion—reinforces the impression there was a fair amount of organization very quickly and that it took place before the Americans arrived. That contradicts Prince's dissertation that suggests it took the Americans to bring order out of chaos. The minister's pass was discovered when his daughter wrote the author after hearing him interviewed on a CBC radio program in Halifax. Johnstone's pass is in the Public Archives of Nova Scotia.

Sometimes by using modern data processing it is possible to get archival records to reveal more than was known when they were created. It took more than three days to type morgue records into a computer file, but the results made that time well spent. One diary mentioned visiting the morgue looking for two different missing persons. The information in the diary was sufficient to establish who these persons were. By using the dates in the diary which show when the bodies were identified it was possible to date those two morgue entries and, from that, to start to figure out how quickly the identification process moved along. Equally important, it showed how the process worked, something not clear from other records. Once the data were on file, it was possible to find an address. By comparing the search records, information about unidentified bodies found at specific locations, and addresses in the city directory, it was possible to figure out the probable names of the dead. For example, one woman's body with a wedding ring was found where the father and children were dead and the mother missing. The body was buried as unidentified. That again helped indicate how the morgue worked. Often data point to new directions. The morgue record says that the body of the captain of the *IMO* was embalmed and shipped home to Norway. That led back to the Sandefjord cemetery where, as it turned out, the author had missed the captain's grave during his first visit.

VALIDITY

People create records, and the same rules apply for testing validity as apply to checking personal stories. Does the material have internal consistency? Is there any corroboration? Is the account something that reasonably could have been known to the person who created the record? If it is not evident, it is important to ask, "How did you know that?" Sometimes persons will provide information both about things they did or saw *and* about what they heard. It is important to separate observations from second-hand accounts. The first are usually accurate, the second are not. While this is harder to do using documents, it is important to ask, when reading a written account, "How would that person have known that?" One man who was on the Boston Express described both what happened on that train and what happened in the heart of the city. His account of what happened on the train fits with other accounts. His account of what happened in the city is unsupported. Since he left on that same train and since the train never made it to the city, it seems reasonable to break his information into two parts, one reliable, one not. It is also important to ask if there is anything which suggests why the person might have been less than truthful or have had a systematic bias. Finally, one must pay some attention to when an account was recorded. As time passes, members of organizations are likely to recall better organization than actually existed. They are also likely to recall that decisions were made at a higher level than was the case.

The statement that the railway telegraph message was relayed by Truro to Moncton is credible because that is how the system worked. If the man who sent it had said he used some other method, it would have been necessary to search for evidence as to why he deviated from standard procedure. Some stories are easy to verify. There are accounts about what happened on the Boston Express from the conductor, passengers on the train, and the physician who came to assist. Accounts also appear in the Truro paper and in Canadian Army records. In addition, there are reports about the yard engine that was traveling back and forth between the Express and Rockingham from the same physician, from a sailor from *Curaca*, and from a crew member off *Hilford*. There is an official damage report from the Intercolonial Railroad. All these fit together.

Other accounts are credible because one meshes with another. The woman who wrote the author after hearing him on CBC Halifax said that her father was a Methodist minister who left for Halifax by train the day of the explosion. She said he met a minister who had lost his wife and a child. There was such a minister. His name was William Swetnam, and he, too, was a Methodist. The author met his surviving daughter, Dorothy, at the Maritime Museum. Similarly, the diary of Archibald MacMeachan, which mentions a naval cadet receiving surgery on his living room couch, connects to the diary of a naval cadet who recorded bringing that cadet to MacMeachan's home. The first diary was in the Dalhousie University archives, the second in private hands in Esquimalt on the West Coast.

Other accounts do not mesh so easily. Shortly after *Mont Blanc* exploded, sailors and soldiers rushed through the city shouting that there was going to be a second explosion. They said the magazine was on fire at the Wellington Barracks. Archibald MacMeachan collected first-hand accounts from a sentry on duty at the magazine and from the young officer in charge there. They said there was no fire. Their version is supported by General Benson, who went to see for himself. At first, this seems to be a conflict. A closer examination shows it is not. Soldiers and sailors did rush through the streets warning of a second explosion, and many persons fled to safety; but there was, in fact, no fire. The fact that soldiers and sailors shouted, "Fire," is not evidence of a fire, only evidence of their shouting that there was. (MacMeachan himself is not a primary source; his personal diary makes clear that he was not present when the warning was given.)

Some material is credible because the source has no apparent or conceivable reason for bias. The artist, Arthur Lismer, records that his arrival at Rockingham occurred when a trainload of injured was heading out. Only one trainload of injured left that day, and the time fits. Sometimes material is useful because it helps establish credibility of other accounts. There are enough stories of telephone messages to indicate that some phones were working. A story that mentions a phone call is, therefore, credible as long as it does not suggest that someone was talking between Halifax and its sister city, Dartmouth. The link across the harbor survived the explosion but was destroyed by ships dragging their anchors in the subsequent storm. The fact that something is not credible does not make it useless. There were rumors after the explosion. Some died quickly, others persisted. Many persons thought that the explosion was an attack by German submarines, a German warship, or German zeppelins. That was not true, but the way these rumors stopped is relevant to a discussion of how rumors die.

IMPORTANCE OF DISASTER RESEARCH

Before this research began, the broad outlines of the story were well established. Two ships collided. One, *Mont Blanc*, caught fire and then blew up. Thousands of persons were injured, dying, or dead. The North End of the city of Halifax was in ruins. There was an enormous relief effort from the surrounding area (that was not well documented) and from the United States, especially Massachusetts. One serious problem was eye injuries caused by flying glass. No evidence was found to discredit this basic story, though it was possible to add a great deal to it. Most published material about the explosion deals with the situation in Halifax and Dartmouth. There are limited accounts of the response from the region, on the role of the two railways, and on the background of the ships involved (e.g., that the *IMO* was normally involved in whaling). There is nothing on how the news spread or how the newspapers, especially the Boston press, treated the story.

The latest research closes some gaps. The story of first day response by rail can now be told almost minute by minute. Some medical data can be fleshed out: the author has a growing list of names of persons whose eyes were removed, a list that already has more names than the number in medical statistics. It is also possible to take existing material and look for a new interpretation. There is a lot of evidence about the two ships blowing their whistles before the collision. The reason why *Mont Blanc*'s whistles were blown was confirmed by the captain and the pilot, but those on the bridge of *IMO* died. The court assumed why *IMO*'s whistles were blown. Were those assumptions wrong? Once that idea surfaces, it is possible to review data, including court transcripts.

Much of the memory of the explosion is anecdotal. There are stories of how a father rescued his children or a mother rescued her child, accounts of the response by one nurse or one physician. These show up as interviews in a newspaper, as letters, or as private notes. They are also part of living memory. In addition, there are accounts of specific organizations, the role of the Massachusetts or Maine relief group, the activities of St. John Ambulance. One framework for tying this material together is the story as told by persons like MacMeachan, Bird, and Kitz. It could be retold with more detail. Another framework is offered by 80 years of sociological research. It was used by the author. That framework, the one developed by Charles Fritz, Henry Quarantelli, and Russell Dynes and their students like Thomas Drabek and William Anderson and enunciated in books such as *Organized Behavior in Disaster* (Dynes 1970) and *Human System Responses to Disaster* (Drabek 1986), did not exist in 1920, and it was not known to those who have written about the explosion. The author could examine both new material and existing data from this new perspective. When some authors reported "panic," it was helpful to know that flight behavior is not panic—and to examine the evidence. Similarly, when it was discovered how well the railways performed the day of the explosion, it was helpful to know that in 1917 there were so many troop trains, supply trains, and munitions trains that running specials was routine. Organizations do better at familiar tasks. One might say there was a "special" subculture on the railroads in 1917.

Is it possible that this framework acted as a blinder, blocking the possibility that Halifax was different, that points were missed because of a mind-set? It seems unlikely. The material does fit the model. When the author went to Darwin after Cyclone Tracy, he went because he was convinced Darwin was different. There, he discovered the data did not fit that theory. Darwin was not different (Scanlon 1979). Halifax was what Henry Quarantelli labels a "catastrophe." The evidence suggests it follows the model built over the past three-quarters of a century. The tragedy of Halifax from a research point of view is that this evidence has been available for decades and that, in the intervening period, some was lost. Particularly sad is that some lessons from Halifax, especially the lessons about emergency medicine, were buried in government files, forcing others to learn from their own experience. However, enough remains still to make this 80-year-old disaster a fruitful topic for research.

NOTE

1 The suggestion for this chapter came from Dr. Henry Quarantelli who said he was intrigued by how some things on Halifax were tracked down. The first version was substantially improved by scores of comments from a long-time ECRU researcher and friend, Scott McClellan. He pointed out that the methods described were developed and tested during 25 years of quick response field research.

PART III
PROSPECTS

The four chapters in this section take up topics that reflect the prospects of future disaster research. While the methods currently in use will continue to be applied, new technologies and new contexts, both in the developed and the developing world, will affect research in the field. Wolf Dombrowsky (Chapter 12) discusses both positive and negative aspects of computer technology in general and of the Internet in particular for future disaster research. Nicole Dash (Chapter 13) describes how Geographic Information Systems (GIS) are currently used in the practice of hazards mitigation and disaster management and suggests a research agenda on GIS that could benefit both practitioners and the research community. Habbibul Haque Khondker (Chapter 14) draws on his experience in doing fieldwork in Bangladesh to identify problems and solutions for future field studies in countries in the developing world. Kathleen Tierney (Chapter 15) identifies changes in American society since the pioneering days of disaster research that will present both challenges and opportunities for future field researchers.

12

METHODOLOGICAL CHANGES AND CHALLENGES IN DISASTER RESEARCH:
Electronic media and the globalization of data collection[1]

Wolf R. Dombrowsky

Within the last decade, computers have conquered and transformed our field. At first they only substituted for our typewriters: then, step by step, we withdrew from the "hard return" at the end of each line and understood what "floating text" means. We qualified in wordprocessing, desktop publishing, spreadsheeting, and the tricks of converting differently formatted text and data files. We spent hours and hours with all that, realizing in the end that we had become quite good at taking over all the professions which were necessary to transform a formerly handwritten manuscript into a ready-to-print copy. However, these aspects of social change almost came about behind our backs. In the foreground, other changes were foreshadowed. Our modes of writing began to change. We discovered the advantages of "modular writing" and "multi-using" of text-blocks, of cut and paste, move and insert, and, above all, of databases: addresses, literature, notes, quotations, excerpts, news, clippings. Our productivity increased as well as the number of our publications. Whether quality increased is harder to say. We all know publications which differ only in their titles, and we also know this diffuse feeling of *déja-vu*: Didn't I see (read) that before once or twice or even more often?

The next improvements entered our studies with scanner technology, imaging, and optical recognition. Newspaper articles no longer needed to be retyped, originals and pictures no longer needed to be stored in space-consuming archives. On the other hand, gigantic hard disks became necessary as well as fast backup media and powerful software to organize and manipulate these image files. Again, our modes of data-handling changed: we learned to present our data better, we created computerized slide shows, and we slowly entered multimedia. With that, teaching changed also, because we learned to utilize all those nice little built-ins in SPSS (Statistical Package for the Social Sciences) and other software to project graphs, tables, and multi-layer maps directly from our computer to the classroom screen.

Most of us had scarcely climbed up to higher productivity when the Internet opened its gates and linked us with electronically multiplied strings of millions of users. Kling (1997) gave a nice example of how the Internet and its services influence our scientific handicraft: e-mailing, electronic discussion groups, Web-conferencing, searching indices, publications, abstracts, conferences and their papers and proceedings. Most of us have put our homepages into the Internet; almost all institutions—universities, labs, companies, churches, agencies, governments—are represented. Within minutes we can click through their branches and offices, make contacts, and get information. We can download and upload files and software, drivers and bug-fixes, updates and viruses—and we leave prints and profiles. The best of all, no doubt, is searching the Web for, let us say information—no matter of what it consists: services, software, images, movies, sounds, etc. Within seconds, or after a coffee break at least, our search engines and, more sophisticated, our Web agents serve us best while they optimize themselves along our preferences. (And some secretly work for others, too.)

Nevertheless, we again spent hours and hours, trying to survive the battle between browsers and their incompatibilities, the horror of encoding and decoding, zipping and unzipping, and converting files and data into "legible" formats. We remember our first impressed amazement at gazing at the hit-list our search engine has cataloged for, say, "Panic"—but also our disconcert when we realized that more than 90 percent of these 233,460 hits (with Alta Vista) were useless for disaster researchers' purposes. Yet, endowing surprises is unavoidable: "Rip Rig And Panic" are, as well as the German "Panik Orchester," nice and cranky musicians.

Unavoidable as well, then, was the need for further qualification. Now, most of us know a lot about Boolean operators, command syntax differences, protocols, networking, Hypertext Markup Language (HTML), cookies, and applets. And again we realize changes behind our backs and of ourselves: Was writing humankind's first intellectual revolution?; if so, then computer literacy may be seen as the second one. Both do sort strictly along access to and dispose of knowledge—which is, of course, not identical with information. However, information is the basis, the material of knowledge. In this respect the Internet is revolutionary because of its accessibility and feasibility. No wonder that dictatorships try to cut the Internet access of their literate people (see www.dfn.org, the Web site for the "Digital Freedom Network"). No wonder too that the Internet becomes a melting-pot of all participants whose traditional, mostly nationally and/or ethnically based "stock" of literacy—language, literature, poetry, narratives, etc.—will be transformed into some sort of "Net-speech computeranto" nobody has heard before.

Successively, a global commune of data accusers and distributors emerge whose links decide about "significance." More than ever, quality becomes key because quantity is overwhelming and, worse, indecisive and indeterminate. The Disaster Research Unit (KFS) I work with, together with significant disaster relief organizations, recently established a Internet-based interactive database (www.soziologie.uni-kiel.de/~wdombro/waktuel.htm) of individuals and organizations that offer psychological assistance for traumatized disaster victims and relief personnel. The main aim is to make the offers public and comparable, to define criteria and standards of appropriate and reliable assistance, and to inform those who seek assistance about qualifications, experience, and references. Many Web sites' offers advertise what nobody can assess or evaluate: Is it serious, reliable, helpful, inventive? Who is behind the site?; somebody highly motivated yet inexperienced, an unemployed psychologist or social worker who seeks an opportunity, or simply free riders who want to exploit the acute topic of post-traumatic stress disorder (PTSD) and crisis intervention? The example and the problem behind it is ubiquitous.

Kling (1997: 438) asked similar questions and raised analogous problems:

> "The Internet generally privileges speed of connecting authors and readers, serendipity, and novelty over epistemological reliability (as mediated by peer review, publishing houses, and curated libraries). Some of the arguments that the Internet enhances democracy rest upon the ability of a wider array of people to post materials for broad readership. But the materials may range from works of genius to foolish ramblings; and it is also possible to easily create spoof sites—such as posting 'translations of the secret correspondence between Durkheim and Marx'!"

But there are not only spoof and indecisive sites, but also spam-mail and hostile activities; there is also time-waste (the "World Wide Wait") and overload with garbage and outdated information as well as misleading links and dead URLs. In the last resort new services become necessary, agents and assistants that help users find, sort, evaluate, and organize the world-wide vagabonding and often vague information, to update URLs and e-mail addresses, and to check links and sites. (See as examples: www.theangle.com; www.ai.mit.edu/people/sodabot; www.agentware.com; rg.media.mit.edu:80/projects; bf.cstar.ac.com/lifestyle/). Consequently, some type of commercialization is unavoidable due to the fact that the Internet is (still) free—free in regard to uploading whatever one wants.

As a matter of fact, the Internet has changed our modes of working, writing, organizing, and teaching. The Internet has accelerated our communication and the interchange of data of all kinds. We can read complete books and articles on-line, and we can download documents right into our own texts. Soon, the knowledge of humankind will be accessible from every desktop terminal, just as virtual visits to museums (e.g., www.isst.fhg.de/ ~lemo), music halls, buildings, and exhibitions already are. The Internet has intensified and globalized our research. Better than ever we can know what our colleagues are doing worldwide. The Internet opened—sometimes burst into—groups and citation cartels because of cataloging sites of scientists from the peripheries in terms of regions, languages, paradigmatic orientations, or resources. We can have broader perspectives, wider arguments, more detailed examples, and a broader empirical basis of on-going and finished projects, results, and applications.

The Internet also has enabled access to institutions, agencies, and governmental and nongovernmental organizations. In my own field, I appreciate the convenience of accessing the sites of FEMA, WHO, DHA, UNHCR, PAHO, IDNDR, and ECHO (to name only some). I also appreciate the useful sites of the Natural Hazards Research and Applications Information Center (www.colorado.edu/ hazards), the Disaster Research Center (www.udel.edu/DRC/), NOAA (www.csc.noaa.gov/), Emergency Management Australia (www.ema.gov.au), and the International Committee of the Red Cross (www.icrc.ch)—to pick only a few (useful overviews may be found in Butler 1997; in Butler's appendix in this volume; and in Gruntfest and Weber 1998). Most of us have experienced the ease of searching for publications, mission reports, materials, and unpublished or unofficial papers. Special services such as listings of worldwide conferences, tables of contents, abstract indices, and sorted events or categorized tables of specific disasters (like volcanoes, floods, earthquakes, etc.) have made research easier and communication more efficient. In many cases, the ease of access has made cooperation and mutual contacts possible and has opened doors to learn about problems and about desirable and possible solutions, about research needs and available funding.

The most exciting aspect of the Web, however, is the integration of text, sound, images, video, radio, TV, and satellite. It is no problem to download the latest CNN video sequences of disasters, wars, or riots, of radio broadcasts or interviews, of Videotext-Information or Intercast-data. Many providers offer a "news reader"—services which provide us daily with specific information; specialized programs like "Paperball" (www.paperball.de) serve with keyword-centered news of German newspapers; most magazines like *Stern* (www.stern.de) and *Focus* (www.focus.de) have opened their archives for research: cut and paste worldwide, no matter of what, no matter for what.

In the meantime, I dispose of 18 gigabytes of exciting pictures, graphics, and videos of volcano eruptions, earthquakes, storms, floods, and air crashes. My students love the collapses of bridges and buildings most; also the jumps to death during high-rise building fires. Teaching is turning into a multimedia Internet session and paperwork into funny eclectic site citing, like pearls on a string, combined to a huge HTML document which runs off-line on my computer but has to be turned on-line when all the cited sites and links should be visited—and checked.

All this is nice and welcomed; some aspects, however, are often overlooked or neglected. After surfing the Internet for hundreds of hours, and, more time consuming, after years of more or less continuous self-teaching, going hand-in-hand with upgradings of the software and hardware (which indirectly did take months and months to earn the money for), I doubt a clear acceleration of or a sum-total increase in productivity. I would not miss either the computer or the Internet; both operate far below their potential. I still miss an address software or module that really works together with all others (with browsers, office suites, databases); I still miss standards in protocols, in text, data, and graphic/picture files; I still miss automatic converters to make encoded attachments legible; I still miss a secure driver management, instead of being forced to backup a stable system before testing a new program or a hardware component. I hate messing around with layouts, fonts, and sizes due to the requirements of different printers (only a few use postscript printers or LaTex). I hate all those needless throw-backs from so-called software "updates" that change key combinations, macro languages, or other central features for organizing one's work. And I still hate to compensate for the lack of quality software by searching the Web for bug fixes, "tricks & tips," and "inside information."

The computer still is not the universal machine it could be, if released from economical, political, and entrepreneurial limitations and competitive interests. From a technological point of view, there is no further need for a DOS-, Win3.x/Win95/98-, NT-, OS/2-, Mac-, Unix/Linux-, Alpha-, and BeOS-World (look at the emulation of all under Linux!) and its software incompatibilities the users are not longing for. Neither is the Internet as universal as it could be. We still suffer from 8-bit limitations and ASCII, albeit the 16-bit-based "unicode" is available. Nevertheless, our global communication still relies on the 8-bit codings of ISO 8859/Latin-1-Standard which is incompatible with Asian fonts (CJK-texts). In contrast to that, the scientists and literate people of the Middle Ages, arrogantly labeled the "Dark Ages," applied a universal language: Latin, their *lingua franca*. Today, with the help of the computer and the WWW, such an universal language is more necessary than ever and the most possible to implement ever. Technologically, we already could have a standardized, universal operating system with a stable, secure kernel, modular driver packages, and safeguarding against hacking (see www.hacked.net) and virus attacks, and globally compatible software. Practically we suffer from the contrary, from computers, software, and Internet services that pretend to be solutions but force us to pay attention to them, although we would not be without the problems which we do not need anyway. (The very best example is the myriad of Windows add-ons and tools which make the sick machine run "better" and crash.)

All this has a lot to do with data collection and electronic media. Globalization, however, remains ephemeral still, a fashionable label. My student's homework, this brilliant "site-and-link-quilt" or "net-patchwork," is suitable to make the point: picking examples from all over the world has nothing to do with globalization, but rather with universal unconcern and shallowness, with naive illustration at best. In the first place, the WWW is such a global illustration, a huge reservoir of signs of life, but without sociological meaning. Most of us know our national statistics and their derivatives: the average man/woman, demographic distributions, stratification. What or who represents the Internet, and of what is it representative in statistical terms? In Germany 2.49 million homes have a PC with a modem, 1.85 million homes access the Internet via the leading providers like T-Online (1.5 million members), AOL (about 300,000), and Compuserve (50,000) (all figures for 1997). A *quantité négligable* compared with the numerical basis of all telephone owners, consisting of approximately 30 million private homes, 42 million registered "main sockets" including business and fax but excluding cellular/mobile/satellite telephones. But what does "registered" really mean? The German computer magazine *PC Praxis* (10/97:52 58; www.pcpraxis.de) has tested six Telefon-Info-CDs and seven Internet-Telefon-Services showing that up to 50 percent of the entries are incorrect. The German social science research and survey institutes withdrew from sampling "telephone owners" because of the incorrectness of entries and because of a growing number of owners who refuse to be publicly registered (estimated at 30 percent). Neither the German Telefon owner nor the Web surfer is representative; especially the latter is particularly biased—but biased for what?

    This question targets the core of globalization and of global data collection. The Internet is judged to be becoming one of the fastest growing global markets. Internet users are the pathfinders and trend scouts of future services, products, and habits. The study, "Computer Kids Tracking," by Roland Berger of the Institute of Youth Research (IJF, Hamburg), has found that the children of the "biased" segment of computer—and modem-owning households (N = 1,204) spent an average of 70 minutes per day with computing, and 15 minutes of this with surfing the Internet.

Most children use the computer almost exclusively for playing games; 48 percent use it for school and other "serious" tasks. Seventy-seven percent of the male adults use the computer, two-thirds for business and job purposes and one-third for gaming and "goofing around." Female users are still a minority, using the computer preponderantly for educational or "serious" purposes. The figures show that in Germany the "knowledge gap" (Donohue et al. 1987) between "networkers" and "computer and Internet junkies" has not really opened up yet.

Most German users are looking for "serious applications" which are, in most cases and for both sexes: home banking, shopping, travel, events, jobs, and fitness and health information. Specialized (German) "on-line magazines" with about 2.5 million readers periodically inform about such sites and services. Mailing lists and newsgroups as well as the "novelties" of the mass providers also show trendy developments; they are the early-warning systems of changing users' habits and needs.

But who wants to know about these changes? Obviously not sociologists—disaster sociologists the least. More radical than ever, the Internet is spied upon by market analysts. Focalink Communications of Palo Alto, California, for example, is advertising its "Smartbanner" Technology with which Web activities can be analyzed and transformed into detailed user profiles showing preferences, habits, and other individual information. Based on these profiles, Focalink sells "psychographical data" with which Internet users can be directly addressed and advertised. The company, Doubleclick, works even more effectively. This firm evaluates the "fingerprints" of all users who address companies which cooperate with Doubleclick. Neither the Web server nor the user realize that they are detoured via the Doubleclick server, where all the informative cookies are collected.

Doubleclick, established in February 1996 by the New York public relations agency Poppe Tyson (PT), specializes in "Cyber PR" and has cooperated closely with Netscape since 1994. PT strongly recommends Netscape's Navigator and offered it for free on its homepage—now Navigator is generally freeware, and one can guess why. PT advertises to its clients:

> "We dispose of the globally largest Database of Internet users and we are able to generate detailed information about single and grouped users. We know the operating systems of our users, installed software, residence or company of the user, his on line habits and preferences and many other interesting data." (cited from the German edition of *PC Professional* 6/1996:25f.; own translation)

PC Professional (6/1996:25) therefore spoke of a "glass user"; the built-in pursuit via cookies became a big issue. (Other mechanisms to surreptitiously collect data or to intrude into one's PC are known—and applied, like Javascript or "action triggers" in Adobe Acrobat PDF; see www.aleph2.com/tracker/tracker.cgi and www.cs.wisc.edu/~ghost/index.html). The German law (Datenschutz Recht) is clear-cut, but difficult to execute abroad. (Even AOL Germany has sold user data to U.S. companies.) The judicial aspects exemplify what globalization also means: the WWW is to a certain degree an extra-territorial, virtual global community, outside, better: beyond nationally binding laws and norms. The emergence of "netiquette" (see www.jura.uni muenster.de/netlaw/) and the attempts by states to regulate the Internet demonstrate, at least rudimentarily and in a nutshell, that globalization always tends to burst smaller-sized institutional aggregates. Seen from the Internet, nation states are historical relics, often competing and contradicting obstacles for more complex and higher integrated aggregates (which is also true for the world market, the international system of finance capital, and globally operating companies). However, the state has been the institutional aggregate for national societies; what will become the institutional aggregate for Web citizens?

Until today, the Web citizen lives (almost) without "informational rights" in the sense of an (internationally legal) inviolable privacy of data. But not only are the individually attributable data of WWW-users questioned. In modern societies, everybody leaves dozens of data tracks behind which create perfect profiles of one's personal, economic, political, social, psychical, sanitary, and sexual situation. Today, every credit card company knows more about individual mobility, shopping habits, market and service preferences, and other ways to spend one's money than sociologists do. Merge together, let us say, the data of censuses, of communal bureaucracies, of banks and credit card companies, of power plants and waterworks, of insurers, social security, health services, and pensions, and one's individual life becomes highly predictable, much more precisely predictable than the estimations of Quételet, the Gluecks, Eysenck, and Herold altogether.

Famous names; sociologists should know them well. They all contributed to a sociology which was strictly based on statistics, not on surveys asking for opinions or personal attitudes. At this point, our thoughts should return to the origins of sociology, and we will be surprised that statistics and probabilities were key instruments. Society and human behavior were seen as results of variables which could be investigated and influenced. The so-called "political arithmetic" of Sir William Petty (1623-1687) was first used for the calculation of an "medium life expectation." Together with John Graunt (1620-1674), he calculated the mortality rates (number of the deceased in 1,000 inhabitants) for London. They recognized oscillations that are the correlations between life expectancy and residence. Edmund Halley (1656-1742) went further. In 1693 he calculated "medium death rates" for Breslau and the chances of survival of newborns, as well as insurance premiums in relation to age and life expectancy. The Prussian army chaplain Johann Peter Süssmilch (1707-1767) applied this knowledge in an effort to prove that infant mortality could be reduced through smallpox vaccinations. He argued that the risk of death from vaccination is essentially less than that of falling ill to the deadly smallpox.

Daniel Bernoulli (1700-1785) based his considerations in the field of public health upon the same arguments. In addition he introduced the economic category of "benefit" (Bernoulli 1896) when he recommended compulsory vaccination to the French king: inoculations, he argued, would be beneficial even if thereby many babies died, if the losses fell "only on the children useless for society, whereas the valuable ages benefit" (cited after Huber 1958:91). The implicit concept was quite simple yet anticipated Social Darwinism and eugenics: who already succumbs the relatively slight demand through the vaccine saves society pains with and costs for a probably weak life anyhow.

In principle, modern societies also have not solved this founding problem of decision-making (see Schrage and Engel 1982). Wherever costs and benefits of measures must be calculated, particularly those which influence ultimately life, health, and property, humankind's supreme interests are turned inescapably to assets (see Kunreuther and Miller 1985; Moore and Okamoto 1985). Bernoulli had made this aspect intentionally the starting point of his "political arithmetic." His probabilistics should lead to a purely mathematical theory of decision-making with which human decisions should be made assessable and capable of evaluation in advance (Bernoulli 1896: 43).

The insurance industry applied Bernoulli's methods and calculus. Moreover, it promoted the development of the calculating machine and modern social statistics. King Fredric II (1712-1786), convinced by the clergyman and statistician Johann Peter Süssmilch, established "historical indices" in which the population of Prussia, the buildings, and the finances of the communities were recorded. In 1805 the Statistical Bureau of Prussia was created.

In France, Adolphe Quételet (1796-1874) conducted the second census, out of which he tried to derive the "average Frenchman" as well as an ideal type of good and beauty and the reasons for deviance and crime. Herein the nucleus of an early diagnostics may be seen, like the one developed by Sheldon and Eleanor Glueck (1959, 1972) and later by Eysenck (1977) and Herold (see Cobler 1980; Simon et al. 1981; Simon and Taeger 1981).

In the course of industrialization the social arithmetic went far beyond census and economic statistics. In the U.S. every ten years since 1790 a census was undertaken to register the voters and, later on, to record industry and commerce, farms, mines, and administrative units. The Tenth Census of 1880 produced so much data that the evaluation of them took seven years. The need to process huge amounts of data once again promoted the development of calculating machines. Without Hermann Hollerith's (1860-1929) data-processing machines, which extended Jacquard's punch cards to tabulation of U.S. Census data, neither a modern, rational industrial planning and logistics nor the "battles of materials" of World War I and II would have been possible (see Beniger 1986).

Consequently, the census of the Third Reich was the most advanced instrument to record the human and economic potential of the society as a whole. The National Socialists designed an ID punch card for each citizen with 28 marks having about 600 punches. The owners of vehicles (cars, trucks, busses), physicians, paramedics and nurses, and people with strategic and other relevant qualifications became specifically registered. The data were well suited for mobilization, for selection based upon qualifications but also based on racial features.

In the face of user profiles through the WWW, this brief historical reminder should make sense as should the question, which has not changed after three hundred years of social arithmetic: What type of data, in which quantity and quality, and in which connection, do we need to enable us to govern our world rationally and appropriately?

At this point methodological, epistemological, and ethical aspects come into play—and, most important, so too the role that disaster research can play. Earlier (see Dombrowsky 1987, 1995) I have suggested that "disaster" be defined as empirical falsifications of our planned and intended activities. Disasters prove the fact that not everything has been under control or in a state of controllability. Most findings of international disaster research demonstrate that failures are the final outcome of mistakes being joined together, of wrong decisions, operator errors, misinterpretations, and misjudgments. Almost all so-called "sudden and unforeseen" outbreaks would have been correctable. Even Chernobyl had a window of return of about 45 minutes to regain control (see Haynes and Bojcun 1988). We have also learned our lessons with respect to so-called "natural" disasters. It is not the earth tremor that makes people suffer and die, it is the way we integrate this hazard into our lives. Most students of disaster agree that the severity of a disaster triggered by an earthquake depends more on cultural, economic, and technical conditions and social adaptations than on its physical intensity. The primary causes lie on one hand in factors such as

- land use
- settlement
- architectural structures
- infrastructures
- energy supply, raw materials, and resources
- and metabolism with our natural basis

On the other hand, they lie in the availability of

- preparedness, warning, and protection measures
- rescue, relief, and mitigation systems
- and recovery and reconstruction capabilities and resources

For all these factors we already possess empirical data based on spatial distribution. Risk maps and vulnerability assessments for almost all natural but also for most man-made hazards have been made. Via the WWW we can trace tornadoes along the continents, the development of desertification, of vegetation, or of water resources. The Munich Ruck Insurance Company periodically publishes the *Worldmap of Natural Risks*; NOAA has reexamined the Yellowstone National Park forest fire on the basis of stapling "thematic maps" derived from radar, temperature, wind, and humidity data as well as from morphological, hydrological, and biological data. Dynamic geographical information systems enable simulations of events such as fires or disasters up to more complex systems such as climate or vegetation (see Andrews 1983; Association of Bay Area Governments 1982; Bremer 1987; Bremer and Gruhn 1988; Gearhart and Pierce 1989; Litjen et al. 1978). The Internet is a treasure vault of approaches, paradigms, and methods with which complex systems and interrelations could be modeled, assessed, and analyzed (see http://sigma.unisg.ch/~sgzz/links/stp/futres/scenplan.htm; http://iswww.bwl.uni-mannheim.de/lehrstuhl/forschung/sd.htm). Together with these methods and approaches, we can develop a universal, standardized scheme of investigation, an Internet questionnaire in which all of us could fill in knowledge of our country, types of disasters, their occurrences, severity, effects, and so forth. In the end we would have created a working *Global Disaster Handbook* which is much better than, say, the *CIA Nation Reports*. This idea may sound naïve; however, the WWW opens up opportunities that we never had before.

That is my central thesis: sociologists are using the Internet just as they (mis)used the computer in the beginning—as a mere typewriter. The Internet is much more than a library of libraries, a gallery of galleries, a videotech of videotechs, or a hyperfast post office with built-in telephone and fax machine. The Internet is the universal machine to generate social arithmetics, the second, virtual globe with which we can simulate a rationally planned and developed world. That is the true methodological change and challenge, and a genuine opportunity, too.

NOTE

[1] Bob Stallings has my gratitude for his patience and linguistic instinct with which he transformed my manuscript into legible English.

13

THE USE OF GEOGRAPHIC INFORMATION SYSTEMS IN DISASTER RESEARCH

Nicole Dash

A little less than a year after Hurricane Andrew (1992), I had the unique opportunity to work on the ground floor of the implementation of the Federal Emergency Management Agency's (FEMA) use of Geographic Information System (GIS) in a disaster field office (DFO). The DFO after Hurricane Andrew was one of the first large-scale uses of GIS by FEMA in the aftermath of a major disaster. While GIS is far from being a new technology, it is only in the last five to ten years that it has become part of the everyday discourse in geography, urban planning, and emergency management, to name a few disciplines. GIS has yet to become a part of the methodological discourse in the social sciences generally, and in social science research on disaster specifically.

When I was first asked to contribute to this volume, I was convinced that I could separate the issues of GIS in disaster research and GIS in emergency/hazard management. As I continued to think about the future of GIS use in disaster research, however, I realized that to a large degree the two issues are somewhat tied together, particularly in the case of applied research. This chapter, then, will use lessons learned from GIS in emergency management to help define the direction of GIS use in disaster research.

Space can be defined in a multitude of ways: social, political, or economic, for example. The power of a GIS is that it allows us to link these different definitions together in one all-encompassing system. In disaster research the ability to link social, economic, and political data with geographic or spatial data is extremely important. Various researchers such as Britton (1986), Dynes (1970), Kreps (1984), Streeter (1991), Pelanda (1982), and Quarantelli (1987a) have struggled to define disaster. Disaster is seen as an unpredictable event which renders a social unit unable to cope with previously normal activities. By nature of this definition, disaster occurs in some type of space. This space can be defined geographically.

Simple geography, however, does not define disaster. Rather, disaster occurs in social, political, and economic space. And it is within this space that individuals, families, businesses, and communities must cope with hazardous events. The implication is that social, political, and economic space have inherent power structures that create conditions of discriminatory and unequal distribution of resources. Both on a micro level and a macro level, the outcome is that some social units have the resources to better mitigate against and prepare for disaster. The power and strength of a Geographic Information System is its ability to integrate geographic, physical, social, political, and economic data.

WHAT IS A GIS?

A GIS is an "organized collection of computer hardware, software, geographic data, and personnel designed to efficiently capture, store, update, manipulate, analyze, and display all forms of geographically referenced information" (ESRI 1994: 1-2). A GIS uses both geographic data (a digital Census tract coverage) and nongeographic data (a spreadsheet with Census tract numbers and populations), and it includes the ability to tie these two datasets together. A GIS integrates geographic or spatial data and other information into a single integrated system. The key feature in a GIS is that it also includes operations which support spatial analysis.

Additionally, it allows for the manipulation and display of geographic knowledge in new ways. These new ways include large-scale maps in addition to standard tables and charts. A GIS, then, can be more broadly defined as "a system of hardware, software, and procedures designed to support the capture, management, manipulation, analysis, modeling, and display of spatially-referenced data for solving complex planning and management problems" (National Center for Geographic Information and Analysis 1990: 1-14). This system taken as a whole can be a tool in disaster research and emergency management by helping users answer various spatial questions.

These questions can be simple or very complex (DeLiberty 1995). A GIS can be used to answer questions of location. Such questions include determining "what is at" a specific location. Determining what house is on the southwestern corner of a specific intersection is one type of location question. Second, a GIS can be used to answer conditional questions. In other words, it can help us determine locations that meet certain criteria, such as how many schools are within 10 miles of a nuclear power plant. Third, a GIS can help us determine trends. Using data from different periods of time, we can determine what has changed over time. For example, by comparing housing coverages from 1990 and 1995, we can determine the housing density change over time. Fourth, spatial patterns can be found by using a GIS. Patterns of damage, for example, can be determined by analyzing the spatial distribution of damage over a specific geographic area. We can determine whether damage is worse in areas of poverty. Finally, a GIS, as its most complex function, can help us model possibilities. For example, if we add a road to a neighborhood, will egress evacuation times be significantly altered? If Hurricane Andrew's track were 20 miles to the north, what type of damage could we expect from wind, rain, and surge?

In order to answer these questions, the geographic data necessary to represent the map features can broadly be in three forms. *Point data* are specific location data such as the coordinates of schools or homes. *Line data* are used to represent information such as roads or narrow rivers. *Polygon data* represent areas such as city boundaries, wetlands areas, or Census tracts. Any given project can include all three types of data. A GIS stores data as individual coverages that are layered to answer the various questions. As we will see, GIS can be a powerful tool in emergency/hazards management and an important future methodological direction for disaster research.

GIS AND HAZARDS

In 1995, Chartand and Punaro argued that in order to further improvements in disaster mitigation, preparedness, response, and recovery, emergency managers need " . . . advanced, technology-supported information handling systems for collecting, indexing, storing, processing, retrieving and disseminating essential data" (Chartand and Punaro 1995: 1). One such system is a GIS. By allowing emergency managers to incorporate spatial analysis and modeling into their everyday decisions, they are better able to plan for potential disaster. In addition, GIS can be implemented in the wake of a disaster. The ability to spatially analyze data allows emergency managers to better plan and coordinate their recovery efforts.

In one of the most recent works on GIS and natural hazards, Coppack (1995) attempts to assess the use and development of GIS in helping to reduce the impact of natural hazards. He finds the literature connecting GIS and natural hazards to be scarce. In part this is due to the multi-disciplinary nature of both GIS and natural hazards. However, even when looking at the two fields independently, the search for literature yields similar findings: it does not exist. Furthermore, Coppack (1995: 26) states that " . . . it is not known how many operational GIS's for mitigating the impact of natural hazards exist." Coppack concludes that GIS does not appear to have lived up to its expectations. In fact the International Decade for Natural Disaster Reduction (IDNDR) Science and Technical Committee called for " . . . better coordination in the technologies such as remote sensing and GIS" (Coppack 1995: 27). However, the failure of GIS to live up to expectations is not a function of GIS itself, but rather a failure to implement a functional system with appropriate data and personnel.

Emergency managers and community leaders are under increasing pressure to be prepared for all types of potential hazards (Wolensky and Wolensky 1990). In recent years, the federal government has pushed states and localities to take a greater role in disaster preparedness and response. One example of GIS use in a coastal disaster context is the State of Florida's Florida Marine Spill Analysis System (FMSAS). Until a short time ago, officials responding to oil spills had to use old paper maps as their primary data source. In order to create a system that would be readily available if a spill occurred, the Florida Marine Research Institute developed a GIS application that would use digital data to give real time current maps and on-line spill response ability. The system was developed by incorporating satellite imagery, basemaps, and intensive field observations to create a system that could produce customized maps identifying vulnerable habitats during the various states of an oil spill. The system was put to a test in August 1993 when 300,000 gallons of fuel spilled into Tampa Bay. The FMSAS system within hours was able to produce maps for Coast Guard officials detailing the extent of the spill. The process was as follows: helicopters monitored the spill, its boundaries were mapped, and maps showing the movement of the spill were furnished to determine the most vulnerable areas. This oil spill model is one small example of how GIS can be used to assist emergency response. A broader example is the use of GIS after Hurricane Andrew. Unlike the oil spill model, the Hurricane Andrew case shows how a GIS can be used when real time information is not as critical.

HURRICANE ANDREW: A CASE STUDY IN THE APPLICATION OF GIS

At the end of August 1992, Hurricane Andrew swept through southern Florida, leaving a path of destruction throughout the region. Along with the physical and emotional devastation to the area and its people came a large, somewhat unmanageable response and recovery phase. In areas such as Florida City, a poor Black neighborhood, not one house was left unscathed. After the initial shock, leaders in federal, state, and local governments converged to begin the process of "Hurricane Andrew Recovery." The initial use of GIS was in mapping damage and analyzing community demographics. Later, as the potential of GIS was better understood, its use grew in areas such as public assistance and hazard mitigation.

The Federal Emergency Management Agency committed to setting up a GIS after Hurricane Andrew. The FEMA system was a multi-platform, Unix-based Apple Computer networked system with a 13 gigabyte hard-drive system partitioned into distinct nodes: a data node that housed all new incoming data; a GIS for working files; and a raster node where all rasterized (digital output file) maps were stored. Map output was not sent directly to the networked plotter, but rather was output as a file that could then be stored for repeated printing. At the Hurricane Andrew DFO, and consequently by FEMA as a whole, MapInfo has been adopted as the GIS software package. In part MapInfo was chosen for its relatively low learning curve. While its analysis features are not as sophisticated as upper-level systems such as Arc/Info, it does have a strong data engine that is able to complete sophisticated and complex data linkages both on spatial and attribute data.

More important than the hardware and software in this GIS were the datasets. The FEMA GIS department attempted to support various departments within FEMA, the State of Florida, and local government; measure impact and recovery; and set up a system that could be in place the next time a disaster happened. Projects included tracking (1) debris and debris removal, (2) clean and secured abandoned damaged homes, and (3) the location of trailers used for temporary housing. Damage assessments were geo-referenced to allow these data to become a digital coverage. In addition to basic basemap data such as street coverages, zip code boundaries, community boundaries, hydrology, and Census tract boundaries, emergency-specific data were gathered. These data coverages included Dade County evacuation zones, hurricane shelters, and Turkey Point Nuclear Power Plant evacuation zones. FEMA's GIS also included data that could assist in measuring impact and tracking recovery. These data included Florida Power and Light electric service disconnect data at four points in time, United States Postal Service (USPS) forwarding address data, USPS undeliverable address data at two points in time, FEMA public assistance data, and the Dade County tax assessment database. It is important to note that none of the data for tracking recovery was in a digital format. Yet, with all these data, the majority of products produced by FEMA GIS after Hurricane Andrew required little spatial analysis.

What did we learn from Hurricane Andrew with regard to GIS, and how can we use it to inform the future direction of GIS as a methodology in disaster research? I think one of the most important lessons that was learned from Hurricane Andrew is that the immediate postimpact period is not the ideal time to set up a GIS to address the needs of an impacted community. The first "GIS" product produced after Hurricane Andrew was not completed for at least four months. The first problem was a data problem. Getting data is much easier said then done. Once the data were received and processed, the first reports were produced. The first report presented damage, assistance, and demographic data for communities in southern Dade County. While the report was informative, it failed to use the power of a GIS system, which is analysis. The actual printed map is a by-product of the analysis, not the main object. However, the majority of requests received after Hurricane Andrew were specifically for maps showing specific data, not maps that answered any of the five questions a GIS can help answer.

The problem is that GIS is a very expensive graphics tool. On the other hand, it is a powerful analysis tool. New technologies can be powerful if they are integrated into the planning of emergency response. Likewise, GIS can be a very effective tool in disaster research. However, to effectively use it requires not only computers and data but, more important, a person actually "thinking" about the issues to properly utilize its integrative powers.

GIS AND DISASTER RESEARCH: FUTURE DIRECTIONS

As discussed earlier, disaster by nature of its definition occurs in social, political, and economic space as well as in geographic space. Geography alone, however, cannot define disaster. The power of GIS in disaster research is its ability to bridge these disparate types of data. In other words, the future of GIS in disaster research is in taking the type of work FEMA was doing after Hurricane Andrew to the next level. That level is one which not only looks at disaster phenomena such as damage but also examines the distribution of phenomena in social, political, and economic space.

In order to understand this social distribution, researchers must look for data sources that can be used in spatial analysis. Understanding that data can always be aggregated up (i.e., to more-inclusive units) but never down, the best data are point-specific data. Tax assessor's databases, for example, are an excellent source of data. These databases include data on land value, real property value, age of home, and size of home. Understanding how damage correlates not only with the eye of a hurricane or the epicenter of an earthquake but also with housing value allows us to look at the effects of stratification on the distribution of damage. Again, GIS allows us to integrate social and geographic data in order to understand disaster as a social phenomena.

However, as simple as this may seem, the most crucial element of a GIS is the data. The data drive the system. Bad data yield bad results. As the saying goes, "Garbage in, garbage out." But even knowing this, processing and ultimately using data are complicated endeavors. More often than not, data are in the wrong form. In order to link or match data, they must be in a form that a GIS can understand. For example, in order to reference an address in real space requires that the address be given an X,Y coordinate. In order to do this, most GIS software packages require that the address be in a standard form. However, this standard form is rarely the form the data are in. Additionally, data need to be at the correct scale and level of aggregation. Most GIS data are from secondary sources such as the Census Bureau or local government; thus, it can be expected that the data will require time-consuming cleaning in order to link or overlay them with data from other sources. Again, data drive a Geographic Information System and need to be the primary concern of researchers.

The future of GIS in disaster research is not one of a linear direction. Rather, disaster researchers must look at GIS as an emerging technology with major implications for not only researchers but also emergency managers. GIS, however, must be understood before it can be utilized. Besides needing to understand GIS in order to integrate it into research projects, researchers need to begin looking at the use of GIS (and other technologies) in disaster management. The following are some future agendas for disaster research, in no particular order.

1. *Disaster researchers need to study the use of GIS in emergency management.*

Too often, GIS is seen as an end and not as a means. GIS is a decision-making tool. We have yet to develop technology that can on its own make decisions for a community. The presence of a GIS does not in and of itself make a community better prepared for a hazardous event. With the heightened emphasis on GIS in emergency management, disaster researchers have an opportunity to better understand technology use within emergency management. Indeed, much has changed since Drabek (1991) investigated the extent to which microcomputers were being used in emergency management.

When Hurricane Felix was approaching the East Coast of the United States during the 1995 hurricane season, I was called by the Emergency Operations Center (EOC) in one of the affected states. I previously had consulted with this state to discuss their GIS system and data. They asked if I would like to come in and observe the EOC. When I got to the EOC, I was led to a computer and was asked to tell them what data they had available. I ended up spending the next two days in the EOC working with their

GIS data. In fact, there was very little I could offer this state, because its system consisted of GIS software and Army Corps of Engineers/ FEMA data. They had no plan for implementation. They had no idea what they even had available. As I sat in the EOC, I realized that, while they thought they had a GIS, they actually did not. Somehow this system, all on its own, was supposed to answer questions that no one knew to ask.

While some states such as California use GIS effectively, researchers clearly do not know enough about the technology or its use. Nor do we understand the possible implications of the technology. Questions that remain are: (1) To what extent are GIS and other "decision-making" technologies being implemented?; (2) How are these systems being used as decision-making or information-dissemination tools?; (3) How are these new technologies being implemented (do they have dedicated employees assigned to them, for example?)?; and (4) Is a "techno-elite" being created on a local or state level creating a new type of technology-based stratification system?

2. *Applied researchers need to begin to look at GIS as a valuable tool to understanding disaster-related phenomena.*

We need to begin to incorporate GIS into our research agendas. After the Kobe earthquake in 1994, researchers in the March 1995 edition of the *Natural Hazards Observer* made 12 observations on what the United States could learn from Kobe's disaster. In one of their observations, they argued that "a GIS-based model can give city officials a rapid sense of the scope of damage and point to neighborhoods or facilities that are most likely to be in need of assistance" (p. 5). In addition, they argued that "cities with a history of seismic activity, or in a recognized seismic hazard zone, should begin to identify hazardous existing buildings and establish a program to strengthen or remove those structures" (p. 5). While they did not sight GIS as a specific tool for this, a GIS is an ideal tool for such inventories.

In addition, remote sensing or satellite data can be used to offer more prediction and planning information. In California, for example, digital data on fault lines with buffer zones around the faults can be used in conjunction with property tax data to not only predict those properties at greatest risk but also to tag empty lots as hazardous. With this information in digital form, builders and zoning departments can begin to plan ahead for buildings most at risk. Additionally, a GIS can be used for modeling. Scenarios can be built to better understand what happens in an earthquake, hurricane, flood, or other hazard. As Streeter (1991) argues, disasters are not random events since some regions are more prone to disaster than others. Hurricanes, for example, do not pose a threat to states in the central plains of the United States nor, for that matter, do they affect inland cities in coastal states to the same degree that they affect coastal cities. Our propensity to develop beaches and floodplains significantly contributes to the progression of weather situations from natural acts to highly destructive human disasters. While a GIS cannot eliminate natural or technological hazards, it can be a valuable tool for understanding the social, political, and economic implications both before and after an event.

3. *Survey researchers should use GIS to test the representativeness of survey results.*

GIS can be used to test the representativeness of survey research responses, particularly mailed surveys. More often than not, survey research is done in areas with cultural diversity. As we know, not all people experience events such as disasters equally. Culture, position in regional stratification systems, and gender, for example, can significantly affect individual, family, or business disaster impact and recovery. In order for us to understand how social position affects disaster experience, it is vital that our samples and responses are representative of the social geography of our population.

A GIS can be an effective tool in tracking survey responses without violating subject anonymity. By creating a geo-referenced database of all mailed surveys, each survey can be tied to demographic data. In addition, the database would simply include a data field that would indicate whether a survey instrument had been returned. This database would be independent of the survey results, used solely as a means for testing result representativeness. A GIS allows researchers to analyze the spatial pattern of responses and nonresponses. For example, in an area such as Los Angeles, Anglos, African-Americans, Hispanics, and Koreans need to be represented in survey results because each group is embedded in a unique culture with unique perspectives and experiences. By tying each address in a sample to its demographic characteristics, we can analyze our responses on any one of a number of dimensions. We can query the dataset to obtain a subset of all surveys returned from block groups that are predominantly African-American, for example. Likewise, we can determine return rates from areas with large proportions of single mothers. Understanding these patterns allows researchers to concentrate follow-up efforts in underrepresented areas. Additionally, researchers can understand the limitations of their work.

All too often, researchers are more concerned with overall return rates rather than the overall representativeness of their returned sample. The use of a GIS allows for the simple tracking of survey returns. Similar methods can be used to determine the representativeness of telephone surveys. Researchers need to incorporate new technologies such as GIS into their overall sample framework allowing for tracking and representativeness of the returned sample.

4. GIS should be used to better understand the social aspects of disaster.

We know disaster occurs in a specific geographic area, but it is not defined by geography. Researchers need to use GIS to analyze the social distribution of disaster effects and recovery. The ability to look at the spatial-social distribution of a disaster allows us to better understand populations at risk. GIS allows us to overlay vastly different datasets that without GIS would require large amounts of time to compare. Damage data by address, for example, can be overlaid with Census block group boundaries. Damage data can be aggregated to the block level, and comparisons can be made to better understand the relationship between damage or recovery and socioeconomic status. While this can be done without GIS, it is labor-intensive, requiring manual look-ups of each address on paper Census maps.

However, we need to remember that GIS should not drive research. Just as a Geographic Information System, in and of itself, does not make an emergency management team better prepared for a hazardous event, it likewise cannot in and of itself generate research questions. Nor for that matter can GIS technicians, who understand the technology but not the theory, drive the use of GIS in disaster research. Rather, the only effective use of GIS is through research teams that include researchers who can generate ideas from theory and previous research and technicians who understand the use of the technology. Our experience needs to drive the use of GIS; it cannot be used merely because the technology exists.

5. *Vulnerability Analysis: A Future Application.*

At best, GIS is being used in terms of hazard and risk mapping. Maps exist that show the location of earthquake faults, for example. Maps are also available that show coastal flood zones and areas at greatest risk from hurricanes. However, we are less likely to use GIS to analyze vulnerability, which is more than a geographic phenomenon. GIS should be used to link the physical attributes of risk, such as earthquake faults and flood zones, with the social, economic, and political attributes that render individuals, families, businesses, and communities most vulnerable. Vulnerability is a function of power hierarchies within the social, political, and economic spheres, not merely proximity to flood zones, fault lines, or hurricane-prone coast lines. Disaster researchers can use GIS as a tool to better understand these complex vulnerabilities.

SUMMARY

While GIS is not a new technology, it is a technology that today is on the cutting edge in emergency management and in the future can be on the cutting edge in disaster research as well. Its use requires that researchers understand disaster in a spatial context, linking social indicators and geography. Researchers need to look at operationalizing variables differently to allow for spatial analysis.

Beyond incorporating GIS into disaster research itself, researchers must begin to look at the use of technology as a whole, concentrating on how it is being implemented and used. The use of GIS is not in and of itself a solution for emergency managers or researchers. Rather, it is a tool that, if used correctly, can offer new understandings of age-old problems.

14

PROBLEMS AND PROSECTS OF DISASTER RESEARCH IN THE DEVELOLPING WORLD:

A case study of Bangladesh

Habibul Haque Khondker

It is not a matter of pure chance that disasters are more prone to strike the developing world than the so-called developed world. The main reasons are demographic and politico-economic rather than "natural." Since the majority of the world's population lives in the developing world, a region mired in widespread poverty relegating a vast number of them to a precariously marginalized existence, disasters become more telling in their consequences. Yet ironically, disaster research in so far as it concerns the disasters in the developing world has remained virtually nonexistent or, at best, underdeveloped. Take the case of Bangladesh, a country which is almost synonymous with natural and man-made disasters of all kinds where disasters vastly outnumber researches on disasters. The report of the Disaster Forum, a Dhaka-based "think-tank," recorded a total of twelve disasters ranging from floods, river erosion, water-logging, cyclones, and droughts to cold waves in 1997 (*Independent*, May 6, 1998 p.4).

In the rainy season (June-August), up to 70 percent of the landmass is submerged (Novak 1994: 22). Bangladesh, largely because of its vulnerable geographical location, happens to be the most disaster-prone country in the world. The deltaic landmass that makes up the bulk of the flatland of Bangladesh is prone to both river flooding as the major rivers make their run through this country into the Bay of Bengal as well as to the tidal upsurge of the bay itself. On top of this, cyclones developed in the bay make periodic landfalls on the coastal region. Disasters strike this country of 125 million people with a per capita income of less than $300 (U.S.) on a regular basis. Yet research on disasters in Bangladesh is, at best, in a fledgling state. This chapter discusses the intellectual and politico-economic conditions that may account for this underdevelopment. This chapter highlights a number of problems as well as lessons based on the author's own experience in disaster research in Bangladesh. In the concluding section some guidelines for the institutionalization of disaster research in Bangladesh are proposed.

THE POLITICO-ECONOMIC CONTEXT OF DISASTER RESEARCH

One of the reasons for the underdevelopment of disaster research could be the difficulty in analytically separating a disaster situation from a normal condition of mass poverty and deprivation. Here an anecdote may throw some light. While proposing to a senior official of the Bangladesh Biman, the national airlines, the need for organizing crisis management courses, the author was told that a permanent crisis exists in the airlines. The Biman official observed: "What is needed is a mechanism to deal with the routine, everyday crisis and once that has been accomplished, we can go for the special type of crisis" (interview conducted in 1996). What is implied here is that before we identify a situation of disaster for the purpose of comparison we need to have a condition of normalcy. The condition of normalcy in various sectors of the society is yet to emerge. At the national level, one can also see a parallel of this situation. The multitude of problems that affect this country are enough to keep the government on its toes on a permanent basis.

The Trotskyite notion of "permanent revolution" has gained a new meaning in contexts such as Bangladesh.

The conditions of mass poverty and malnutrition are so endemic that from the point of view of the suffering multitude a condition of permanent disaster is prevalent. Yet, it can be argued that a distinction can be made between the routine crisis and a sudden amplification of that crisis. In this regard, disaster agents can be seen as triggers since they spark a crisis in a situation where a vast number of people somehow learn to cope with their day-to-day troubles. In a disaster their coping strategies fail because the circumstances change. There is a change in the scale and magnitude of the crisis. The routine problems of the airlines, which are indeed plentiful and sometimes quite paralyzing, are not the same as an airline disaster. Similarly, a condition of mass poverty can be analytically separated from a famine, although the prevalent condition of generalized deprivation, precarious entitlement arrangements, and an inadequate safety net contribute to the making of a famine situation.

The infancy in analytical and critical understanding of disasters and their consequences is, however, symptomatic of the sluggish growth in sociological research in nontraditional fields in Bangladesh. The research focus in sociology still revolves around traditional concerns such as village studies and social and cultural consequences of modernity and economic change. The heavy influence of the social anthropological tradition in Bangladesh sociology is still very much evident. The only new ground in social research that has been broken in recent years is in the area of gender studies. Another related theme has been nongovernmental organizations (NGOs), which has seen a proliferation of research (Chowdhury 1989).

Research on disaster has been growing, albeit sluggishly, since the 1970s following the independence of Bangladesh, which left a trail of disastrous consequences. The study by Lincoln Chen (1973) probably was the first of its kind in the postindependence Bangladesh. The works of Bruce Currey (1978, 1980) also made an important contribution in this regard. For pioneering status, however, one has to consider the works of Ram Krishna Mukherjee (1944) which resulted from the survey conducted in the aftermath of the great Bengal famine of 1942-43. That study was initiated with the support of Professor Mahalanabish of the Indian Statistical Institute. Research on famine and food crises poses special problems because of the political sensitivity associated with these crises. The poor quality of data recording poses an additional problem. As Currey (1978) observed, "Agricultural statistics still remain amongst the nation's most serious data deficiencies." One can easily comprehend the implication of this deficiency not only on famine research but also on famine management and forecasting. Sometimes, data collected at the village level are not properly stored. Data loss due to natural calamity or to termites is fairly common, as Currey points out. Although in recent years quality of data management has improved to some extent—thanks to donor assistance and the political will of the government—, there are still considerable gaps.

The deficiency in terms of quality and reliability of data make it doubly necessary to conduct fieldwork in Bangladesh. W. van Schendel, a Dutch scholar on Bangladesh, observes: "Social research in Bengal cannot do without the insights that result from good fieldwork. It is well known that the reliability of statistical material on Bengal leaves much to be desired and that it is hazardous to base analyses on official or semi-official data" (van Schendel 1985: 66). This warning applies very well to data on various disasters. Besides, official data on various aspects of disasters are simply nonexistent, either because these data were never collected or because what little exist are highly inadequate.

While conducting research on the governmental response to the 1974 famine (Khondker 1984) in the early 1980s, the author was told by officials of the Ministry of Relief and Rehabilitation that many of the official records had been destroyed on order by the then President Ershad, apparently to create more office space. However, no official record in terms of orders or memos to this effect exists. The destruction of the so-called "old papers" was carried out not to cover up the famine, but it was done across the ministries and departments in the Secretariat, the headquarters of several ministries, which actually faced a serious shortage of space. How much space was gained is not known, but the loss for researchers is, no doubt, irreparable.

The specific problems associated with disaster research can be viewed in terms of the overall sluggishness of sociological research in Bangladesh. Slow growth in sociological research in Bangladesh itself can be explained in terms of politico-economic circumstances and is a subject worthy of attention in its own right. Since the inception of sociology in Bangladesh (formerly, East Pakistan) in 1957 under the auspices of UNESCO, methodological issues have been discussed. The first generation of sociologists such as Karim (1960), who was the first local head of sociology at the University of Dhaka, was in favor of historical and social anthropological methods. A tradition of empirical research did not gain ground in Bangladesh sociology. This has serious consequences for the growth of research in disaster sociology. Most of the research worth mentioning in the area of disasters has been conducted by foreign experts. Most of these studies are in the genre of the ethnographic tradition of social anthropology. Very few studies have been conducted by local social scientists. The presence of local scholars became visible only from the 1980s onwards.

One of the important features of disaster research in particular and social research in general in Bangladesh is that it is donor-driven. In the 1970s, a good deal of intellectual and other resources were devoted to research related to demography and fertility control. The control of social research lies firmly in the hands of the providers of research grants. This experience has resonance in the developing world as a whole. The funders of research in Bangladesh did not show interest in disaster research until late 1980s.

One of the unintended consequences of the disastrous floods of 1987 and 1988 has been in the area of disaster research. Following the calamity, a number of studies were carried out on the causes and consequences of floods. The works of Adnan (1991), Hossain et al. (1992), and Khondker (1992) among others are examples. The floods also led to a massive Flood Action Plan (FAP) drawn up by the World Bank, which prompted a slue of feasibility studies. A total of 26 studies at the cost of an estimated $150 million (U.S.) were proposed (Chowdhury 1992: 3). The World Bank-initiated feasibility studies have drawn a good deal of intellectual attention to the area of disaster research.

A STUDY OF TWO VILLAGES: PHALIA DIGHAR AND CHOTO CHONUA

Administratively, Bangladesh is presently divided into six divisions: Dhaka, Rajshahi, Chittagong, Sylhet, Barisal, and Khulna. Each division is divided into an unequal number of *Zila*, or districts, that are further divided into *upazila*, or subdistricts. These *upazilas* are comprised of unions, which in turn are composed of villages. Although Bangladesh may appear to be a very homogeneous country in both social and geographical terms, there are important geographical and other social-structural differences. A field study was conducted in two disaster-prone villages from two geographically distinct regions of Bangladesh. In both physical and social terms the data from Phalia Dighar, a village in a northern district, present certain contrasting features with those drawn from the southern region of Chittagong. Yet there are certain similarities in the way people in these two regions respond to disasters, reflecting certain social-structural and cultural uniformities.

Two villages were selected based on the criterion of their exposure to natural disasters. To study the impact of river floods in the northern part of Bangladesh in 1992-93, a village from the northern district of Gaibandah was selected. The village in the north is prone to river flooding on a fairly regular basis. To assess the impact of tidal waves, a village from the southern district of Chittagong was selected. The village selected from Chittagong, a district fronting the Bay of Bengal, was most recently battered by a cyclone in May 1998. In view of the paucity of data relating to disasters, we sought to collect field-level data to assess the differential impact of disasters and gender-specific differences in coping strategies. A comparative study of the impact of floods on women in rural society was proposed. It was deemed that comparative research would give us in-depth understanding of the disaster response.

The villages were selected in consultation with the author's friends and contacts in the NGO sectors in Bangladesh. The research was supported by a small grant from UNDP/DHA-UNDRO Disaster Management Training Program. The sponsors were preparing original research material for use in their training programs on disaster management.

FIELDWORK

Enlisting the support of the NGOs was crucial. The author had friends in *Nijera Kori* and *ARBAN,* two very active NGOs in Bangladesh. While *Nijera Kori* had a project near the study village in the north, *ARBAN* was involved in a development project in the study village in the south. The help received from Ms. Khushi Kabir, the founder of *Nijera Kori*, and Mr. Kamaluddin, the main organizer of *ARBAN*, was immense. Both of them are leading lights of the NGO movement in Bangladesh. Despite their very busy schedules, they extended support and cooperation in organizing the fieldwork. The author was particularly privileged in being able to count these two very dedicated persons as his friends. Use of social networking and friendships made the fieldwork reasonably easy and an enjoyable experience.

The realities of Bangladesh are such that even to conduct social research in an unfamiliar area one must enjoin the support of the influential members of that locality, especially those with political connections. One is advised to contact the village elders and other influential people in the village, keeping them abreast with the nature of the research project. Again networking was necessary in this regard. The author had a cordial relationship with one of the influential businessmen of Rangpur who was also a leader of the Bangladesh Nationalist Party (BNP) at the local level. Apart from providing logistical support, this person took a keen interest in the study.

As researchers of rural life in Bangladesh know very well, one has to have knowledge of the dynamics of rural society to conduct fieldwork in the rural setting. The NGO workers had very good rapport with the villagers and were able to collect data without arousing anyone's suspicion.

RESEARCH DESIGN AND METHODOLOGY

In examining how disasters undermine the economic position of rural women, how their income-generating activities are disrupted, and how rural women cope with disasters, we decided to interview them. We also asked: Do women have equal access to postdisaster relief? If not, what are the prevailing social-structural conditions and cultural norms that perpetuate such gender inequalities? What are the factors involved in the process of recovery for these women? We sought to explore these questions. Given the limitation of time, we decided to select only two villages.

Data were collected with the help of a team of interviewers hired from among the NGO workers. For the village in the north, there were eight interviewers, four male and four female. For the village in the south, the interview team was comprised of three females and two males. Data were collected through the use of a questionnaire which had three parts. The first part (Questionnaire "A") was used in the intensive interviews conducted with people (mostly men) in positions of leadership in the village. These interviews were conducted by the male members of the interviewing team. The second part of the questionnaire (Questionnaire "B") was used to collect data from heads of households. Usually, the male interviewers conducted these interviews with the male head of the household. In cases of female-headed households, female interviewers were used. Female members of the household were interviewed using the third part of the questionnaire (Questionnaire "C"). These interviews were conducted by the female interviewers. It was necessary to match the gender of the interviewer with that of the respondent in view of the local cultural norms of sex segregation. In one instance, however, a male interviewer interviewed a female community leader with no perceptible variation in the quality of data. Discussions were also held with some of the relief workers and organizers of various NGOs who have knowledge of disaster impact on the rural society.

A selective sample survey of households was conducted to find out the existing socioeconomic conditions such as income, education, religion, occupation, condition of housing, availability of flood shelters, and family structure. Information on head of household (whether it is female-headed or not) as well as demographic characteristics, which included information on family size and age-composition of the family, were collected in the two selected villages.

To find out additional information on the disaster coping strategies of women, we asked questions about the quantity as well as quality of relief materials. We also asked questions on diseases and loss of lives as a consequence of disasters. On this, however, we have certain compunctions. We needed the information on loss of lives in the family, but by bringing up this issue we reminded them of a tragedy they would rather forget. No systematic data on mortality or epidemiological patterns were collected. We also gathered information on the women's access to relief and rehabilitation and on incorporation of women in postdisaster development initiatives.

Background information on the physical characteristics of the villages were collected from existing sources. Ideally, a complete household survey would have yielded valuable results. Since the study was done under constraints of time and resources, such a survey could not be conducted.

Questions were asked about the production of food crops, cash crops, occupation, and work in general. The majority of the villagers can be classified as poor, yet information on the nature, depth, and distribution of poverty in the study villages was collected.

In terms of disaster experience, questions were asked on their ability to recall past floods and how they used those lessons in coping with the recent floods. We were also interested in finding out about the sources of information about disaster warnings, evacuation plans, and disaster shelter. The villagers, it turned out, were not prepared for the sharp rise in the water level that would submerge their temporary shelter.

The study sought to assess the impact of floods and cyclones, the two most common natural disasters in Bangladesh on the rural women. Although in terms of loss of lives and property natural calamities in Bangladesh seem to affect groups and sections of the population irrespective of gender and class, nevertheless certain interesting gender variations can be discerned. Based on the findings it can be mentioned that one should not exaggerate the loss of lives due to the floods, especially in the northern part of the country. The cyclone that affected the southern coastal districts took a much heavier toll of human lives compared to floods in the north. Floods sweep away the food crop and other resources, thus shackling the vast majority of the people in the northern part of Bangladesh to grinding poverty. It is often the diseases that follow the floods which prove fatal. After the recent flood, according to our respondents, medicines and water purifying tablets were made available through the channels of official relief as well as through nongovernmental organizations, which helped contain the outbreak of epidemics. Here is an instance of utilization of knowledge learned from previous experiences with calamities. It has become common knowledge in the flood plains of Bangladesh that epidemics follow floods, and thus appropriate actions were taken.

The overall impact of the calamity on the lives of the villagers was indiscriminate; in this sense, flood is an "equal opportunity" disaster. As the entire village was submerged, all the inhabitants of the village—the rich and the poor; males, females, and children—were all forced out of their submerged houses and were desperate for shelter. It was difficult to identify the special impact of the calamity on the women. Respondents were concerned over the losses that the families and households suffered. They were not able to recount the gender-specific impact. Gender sensitivity became clear when we examined the economic losses, especially the effects on income-generating activities of the women. For example, a large number of the cattle herds were swept away at the first sweep of the flood. Many village women were dependent on cattle for their income. The task of separating the collective effects from the specific impact on women was a challenging task.

To assess the unevenness of the impact of flood along gender lines was complicated by the fact that most studies as well as administrators in charge of relief distribution take the household as their unit. From the point of view of relief administration there is certain merit to it. Floods affect the households and the families who comprise these households. In the distribution of relief the needs of the household as a whole are often taken into account. For example, provision of building materials for houses benefit the entire household and not just the gender groups.

Moreover, the majority of the women in our sample were not "working women" in the sense of having employment outside the household. However, they were active in contributing to the household economy by raising poultry and cattle herds (mostly goat). Many of them also took up vegetable gardening in land adjoining their homestead. In most cases women who went outside the perimeter of their household did so in order to fetch firewood and water. A number of women reported that they accompany their children to local schools. Although the religious and cultural values in the village still weigh against the outdoor activities of women and their movements, economic necessities have already eroded the cultural values. The disaggregation of disaster effects by gender-group could have been done more satisfactorily had the researcher spent a longer period of time in the village observing and interacting with the villagers. This could not be done. More research on women's contributions to the household economy is needed. The fact that disasters negatively affect the employment opportunities of women is borne out by other studies as well (Adnan 1991: 65-66).

The study of Choto Chonua revealed that it was an economy where rice cultivation and salt making were the two major economic activities. Most of the respondents viewed salt production as an agricultural activity. High landlessness and unemployment were also common.

Cyclones and tidal waves are not too remote from the experiences of the villagers. In November of 1970, one of history's major disasters affected parts of the coastal district of Chittagong. Since then a flood protection embankment has been built which was, however, overrun by the tidal wave in 1991. Cyclones of lesser intensity are almost annual events to which the villagers are quite accustomed.

The cyclone of 1991 was tracked by satellite at least three days before landfall. The warning was quite adequate. Grave danger signal 8 and later 10 were posted. The warnings and special weather bulletins were broadcast over radio. We wanted to find out how many of our respondents actually had heard these warnings. We found out that 35 of our 40 respondents did hear the warnings. We also asked them the source through which they heard warnings. Radio was the main source of information about the disaster.

Most of our respondents were aware of the impending cyclone, but few of them could foresee the actual extent of devastation. Of the 35 respondents who knew of the cyclone warning, 23 of them considered moving to a safer place. There was, however, no evacuation plan in place, and the cyclone shelters were simply inadequate. Besides, a number of our respondents (about 15 percent) said they did not take the warnings seriously. Such an attitude can be explained by the fact that previously on a number of occasions there were "false warnings," and the people were not educated on the change of direction of cyclones. Educating the people on disaster warning is important. This is an area that needs immediate attention in counter-disaster planning

SOME SIMILARITIES IN RESEARCH EXPERIENCE

The nature of natural disasters is different in these two regions, which poses a serious problem in comparing the two disasters. However, since the focus was on the effects of the disaster rather than the disaster agent, the problem was minimized. Respondents in the north mentioned a variety of natural disasters such as flood, too much rain (which itself may contribute to flooding or damage the agricultural crop otherwise), and drought. In the south the only natural disaster they are familiar with and constantly worried about was cyclone. These cyclones are formed in the Bay of Bengal and lash the villages with destructive consequences.

Given the differences in the types of natural disasters, the socioeconomic consequences as well as the coping strategies were also quite different. All our respondents in the village of *Choto Chonua* in Chittagong recounted the April 29, 1991, cyclone. The death toll from this cyclone alone was much higher than all the floods in the recent history of the northern districts put together. The dramatic nature of the disaster left a permanent impression on the minds of the victims. As such their recall of the catastrophic event was more reliable. The highly destructive cyclone in the south, apart from taking a heavy toll of human lives, also caused severe damage to livestock and agriculture. Many of our respondents relied on agriculture. Their crops were also wiped out in the 1991 cyclone. The differential impact on gender of such a sudden and calamitous disaster was even more difficult to gauge.

In *Choto Chonua*, the visit by the central communication minister at that time, Mr. Oli Ahmed, expedited the delivery of relief materials. Ironically, it is only a disaster event that brings the attention of the political elites and the higher officials of the government to the conditions of the villages. Politicians in Bangladesh over the years have learned to manipulate and take advantage of these "natural" calamities to their political advantage. The leaders can project themselves—thanks to the spread of television coverage—as saviors. National leaders and high government officials compete for television time, showing their direct involvement in relief operations. The symbolic use of disasters and crises in Bangladesh can be the subject of a future research project. Once the disaster is overcome, the plight of the villages is conveniently forgotten. Thanks to the presence of democracy and competitive politics, however, at least the victims of disasters get some attention, albeit short-term, in moments of crisis.

LESSONS LEARNED AND MISSING FINDINGS

1. First, in both villages the worst sufferers were the poor women. Collecting data *from* the poor—as opposed to *on* the poor—is difficult for a variety of reasons.
2. The coping strategies of the women could be better understood if research took place immediately following impact. We could not ascertain the exact nature of population movement following the flood. Usually, the male members of the household are more migratory. As husbands leave behind their families, the women are left to fend for themselves in the event of future disasters.
3. Although according to our respondents the pilferage of relief goods was not a major problem, we were unable to ascertain the degree of nepotism and favoritism in the distribution of relief materials except to see the influence of political patronage.
4. It is very difficult, in the absence of transparency, to be absolutely sure of the amount of relief supplies available and the amount distributed. We had to settle for estimates given by the village leaders and the local administrators. Here, newspapers play an important role in reporting any perceived malpractices. However, whether newspaper reporting could be used as reliable research data is something that remains arguable.
5. We are also somewhat queasy about the ethics of disaster research. Research on tragic events is always problematic. Disasters in Bangladesh involve death and destruction of property and income. By asking the respondent to recount these sad memories, the researcher risks being invasive. Yet at the same time, research on disasters can help us formulate better counter-disaster plans and policies that might help in preventing future tragedies. This poses a serious dilemma for the researcher.

CONCLUSIONS

A list of guidelines for the promotion of disaster research in Bangladesh are as follows:

1. More research grants are needed in the area of disaster research.
2. More cooperation between local and international experts is needed.
3. More studies on the consequences of disasters are needed.
4. The political overtones and sensitivities of disaster research should be de-emphasized. The political use of disasters must give way to objective assessment.
5. Studies of local power structures and of social inequalities are needed. Pre- and postdisaster studies should be made.
6. Studies of corruption are needed, especially on the distribution of relief goods.
7. There needs to be an increased focus on policy research and on evaluation research of policies developed by the government as well as by the NGOs.
8. Courses in disaster sociology in the universities should be introduced.
9. Training of disaster professionals is needed.
10. An international center for disaster research should be created to attract researchers from around the world as well as to provide research facilities to local researchers with the aim that expertise and research findings could be shared with other countries in the region and beyond.

This is a tall order, but we must begin somewhere. The time is now.

15

THE FIELD TURNS FIFTY:

Social change and the practice of disaster fieldwork

Kathleen J. Tierney

Fieldwork has been the most important data collection strategy used in disaster research since its earliest days. Perhaps because of its Chicago School, symbolic interactionist roots and its applied focus, the field has always gravitated toward naturalistic studies of the ways groups, organizations, and communities respond in actual disaster and threat situations, as opposed to conducting experiments or large-scale, quantitatively-oriented surveys.[1] As Brenda Phillips shows elsewhere in this volume, there has historically been a close affinity between disaster research and qualitative data collection and analysis, particularly field-oriented data-collection strategies. The first field studies, which were conducted by the National Opinion Research Center (NORC), the University of Maryland, the University of Oklahoma, and the National Academy of Sciences (NAS) during the 1950s and early 1960s,[2] involved the collection of data on a wide range of natural and technological disasters and emergencies including tornadoes, floods, plane crashes, chemical emergencies, explosions, fires, and an earthquake (Quarantelli 1987b). Those early studies became the model for subsequent empirical research which continued to emphasize the exploration of disasters and their social impacts through quick-response research and direct observation (whenever possible), qualitative interviewing, and related fieldwork techniques. The Disaster Research Center, which was founded at the Ohio State University in 1963 and which has now conducted more than 600 different field studies both in the U.S. and in other countries, is probably the best-known exemplar of the disaster research fieldwork tradition.

Both published work (see, for example, Quarantelli's discussion of DRC's field methodology in this volume) and informal discussions with pioneering disaster researchers convey the idea that carrying out field research in the early days was generally unproblematic. From all accounts, entrée and access to data sources did not present problems for early disaster researchers working in the field. Funding was not always abundant, and fieldworkers may have had to adjust to physically uncomfortable situations in conducting research immediately after disasters, but they generally did not encounter significant barriers or resistance from the groups they studied. Even though the field was initiated with funding from agencies such as the Department of the Army and the Office of Civil Defense, and even though those agencies were seeking defense-related insights into human behavior in highly stressful situations, there was very little outside interference, either with field operations or with how data were analyzed and released (Quarantelli 1987b). Promises of confidentiality and anonymity reassured contacts in the field that they could be frank and open with their ideas even if they were being tape recorded, and standardized methods were developed for protecting the privacy of research participants during the data analysis and reporting phases of research. Prospective fieldworkers learned how to operate in both the postimpact, quick-response context and in preplanned fieldwork activities, and those strategies generally worked well. The fieldwork tradition has yielded numerous insights and an impressive literature on human and organizational behavior in disaster situations.

As disaster research reaches the end of its first half-century, fieldwork is still the most common data-collection approach, and fieldworkers are still remarkably successful in gaining access to people, activities, and information sources. People continue to show a willingness to talk candidly with researchers in the field, and they are often extraordinarily cooperative and helpful. Some of the best recent research (see, for example, Peacock et al. 1997) effectively blends survey and ethnographic approaches, demonstrating that skillful researchers can gain access to many different types of settings, from the offices of high-level decision-makers to the temporary living quarters of displaced disaster victims. At the same time, the context and environment in which fieldwork is conducted have changed over the years, mostly in ways that make that work more challenging, but sometimes in ways that make it easier. In some cases, the changing climate affects not only disaster researchers but also other social scientists and scientific research in general. In others, the issues disaster researchers face are tied more closely to the nature of their work.

Having been involved in disaster fieldwork since the mid-1970s, first as a graduate student "apprentice" and later as a researcher, field-team leader, and supervisor and trainer of another generation of fieldworkers, I have experienced many of these changes firsthand. This chapter contains observations on how the environment affecting research has changed since my early days in the field and some cautionary words on what the future may hold. Because my experience has been confined almost exclusively to work in the U.S., my comments apply only to that particular setting. The practice of fieldwork doubtless presents very different challenges for individuals and groups conducting studies in other societies.

Many new developments have influenced our ability to carry out work in the field, but in this essay I will discuss the six trends that I consider the most significant: human subjects regulations and their interpretation; the U.S. "litigation explosion"; agency and organizational orientations towards research and researchers; the dramatic expansion in postdisaster field research activities; increasing diversity, both among fieldworkers and among those contacted in the field; and the professionalization of emergency management.

HUMAN SUBJECTS PROTECTION REQUIREMENTS

From its very earliest days, those conducting field research in disaster settings promised confidentiality to the individuals, organizations, and communities that were studied. Researchers in the disaster area have always been subject to the same ethical guidelines as other social scientists working in the field, and the procedures used to ensure confidentiality have also been no different. Like their social science colleagues, early disaster researchers did not have to worry about human subjects issues and institutional review boards, because there were none. Doing research in disaster settings was challenging, of course, in the same sense that all field research is challenging, but it was not seen as presenting extraordinary ethical dilemmas or risks to those who were studied. Postimpact reconnaissance work was considered a form of participant observation in which a field researcher, once allowed into a particular setting such as an emergency operations center (EOC), had at least tacit if not explicit consent to collect field data. Indeed, getting access to such settings was considered critical for effective research.

Because fieldwork involves the collection of various types of data and different data-gathering strategies, the human subjects issues involved are not straightforward. Some of the data collection that occurs in the field involves the observation of public behavior and anonymous actors. More commonly, fieldwork involves gathering confidential information from known individuals. Most field studies, particularly those undertaken immediately following disasters, consist of a blend of data-collection approaches. As discussed in the chapters by Quarantelli and Phillips, fieldworkers must be able to operate flexibly in order to take advantage of research opportunities that emerge in field situations. For example, after the 1994 Northridge earthquake, our Disaster Research Center field team observed meetings of the Los Angeles Emergency Operations Board that were open to the public and that were attended by a variety of observers, including the press. After those meetings, the team remained in the city's emergency operations center with the permission of supervising officers and with the understanding that we could observe as long as we did not interfere with on-going emergency activities. As is typical in these kinds of situations, our presence in the EOC also made it possible to conduct informal interviews with people who were in the setting. Following standard fieldwork practice, these informal contacts and interviews typically led to lengthier, more structured interviews at a later time.

In today's research environment, the entire range of activities—from observation of people carrying out their public duties through formal interviewing—is subject to review for its appropriateness under guidelines governing research involving human subjects. And while the observation of anonymous individuals in purely public settings is considered the least problematic by institutional review boards, most other participation in research is generally interpreted as requiring some type of formal "informed consent" procedure. In earlier times, it was sufficient for disaster researchers seeking interviews in the field to explain the purposes of their research, promise confidentiality, ask for permission to tape, and answer any questions prospective interviewees may have had. Today, unless they are extremely skilled in negotiating the institutional review process, extremely fortunate, or both, they are likely to be required to present written documentation explaining their research in detail when seeking all but the most informal, casual interviews and to obtain written consent to participate from their research "subjects." The trend is moving in the direction of defining most contacts in the field as requiring these kinds of procedures. This complicates the process of fieldwork since highly formalized approaches to informed consent are inconsistent with the fluid, informal data-collection strategies and techniques that are required in postdisaster reconnaissance studies. More broadly, it is questionable whether the standard approach to obtaining consent, which is geared toward experimental research and studies on "at risk" populations, is necessary or appropriate in most disaster fieldwork situations.

Related to this, issues of informed consent are closely linked to notions about the risks research participants face. Like other social science research, disaster fieldwork has generally proceeded on the assumption that research has very little potential for injuring the people and organizations that are studied and on the hope that it may ultimately actually do some good. Some of us even believe, perhaps naively, that talking to an outsider on a confidential basis might actually provide some direct relief to overworked officials, those who are trying to help community residents, and the disaster survivors themselves. However, the overall trend in the regulation of research has been to see those who are studied as almost invariably at risk during the research process and to require ever stronger protections to ensure that they are shielded from harm. This tendency may be particularly strong for research on disasters, since such events are by definition painful and tragic. The human subjects review board at my own university scrutinizes our fieldwork activities closely, and the board members appear to assume that both residents of disaster-stricken communities and the personnel who are mobilized to aid them should be considered disaster "victims" who require special protection. Agency officials discussing the performance of their official duties under guarantees of confidentiality are also seen as at risk for "reprisals" from their superiors based on what they tell researchers, even though there is no evidence that such interviews have ever resulted in harm. The board's requirements have become increasingly strict over the years. For example, in addition to obtaining written consent for all interviewers, DRC has been asked to provide ever more detailed assurances and cautionary messages to interviewees. Besides offering the typical information on the study, the confidentiality policy, the funding source, and the right to refuse to participate, we must now also tell interviewees that we keep our interview transcripts, that other researchers may want to use the transcripts at some point in the future, and that they can refuse to permit that access. These kinds of requirements are of course no more stringent than those imposed on other kinds of social science research, and fieldworkers in various settings are subject to the same constraints. However, they do constitute additional burdens both for fieldworkers and, I would contend, for the people they study.

Indeed, after being informed about all the factors they need to weigh in deciding whether to participate, our "subjects" may understandably be much more anxious and concerned about taking part in research than they would have been otherwise.[3]

FIELDWORK IN A LITIGIOUS ERA

Nowadays, when we obtain consent and promise confidentiality to the people we study, our verbal and written assurances must also include that ominous tag line "unless pursuant to a court order." Unlike our forerunners in disaster field studies, researchers working today can no longer offer a blanket guarantee of privacy and anonymity, primarily because our academic institutions will not allow it, but also because the reality is that a lawsuit could conceivably require the release of information obtained from those who participate in our research.

In the U.S., disputes between contending parties increasingly tend to end up in court. Lawsuits are also a common strategy used by those who possess the resources to silence their opponents. Researchers get drawn into the legal process as expert witnesses, possessors of relevant data, and occasionally as defendants. In this litigious environment, courts are increasingly faced with balancing the privilege offered to researchers and research participants with the needs of litigants, often to the detriment of the former. By now we are all aware of the difficult circumstances faced by researchers whose confidential data have been sought by outside parties in civil or criminal cases. Medical researchers studying how small children were affected by the Joe Camel advertising campaign found their raw data subpoenaed by R. J. Reynolds Tobacco Company, which charged that their work was flawed and biased against the industry (Baringa 1992). Mario Brajuha, a graduate student who was conducting sociological field research on the restaurant business, had to fight a subpoena seeking the release of his data for more than two years before the legal action against him was finally dropped (Brajuha and Hallowell 1986). Rik Scarce, who was also a sociology graduate student at the time, went to jail for 159 days rather than turn over data obtained from research sources under assurances of confidentiality (Scarce 1994).

The disaster area is among a number of fields of investigation that are becoming increasingly litigation sensitive—often with catastrophic results for the researchers involved. Several researchers who were involved in studying the social impacts of the *Exxon Valdez* oil spill became embroiled in extensive litigation to protect the participants in their research when Exxon demanded disclosure. Steven Picou was among a group of social scientists who studied how the spill affected communities and households in the impact region. His work was funded by a quick response grant from the Natural Hazards Research and Applications Information Center at the University of Colorado and by grants from the National Science Foundation and Earthwatch. While he was still collecting data, the president of Exxon wrote a letter to the director of the National Science Foundation (NSF) protesting NSF's funding of the study. Later, Exxon attempted to subpoena all the data Picou and his team had collected, including material that had not yet resulted in publications, as well as all his personal and financial records involving the Exxon study. With help from the attorneys at his university, Picou vigorously resisted Exxon's efforts to obtain data that had been collected under assurances of confidentiality, and eventually a compromise was reached that involved the release of data, with identifiers removed, to sociologist Richard Berke, the consultant Exxon had hired to review the research for its methodological soundness (Marshall 1993; Picou 1996a, 1996b).

John Petterson, whose research company, Impact Assessment, Inc., conducted studies on affected communities with funding from a group that later decided to sue Exxon for damages, faced a similar fate. In this case, both sides in the lawsuit wanted access to his raw data. Petterson fought attempts to obtain the data at great personal and financial cost, but eventually he was forced to permit access to experts. Steven McNabb, an anthropologist who studied the impact of the oil spill under a contract with the Minerals Management Service, found himself subject to a blanket Exxon subpoena that asked not only for his research data, but also for material related to virtually every aspect of his professional life and personal finances. McNabb relates that when he received the subpoena,

> I engaged an attorney and learned that a comprehensive response would mean that I would deliver everything I had written since 1980; every source I consulted since 1980 (which meant virtually my entire personal library and all files, and an equivalent volume of paper and books from other archives and libraries); and financial records, including IRS returns, invoices, telephone bills, and assorted receipts since 1980 . . . a strict reading of the subpoena would require me to turn over syllabi, course materials, and even grades and student evaluations from the courses I taught that were unrelated to the oil spill and that were dated earlier than the spill. (McNabb 1995: 332)

Exxon was unable to get access to the data it wanted—that is, information on individual respondents, raw field notes, and other documents that could link data with specific research subjects—because McNabb had already made that impossible through a systematic purging of all his files. He now cautions that all researchers who promise confidentiality to their interviewees should be prepared to undertake similar measures and even face jail in order to live up to that promise.

Exxon's attack on social scientists because of the threat their research represented are related to a broader trend involving the use of so-called SLAPP suits, an acronym that stands for "strategic lawsuits against public participation." SLAPP actions are brought by litigants for a variety of reasons, but their main objectives are to intimidate, silence, and financially burden their critics. Typically, such suits accuse those who attempt to exercise their free speech and petition rights of conspiracy, defamation, or intent to cause economic injury (Canan and Pring 1988; Canan et al. 1990; Pring and Canan 1996). As might be expected, suits involving hazards and environmental damage constitute an important category of SLAPP actions. Organizations in the environmental and antinuclear movement and citizens who speak out against polluters and locally-unwanted land uses (LULUs) are common targets of strategically-motivated suits (Pring and Canan 1996).

Researchers are also at risk of being SLAPPed if some powerful party comes to see their work as threatening or troublesome. Earlier this year, for example, Cornell University researcher Kate Bronfenbrenner was sued for libel by Beverly Enterprises, one of the country's major providers of nursing home care, for statements she had made at a town hall meeting attended by several members of Congress. Her research had shown that Beverly was a consistent labor law violator that had tried various strategies to interfere with union organizing at its facilities. Beverly's $250,000 SLAPP suit seeks release of Bronfenbrenner's confidential data and other research materials. (For more information on this case and its implications for researchers, see National Public Radio's "All Things Considered" for April 27, 1998.)

While guidelines such as the American Sociological Association's Code of Ethics are quite clear on the rights and responsibilities of researchers, recent cases and court judgments are anything but reassuring. In several important decisions, courts have recognized the need for protecting the privacy and confidentiality of individual research participants. However, there is currently general agreement that no broad "scholar's privilege" exists that can shield data from subpoena and that researchers must take proactive steps to ensure that they can protect their data sources should the need arise. And the sad fact is that, even when they win in court or are eventually vindicated in other ways, researchers pay a tremendous price in fighting these cases, and they often end up embittered and impoverished as a result of the process.

Some researchers (see, for example, Clarke 1995) argue that in the current legally ambiguous and highly litigious climate it is not appropriate for any researcher to claim that research data are confidential or privileged. Others believe that researchers can still protect their data and sources if they plan ahead and use appropriate safeguards. Everyone agrees that the litigation explosion has introduced a new set of complications into the research process.

(For other discussions on ethics, litigation, and fieldwork and on protecting research data, see Cecil and Boruch 1988; Presser 1994; Erikson 1995; and Picou 1996b.)

AGENCY PERSPECTIVES ON RESEARCH

It is axiomatic that organizations do not like to be studied—either during normal times or in disaster situations. Even when a researcher is invited into an organization to conduct a study, the sponsor is often less than delighted with the results. Virtually all organizations, both public and private, seek a favorable public image, and one means to accomplish this aim is to exercise control over information, including the kinds of information researchers seek. The need for organizational impression management is probably even more marked in disaster situations than during normal times, since crises open up the organizations involved to heightened scrutiny and since any mistakes they make may have grave consequences (Tierney and Webb 1995). As the mass media have become ever more pervasive and disasters loom ever larger as news stories, agencies that deal with disasters have become ever more sensitive to the possibility of adverse publicity. After all, if snow cannot be removed after a blizzard in Chicago, if sea birds and otters are seen on the evening news slick with spilled oil, or if the cavalry is not ready to charge immediately after a hurricane strikes, heads may roll. Disasters can make or break careers and boost or damage organizational prestige. Under such circumstances, it is understandable that the organizations and individuals involved try to exert control over information.

Over time, government agencies and crisis-relevant organizations in particular have devised various ways of more effectively managing impressions and heading off adverse publicity, both during normal times and in disasters. One way has been to professionalize the providing information through the creation of the position of public information officer (PIO). The PIO position—or at least the function—is now institutionalized in most governmental and many private-sector organizations. In addition to disseminating information to the public, a key role of the PIO is to deal with the media in order to obtain favorable press coverage and more generally to ward off threats to the organization's image. To these same ends, another common agency strategy is to hold frequent press conferences and to use various other media, ranging from television to the Internet and satellite communication, to release and continually update information. A third approach is to control outsider access to the kinds of information organizations do not choose to share—including researcher access.

This is not meant to suggest that crisis management, governmental, and other organizations are doing anything out of the ordinary or sinister when they engage in these kinds of activities. Rather, they are merely doing what all organizations do, which is to attempt to manage transactions with their environments, including the flow of information. If a disaster represents a threat to that control, then from an organizational perspective that calls for even stronger evasive and defensive action.

However, as a consequence of these new impression management strategies, the fieldworker's role has become more difficult. When researchers arrive at a disaster site seeking information, for example, the officials they initially contact may try to send them to public information officers or to press conferences rather than granting interviews. Security at emergency operations centers and other places where disaster-related activities are carried out is generally quite tight, with access controlled by badges and other forms of identification. Researchers seeking access to those places may be shunted off to the "public" areas that have been set aside for presentations and for the press. Instead of having the opportunity to observe disaster operations directly and ask questions freely, the fieldworker may instead be handed a packet of preprinted information. Agency officials are increasingly guarded with researchers, treating them like members of the press even after receiving assurances of confidentiality. Frontline disaster workers may worry about sharing information with field researchers, for fear that that could anger their superiors and lead to sanctions.

The trend toward increasingly centralized information control has several consequences for the conduct of research on disasters. One is to promote a "command post" point of view that privileges the official information-dissemination function over the perspectives represented by other elements in the disaster management network. Another is to present a unitary or monolithic view of disaster-related activities rather than one that allows for multiple interpretations.

All these barriers to the free flow of information can be overcome through skillful fieldwork, but they obviously represent additional challenges. Valid research depends on the ability to have entrée to information sources that are willing to be candid and on the ability to observe emergency operations and other activities of interest. It also depends on being able to explore a research question from a variety of perspectives, not just from the official one. Being able to obtain information on events independently as they unfold is particularly critical in crisis situations, when officials and responders are often under so much pressure that they may later lose track of sequences of events or fail to remember when certain decisions were made. Researchers lose a great deal when, rather than direct access to events and people, they instead receive reconstructions and packaged narratives.

Another trend that affects fieldwork in disaster research is the increasing tendency for disaster-related agencies and governmental entities to carry out their own research and fact-finding activities. For example, the Federal Emergency Management Agency routinely administers surveys on "customer satisfaction" containing questions that the agency wants answered. Highly-committed governmental jurisdictions like the City of Los Angeles carry out their own postdisaster reconnaissance activities and hold special workshops to find out how other communities have handled major disasters and to identify lessons learned.[4] This kind of research, which is often done on a rapid-response basis, tends to focus on very specific, practical issues. Its goal is to help organizations solve problems, not to make more general contributions to disaster-related knowledge. Clearly, such studies can have very positive outcomes, particularly when they produce genuinely valid lessons and insights, give the agencies and organizations that carry them out a sense of ownership over the findings, and spur organizational change. However, this trend can have negative effects on the overall research enterprise if it leads organizations to eschew involvement in curiosity-driven research or to believe that more systematic research is irrelevant to their needs. As they increase their capacity to address the practical questions they consider important, either through conducting their own studies or through contracting for specific research products, governments and disaster-related agencies may eventually come to view academic and other "outside" research as little more than a nuisance. This would clearly have a negative effect on the social science knowledge base.

A CROWDED FIELD

During the early days of disaster research, a relatively small number of individuals and organizations were engaged in disaster-related field studies. Pioneering groups like the NORC, the NAS, and the early DRC teams had the field pretty much to themselves. Since social science fieldwork was still so uncommon, the main challenges for researchers in the 1950s, 1960s, and even the 1970s were to establish legitimacy, communicate the objectives of research, and achieve access to sources of data. Since the inception of the field, opportunities for conducting research have expanded greatly. Research budgets have increased, a wider range of field activities are receiving support, and a much larger number of organizations and investigators are actively involved in either conducting or sponsoring data-collection efforts in the field. Fieldwork receives support from a wider spectrum of sources than ever before, including grants to individual researchers; small "quick response" grants from organizations such as the Natural Hazards Research and Applications Information Center at the University of Colorado, funding from professional organizations such as the Earthquake Engineering Research Institute (EERI), and support from research consortia such as the Multidisciplinary Center for Earthquake

Engineering Research (MCEER).[5] Government agencies organize their own postdisaster reconnaissance teams, and international collaborative fieldwork is becoming more common. Major U.S. disasters also attract researchers and research teams from other countries. Rather than being alone or having to struggle to explain why they are there, today's disaster researchers are more likely to find themselves jockeying for position in an increasingly crowded field.

From the very beginning, disaster scholars have recognized convergence as a common problem in disaster situations. However, what we did not envision was the extent to which *researchers* would also converge in disaster situations. Whereas researchers in the early days could more or less assume that they would be the only ones conducting fieldwork in a disaster-stricken area, today's researchers can be equally sure that they will not be alone. Like the convergence phenomenon generally, researcher convergence is most marked in the immediate aftermath of a major disaster. A large, damaging urban earthquake, the "mother" of all fieldwork opportunities for many disaster researchers, is the type of disaster in which massive convergence by researchers of all kinds is a virtual certainty, in part because such events are so rare, but also because a major share of social science research on disasters is funded through the National Earthquake Hazards Reduction Program, which funds National Science Foundation research. Following the 1994 Northridge earthquake, hundreds of researchers representing a wide range of earth science, engineering, and social science disciplines were in the field immediately after the event, and a large number of longer-term studies were subsequently funded. So much data collection was being undertaken by so many different researchers that it became necessary to develop mechanisms to facilitate coordination and cut down on duplication. The Earthquake Engineering Research Institute and the California Governor's Office of Emergency Services established a clearinghouse where researchers could go to obtain information, make contacts, and attend daily briefings. The National Science Foundation subsequently gave a grant to the California Universities for Research in Earthquake Engineering

(CUREe) specifically to provide information to the research community on what studies were being conducted and who was involved. CUREe held a workshop that was attended by several hundred researchers, published a directory of the studies that were being undertaken and the investigators involved, and organized a major conference at which research findings were presented (California Universities for Research in Earthquake Engineering 1995, 1997).

The fact that the field is increasingly crowded, particularly immediately after disasters, is a very good thing. It means that funding for disaster research is robust, that larger numbers of researchers find disaster-related problems intriguing, and that disaster research is becoming increasingly institutionalized. However, this welcome trend can have negative consequences if the large amount of field activity makes researchers more intent on competing with one another in the field than on cooperating. Since careers and professional standing are tied to research performance and since a well-done quick-response study can give a researcher a clear advantage in competing for larger grants, the competitive pressures can be intense. Under these circumstances, those of us who are active in the field, especially immediately after disasters, need to exercise caution lest we come to resemble rival reporters vying for an "exclusive" rather than colleagues and members of a scientific community.

The heightened intensity of field activity can also have detrimental effects if providing information for multiple research efforts becomes overly burdensome for the communities and organizations affected. Communities experiencing major disasters increasingly must cope not only with researchers, but also with numerous other interested parties who also converge seeking information. Large disasters attract the media in droves. Politicians are drawn to disasters, both to offer assistance and for the photo opportunities they provide. As I noted earlier, representatives of other jurisdictions routinely visit disaster areas to obtain information that will help them better prepare for their next emergency. People in specific roles—city managers, emergency managers, public works and water department officials—are contacted by their counterparts from around the country and are asked to arrange meetings and tours of damaged areas. If the disaster is big enough, international delegations can be expected. This intense desire to visit disaster sites, which sometimes verges on "disaster tourism," is so strong that emergency management agencies have occasionally found it necessary to have formally-designated protocol officers on staff to manage visitor-related issues. The challenge for researchers facing this increasingly crowded field is twofold: to clearly communicate the distinctiveness of research activity—as contrasted with journalism, fact-finding, and the search for on-the-record information—and to avoid adding to the demands and problems stricken communities face.

FIELDWORK AND DIVERSITY

Like other social science research and academic research in general, the field of disaster research was closed to women and minority groups throughout much of its history. Indeed, perhaps to an even greater extent than a number of other specialties, disaster research began as a virtually all-male, all-white field and remained so for decades. Over time, mirroring broader changes in the social science disciplines, the field has become more diverse, incorporating a broader range of groups and perspectives. This change has been most marked with respect to gender. More women have entered the field over time, and, concurrently, the increasing emphasis in the social sciences on gender and its ramifications has been reflected in the work of disaster researchers.

Although a case can be made that gender affects the entire range of data collection and analysis strategies in the social sciences and science generally (see, for example, Harding 1987, 1991; Nielsen 1990), that influence is probably most marked in qualitative research (for discussions, see Warren 1988; Fonow and Cook 1991). Gender issues are extremely relevant to the conduct of field research, because the researcher's gender affects the ability to gain access to research settings, the roles in which the fieldworker is cast, and how the fieldworker is perceived and treated by those who are studied. Gender also shapes the manner in which the researcher collects and analyzes data, from the strategies and tactics used in fieldwork to the manner in which field experiences and data are interpreted. Gender conveys both advantages and disadvantages in the field. It provides both a lens for viewing social life and a filter that blocks access to information.

For the first twenty years, the conduct of disaster-related fieldwork was almost exclusively a male province. Women were involved in early disaster field studies (for example, see Bucher 1957), but in vanishingly small numbers, and, prior to the 1970s, the number of women who were active in studying and writing about disasters could be counted on the fingers of one hand. The Disaster Research Center, for example, was virtually an all-male research unit for the first ten years of its existence, and it was not until the mid-1970s that the gender composition of research teams began to become more balanced.[6]

In considering gender issues in disaster research, it is also important to note that, for the first generation during which studies were undertaken, the *informants* and *interviewees* who were contacted in the course of research were also overwhelmingly male. This was a reflection both of male domination of leadership positions within organizations and of the types of organizations (e.g., civil defense offices, fire and police departments) that were the focus of the early disaster projects. Indeed, it can be argued that one of the reasons the early field studies proceeded with relative ease was that the fieldworkers and those they interviewed tended to resemble one another. This was particularly true in the earliest days of disaster field research, when the researchers tended to be males with military experience who were older than today's typical graduate students. Thus, the initial disaster literature was based almost exclusively on research contacts among males.

When women first became involved in disaster research, they faced twin challenges. As newcomers in male-dominated research organizations, they had to prove to their male counterparts and to their superiors that they could operate in the field as successfully as men. At the same time, they had to establish their legitimacy and credibility with the male-dominated and largely male organizations they were studying. As is typically the case for women in field situations, boundaries needed to be re-established when coworkers or research participants veered into treating female fieldworkers in ways that were not consistent with the researcher role (e.g., as social companions or objects of sexual interest).[7] As former DRC fieldworker Joan Neff Gurney has pointed out (1985), while women in male-centered field situations experience pressures that men do not, at the same time they may feel very uncomfortable about acknowledging those pressures for fear that might compromise the credibility of their work.

The gender-balance situation has changed markedly in the nearly five decades since disaster field research began. Although men still outnumber women by a considerable margin, there is no question that the field has produced a number of very active women researchers. The organizations typically studied in disaster-related projects are also much less likely to be all-male bastions. Reflecting changes in the larger society, more and more women are entering positions of responsibility and leadership both in disaster-related agencies and in other agencies that are typical points of contact in field research. In this respect at least, the fieldwork enterprise is less strictly "gendered" than it was in the past.

To what extent did these gender-related issues influence the conduct and results of earlier research? Clearly, numerous factors have shaped the development of the field and the contents of the literature, and a thorough sociology of knowledge analysis of how the field developed has yet to be written. At the same time, however, recognition is growing that early field research (and disaster research in general) was colored by unconscious gender biases. Arguably, the absence of a gendered perspective, together with the tendency to focus on the activities of officially-designated organizations, helped create a skewed picture of disaster response and recovery in the early literature.[8] Disaster response was equated with what public organizations did—and what they did to cope specifically with disaster effects (see, for example, discussions on the concepts of "agent-related" and "response-related" demands in early postimpact response studies). Missing was a more holistic perspective that would take into account the contributions made by a broader range of community groups and the less-visible contribution made by informal helping networks, volunteers, and work carried out inside the home—spheres in which women tend to be more involved. In her analysis of gender issues in disaster research, Fothergill (1996: 44) also observes that early study findings tended to characterize men as helpers and women as needing help, revealing a "normative bias in disaster research concerning 'appropriate' gender role behavior."

Because of the broader paradigm shift that has occurred in the social sciences and because of the gender-related interests of many researchers who are currently working in the disaster area, the recognition is growing that gender stratification and gender relations, which are pervasive throughout social life, affect the entire spectrum of hazard-related behavior, from risk perception through disaster preparedness, response, and recovery (for good overview discussions, see Morrow and Enarson 1994; Fothergill 1996; Enarson and Morrow 1998). It is now becoming understood that men and women experience disaster impacts differently and that these differences are legitimate topics for investigation. For example, in work that presents ideas that are new to the field, Morrow and Enarson (1996) note many distinctive features of women's disaster-related experiences, among which are that women's ordinary caregiving activities expand and become more intense following disasters; that existing disaster assistance policies and programs fail to address the needs of women, particularly poor and minority women, who often experience the most difficulty recovering; that women play a very significant though often unrecognized role in disaster response activities; that women tend to be shut out of official response and recovery efforts; and that following disasters women face risks to their safety and security that men do not.

Despite these kinds of changes, disaster studies continue to be carried out in highly gendered terrain in that field researchers—

both male and female—still conduct the bulk of their work in organizations and situations that are "male gendered" (Acker 1991, 1992). Not only are crisis-relevant organizations historically male-dominated, but *as organizations* they exhibit and extol qualities that are culturally associated with maleness, such as hierarchy, decisiveness, quick action, strength, and risk-taking. As recent field research has shown (see, for example, Chetkovich 1997 on the fire service), these features, along with an attendant hostility to "womanish" perspectives and modes of action, persist despite changes in organizational gender composition. Indeed, the gendered quality of crisis-related agencies and institutions is a topic that is ripe for further research.

I have chosen to focus my discussion on gender issues in disaster research because the field has probably made the most progress in the area of gender diversity. It remains relatively homogeneous in other respects. Regarding the ethnic composition of the disaster research community, with certain notable exceptions the field has failed to attract people of color. This has no doubt hampered its ability to gain access to nonwhite community groups and to nonmajority perspectives on disaster issues. While bringing more members of historically underrepresented groups into the disaster field will not automatically solve these kinds of problems, it would definitely be a good start. With respect to the content of the studies that have been undertaken, historically very little research has focused on race, ethnicity, and social class as factors that structure both the human response to hazards and postdisaster outcomes. When such topics have been made a focus of research (see, for example, Bolin and Bolton 1986; Perry and Mushkatel 1986; Bolin and Stanford 1991; Simile 1995), findings show that different sociodemographic and sociocultural groups experience disasters in very different ways. The field's new emphasis on social inequality, diversity, and related issues is contributing to the development of a new paradigm for disaster studies that links disasters to broader social-structural and political-ecological factors (Blaikie et al. 1994; Peacock, Morrow, and Gladwin 1997).

THE PROFESSIONALIZATION OF THE EMERGENCY MANAGEMENT FIELD

When the field of disaster research began in the early 1950s, the local civil defense director was likely to be a retired military man operating out of a one- or two-person office that was both physically and functionally removed from the locus of community decision-making. The civil defense office typically lacked both resources and ties to other governmental units, and civil defense directors and their activities—which, in the early days, focused almost exclusively on war planning rather than on disasters—were accorded little prestige or community visibility. The position did not tend to attract either the best and the brightest or young, ambitious people wishing to move up in government. Within other crisis-relevant organizations such as fire and police departments and hospitals, disaster-related activities were given a low priority, except when disaster did strike. Managing disasters was not seen as important to the missions of these kinds of organizations, and people given those responsibilities probably felt justified in not taking them too seriously. Starting around the early 1980s, that situation began to change, and the notion of emergency management as a specialized discipline, significant responsibility within government, and important organizational activity began to emerge.

Although circumstances vary around the country, there is no question that today the field of emergency management is well on the way toward professionalization and that the prestige of the field has grown. At the federal level, the director of FEMA has been given cabinet rank. Locally, the emergency manager now tends to report directly to the mayor or the city manager, rather than finding herself buried somewhere down in the organizational chart. Instead of having to make do in an office in the basement of the fire department, today's big-city emergency managers are likely to find themselves presiding over a state-of-the-art emergency operations center and a sizeable staff. It is now possible for a bright, well-trained, talented, and politically savvy individual to have a very good career in emergency management.

This trend toward greater prestige and professionalization has been accompanied by higher educational expectations for emergency management practitioners. Emergency managers are much more likely than before to have baccalaureate or advanced degrees, and emergency management is a growing area of specialization within fields like public administration. Disaster professionals also continue their training and education by participating in professional associations such as the Emergency Management Section of the American Society of Public Administration and the National Coordinating Council on Emergency Management and by attending courses at FEMA's Emergency Management Institute and specialty conferences like the annual National Hurricane Conference and the Natural Hazards Workshop.

Through these kinds of activities, emergency managers have become more familiar not only with research on disasters but also with members of the research community. Today's emergency managers appreciate and use research, and they are accustomed to interacting with researchers in the course of their work. It is becoming routine for emergency managers to ask researchers to serve as consultants, advisors, or staff on projects they are undertaking, or for their agencies to directly fund research. For their part, researchers also have more of an opportunity to get to know emergency managers on both a professional and personal level. As researchers and practitioners serve together on panels and meet in both professional and social situations, the opportunities for dialogue are expanded.

These kinds of changes obviously help disaster researchers in the conduct of their work. If those who are being asked to provide access and information understand the research process, support the need for research, and know (or at least know something about) the people who are carrying out the work, this clearly helps field operations run more smoothly. Mutual trust and respect encourage disaster professionals to assist researchers with their work and at the same time help researchers gain a deeper understanding of the people they are studying.

CONCLUDING COMMENTS

Like all other social activities, disaster fieldwork is situated in a larger social context. That social context has changed greatly over the decades, and those changes are reflected in the way we operate in the field today. In reviewing the ways in which social and organizational change have affected the conduct of disaster research, I have also tried to show that change brings both challenges and opportunities. Our task is to devise more creative ways of overcoming the challenges while exploiting the opportunities that the new research environment provides.

The recent explosion of interest in field methods and qualitative research generally is leading to a closer examination of what we as social scientists know—or think we know—about the process of fieldwork (for a good overview of the state of the art, see Denzin and Lincoln 1994). As the Phillips chapter in this volume suggests, disaster researchers should try to make the most of these methodological advances and research lessons in their work. At the same time, because of the unique nature of the work we do, we are also in a position to make a real contribution to the methodological literature. As I have tried to show in this essay, doing disaster fieldwork means grappling not only with all the issues fieldworkers have traditionally had to face, but also with an entirely new set of complications. Researchers in other fields will benefit from our experience as we go about working out those puzzles.

NOTES

1 I do not mean to imply, however, that these methods have never been used. Early studies on disaster impacts did include population surveys (Quarantelli 1987b), and laboratory and simulation methods were used in some of the most important early studies on organizational responses under stress (see Drabek 1965; Drabek and Haas 1969; and Drabek's chapter in this volume).

2   The NORC and Maryland studies ran from 1950 to 1954, while the Oklahoma research was conducted from 1950 to 1952. The NAS Committee on Disaster Studies operated between 1951 and 1957; its work was continued by the Disaster Research Group until 1962 (Quarantelli 1987b).

3   A recent DRC field study involved prearranged interviews with community informants. Prior to the interview, after being read and asked to sign the required confidentiality and consent form by the DRC fieldworker, one of the interviewees stated with alarm, "Oh, I didn't realize I'd be asked to give consent," then agreed to be interviewed, but refused to be taped. The individual had already been informed about the study earlier, had agreed to participate, and obviously wanted to talk, but balked when faced with the legal complexities of participation.

4   For example, Los Angeles has sent teams into the field following the 1985 Mexico City earthquake, the 1989 Loma Prieta earthquake, Hurricane Andrew in 1992, and the 1995 Kobe earthquake.

5   MCEER, which was formerly the National Center for Earthquake Engineering Research (NCEER), is headquartered at the State University of New York at Buffalo. It was supported by a major grant from the National Science Foundation between 1986 and 1996. In 1997, the center received a second grant from NSF, and two other centers, the Mid-America Earthquake Center and the Pacific Earthquake Engineering Research Center, were also funded. All three centers will likely field reconnaissance teams in future damaging earthquakes.

6   DRC cofounder Henry Quarantelli provided the following background information on women's representation in early DRC field teams. From the center's inception in 1963 through mid-1974, 56 graduate students took part in field studies. Of that number, five were women. The first woman to take part in a DRC field team did so in 1970. Beginning around 1974, the year I started working at DRC, more women began to be added to the field staff. There were other women involved in disaster field research prior to that time, working both as individual researchers and as members of the early disaster field teams, but their numbers were exceedingly small.

7   In my early days as a fieldworker, I highly resented the fact that the male members of the field team seemed to want to adhere to the "male role" during our trips. For example, they would always insist on driving the car in the field, as if allowing a woman team member to drive was somehow inappropriate. Female team members would remark that the men seemed to view field trips as week-long dates that gave them an excuse to spend time with women other than their wives or girlfriends. Undoubtedly the men also harbored gender-based feelings about the women with whom they worked.

8   Early research did focus on the activities of emergent groups as well as on existing organizations (see Stallings and Quarantelli 1985 for discussions), but by far the strongest emphasis in that work was on the activities officially-designated organizations. The emergent groups studied were those whose activities contributed to the response effort, which was defined as involving "disaster-related" tasks. We see no mention in the classic literature, for example, of emergent groups focusing on child care or of the informal provision of social and emotional support in the aftermath of disasters. Such topics only began to enter the literature in the mid-1970s (see Taylor et al. 1976; Taylor 1976).

PART IV
POSTSCRIPT

Previous chapters have described a range of methods for studying future disasters and disaster-related phenomena. They describe some of the tools available. Ollie Davidson, a career disaster practitioner with both national and international experience, has been at the forefront in creating public-private disaster partnerships. In this final chapter, he identifies many questions and issues confronted daily by disaster professionals to which several of the methods described in this volume could be applied. Davidson focuses on the future. In disaster management practice in the U.S., this means focusing on mitigation. Current policy stresses public-private partnerships as the leading tactic for achieving this goal. It will be up to future researchers to provide the analyses that can bring about the success of this strategy.

16

FUTURE DISASTER RESEARCH:

A practitioner's viewpoint

on public-private partnerships

Ollie Davidson

Emergency management has become almost a discipline, thanks to acts of God and some leadership at the federal, state, and local levels. Former U.S. Federal Emergency Management Agency (FEMA) director James Lee Witt's concern for people and his close relationship with former President Bill Clinton were an outstanding example of such leadership.[1] Whether this transcends the Clinton administration is still an open question. What will transcend is that people's expectations of government's role in disasters have been elevated to a level which may never be fulfilled, especially after a major disaster.

Several state directors of emergency services have taken advantage of past disasters and FEMA's funding to develop partnerships with business and industry, commonly called public-private partnerships designed to reduce state and local disaster losses. These local relationships and the resultant activities may be sustained. FEMA's Project Impact may be the most important legacy of the Clinton administration's disaster work as it attempts to mobilize all elements of a community to take mitigation actions to protect itself. Although more and more companies are participating in Project Impact communities, many still have not recognized that this disaster activity could save their businesses.

FLIPPING THE EQUATION FROM PUBLIC LEADING PRIVATE SECTOR TO PRIVATE LEADING PUBLIC SECTOR

As the number and magnitude of disasters increase, careful thinking needs to focus on what is at risk and how to protect our most vulnerable assets. Recent disaster experience demonstrates that the loss of jobs or the disruption of the work force (e.g., the loss of a workplace, the destruction of workers' homes, or the disruption of their families) is a very serious threat to any country's economic viability. Once the private sector recognizes what it has at risk, it should become the leader in disaster prevention and mitigation. We will recognize success by the advertisements about disaster-resistant home construction and employee benefits which include loans to strengthen their homes against disasters. Disaster awareness will be similar to wearing one's automobile seat belt, having a fire/smoke detector, or installing a home alarm system. Unfortunately, the question of how long it will take to achieve this heightened state of disaster resilience may be answered by the question: How many more destructive disasters will we have?

Business/industry disaster readiness should begin in the workplace, protecting the means of production including the workforce. Former Mayor Tom Bradley of Los Angeles was a visionary when he called local business leaders to a meeting to discuss disaster vulnerability. His message was simple, and it should be heeded by business leaders today: if you want your company to survive a major Los Angeles earthquake, you must prepare yourselves. He continued: government emergency response resources will be fully occupied at critical public facilities (schools, hospitals, etc.) for many days after such an earthquake; businesses must fend for themselves or perish.

Bradley's message stimulated the formation of the Los Angeles Business and Industry Council for Emergency Preparedness and Planning (BICEPP). Today there are BICEPP clones in many cities and a new public-private sector disaster organization, the Disaster Recovery Business Alliance (DRBA), which has taken this concept further in several communities. The real reasons for business and government to cooperate for disaster mitigation need more research, and DRBA communities are a place to begin looking. Why did businesses and government come together? What was accomplished in this relationship? How can these lessons be transferred to other disaster-prone communities? Most important for businesses, how were economic losses reduced because of the relationships built with government and nongovernmental organizations?

A BLUEPRINT FOR PRIVATE SECTOR AND GOVERNMENT COOPERATION

There is still an understanding gap and an attitude gap between public agencies and the business/industry sector. Although more and more examples of cooperation are coming to public attention, the major question of why each group needs the other has not been answered. A volume of public-private case studies, or "Best Practices," would be very helpful.

A frequent question from the public sector at public-private disaster meetings is: How do we start this cooperative process? A few state emergency management directors have suggested that the private sector should develop a strategy for the integration of the nation's private-sector resources into U.S. emergency management. The Private Sector Committee of the National Emergency Management Association (NEMA), the professional association of state emergency managers, has started to develop such a strategy. The ultimate objective is to design a model for private-sector resources to become "integrated" into a national emergency management plan which would augment FEMA's Federal Response Plan and be a major section of any state disaster plan.

In order for true partnerships or integration of private and public resources to occur, much more research needs to be done on the potential benefits of cooperation and the negative results of "business as usual." Once the benefits are better known and this public-private process matures, attitudes will have changed on both sides so that this cooperation will be much more widely accepted.

CUTTING-EDGE RESEARCH NEEDS

Government-industry legal and perceived legal constraints

Despite gains in public-private partnerships for disaster reduction, there are major cultural as well as legal and perceived legal constraints between government and industry organizations. Cultural differences include the strongly held belief by business and industry that government employees don't or can't work. The image of "government as regulator" still dominates much of industry's negative attitude toward government. The Clinton administration's push for a new customer service attitude in government helped reduce this cultural gap.

The idea of businesses as "profit mongers" is an attitude among some government employees which is more difficult to break. Many believe that industry is only out to make a profit and that any help from government should be to all industry, not just one business. Among many government employees there is an almost religious belief that one should remain distant from any industry effort to learn more about government activities, priorities, and strategies. Most government employees do not distinguish between potential corporate partners and businesses that are vendors who want to become government contractors. The most likely corporate partners are those with a significant manufacturing and employee presence in a community that share the need for hazard information and emergency planning to become prepared for a disastrous event.

As more public-private partnerships grow deeper roots, these attitudes will dissipate. However, researchers may become the new "intermediary" or catalyst for interaction between government and businesses that would like to cooperate but do not understand each other well enough to be successful. Research that demonstrates how products will impact on people and how they can be useful in disaster mitigation is needed. Researchers who understand the value of industry participation can become valuable interlocutors, especially bringing experience from one geographic area to another.

Legal and perceived legal barriers inhibit disaster cooperation. While assigned to one federal agency, I asked for advice from its senior career legal counsel. I was shocked by the intense berating I received from this old-line government lawyer. He said that he was against this public-private partnership "stuff" as it was "letting the fox into the hen house," etc., etc., etc. When I asked if working with an industry association was preferable to dealing directly with each company, he responded that such associations "were anti-trust actions waiting to happen." Ironically, today that same federal agency's highest priority is public-private partnerships!

The advisory committee law could be a major resource to facilitate government-industry cooperation. However, many agencies believe that this law is a constraint rather than a facilitator. FEMA abandoned its formal advisory committee, and the U.S. Agency for International Development (USAID) abolished its International Disaster Advisory Committee (IDAC), in part because it was chaired by Marilyn Quayle, wife of the then U.S. Vice President Dan Quayle. During its short tenure, IDAC brought together the most influential companies representing the disaster-causing industrial sectors (chemical, oil, transportation) and the sectors which could cooperate to mitigate disasters (research universities, information, communication, transportation, insurance, etc.). These companies and other participating organizations supported many international and domestic disaster projects and pioneered the concept of public-private partnerships.

All-hazards approach and consolidated information

Although the all-hazards approach has been generally accepted intellectually, it is still difficult to find all-hazards information available to the practitioners who need the information in a useable form. In practical terms we need to know what the hazard will do to the people and the structures in its path. Anecdotal information is abundant among the "war stories" that old-time disaster specialists love to relate. However, solid case studies and documented results are needed to convince management that mitigation measures will save money and lives. Hazard information and new technology without an understanding of how it will impact people and the local economy is difficult for disaster managers to understand and to use.

The latest race to the future is in information technology and the sharing or exchanging of disaster information. Web sites with news and disaster information abound, but few contain documented information to identify the most successful means of protecting people, their jobs, and our communities. The proliferation of Geographic Information Systems and technology from the defense industry may have added to the confusion, not reduced it. Competing systems are difficult for disaster managers to evaluate, and many managers are reluctant to become locked into a system which may not be "the best" or which is produced by a company that may not be in business next year.

Information sources on the Internet have been promoted; however, bureaucratic structures, privacy/proprietary concerns, and "turf" issues continue to impede the necessary flow of information among emergency managers. The Global Disaster Information Network (GDIN) may break through some of these barriers. However, the internal struggle among federal agencies and the difficulty of defining a relationship with corporate partners will continue until clear leadership emerges and funding becomes available.

The "wrong end of the animal"

Big business, like big government, is multifaceted but not monolithic. Finding the right place or appropriate contact in a company is often time-consuming. Identifying the "common business interest" between a nonprofit organization and a corporation is often difficult. Researchers who are familiar with an industry and its members could assist disaster organizations in identifying the right connection. Interviews with business executives about their major concerns and how they would like to relate to government or nonprofit organizations would be valuable to mitigation and disaster operations. The more precise the refinement of common interests, the faster cooperation can be established. Talking to the wrong end of the animal slows progress and, when unsuccessful, often leaves a bad taste on both sides.

The insurance industry: a real partner?

Most public-private partnership presentations eventually lead to the insurance industry and to an example which seems so simple and implementable that one could ask, why is it not already done? When will there be discounts for mitigation measures, the same type of discounts as allowed for residential smoke/fire detectors, alarm systems, and using vehicle seat belts? Jerry Kanter, chief executive officer (CEO) for SCOR Reinsurance, at a Public-Private Partnership Forum 2000 provided a lot of information and answered the above question. Mitigation discounts are not available because insurance premiums are so artificially low that companies cannot discount them. In response to a question about partnering with insurance companies, he confirmed that individual insurance companies are only supporting activities which will most immediately result in reducing their losses or their costs.

Some insurance companies make major contributions to disaster relief work through their philanthropic elements; such donations bring them positive publicity for their generosity. Several companies have excellent disaster preparedness and family education brochures such as the Metropolitan Life Company's "Life Advice" series. More than 80 brochures have been published, and measuring their effectiveness in the community would be a valuable service to Met Life and to community educators attempting to design better means of educating families and communities.

The insurance industry's Institute for Business and Home Safety (IBHS), formerly the Institute for Property Loss Reduction (IIPLR), is recognized as the industry's mitigation leader. IBHS is the innovative "nonprofit arm" of the industry which initiates and supports creative demonstration projects. Partnerships with state and local jurisdictions, particularly in Florida, will become fertile research ground as these projects in community education, home strengthening, and other mitigation measures are implemented. Harvey Ryland, CEO of IBHS, has rallied an alliance of insurance and other industry leaders and creative emergency managers to test the best disaster mitigation ideas in the country.

The projects of the Institute for Business and Home Safety, just starting up at the end of the twentieth century, should be carefully evaluated, because they have the potential to revolutionize how corporations work with governments and communities. IBHS Showcase Communities and IBHS's "seal of approval" for mitigation measures are solid examples of how collaborative activities should be designed. Once implemented and evaluated, many lessons can be identified and applied to other communities.

Will high vulnerability lead to no insurance?

Initial research has identified places where additional vulnerability information could lead to higher insurance rates and fewer policies. Caribbean islands are the most prone to this possibility. When surveyed, some Caribbean insurance agents admitted that they underestimated the risks from hurricanes and earthquakes so that they could sell off most of their risk to reinsurers. Several insurance and reinsurance industry representatives admitted that, if the true risks of Caribbean disasters were known, either they would raise their rates so high that few could afford insurance or they would write no policies there. Some even speculated that no mitigation measures could alter vulnerability enough to encourage insurers to write more policies.

The Caribbean Disaster Mitigation Project (CDMP), created and managed by the Organization of American States (OAS) and funded by the U.S. Office of Disaster Assistance (OFDA) of USAID, should yield valuable insights about this subject. This innovative project brings industry, government, and community-based organizations together to develop ways to mitigate potential losses. As projects are developed and implemented, researchers can evaluate how these activities have helped people. Has this project reduced premiums and increased the availability of insurance, or has it revealed the true vulnerability which resulted in less insurance availability?

Business/industry losses: what is the real impact?

Economic analysis of business losses and potential losses could become a primary motivation for business and government mitigation and disaster cooperation. The Disaster Research Center (DRC) at the University of Delaware has conducted research on business loses from many U.S. disasters. Without this detailed and credible information about losses, it is difficult to motivate "bottom line" focused CEOs to look at their real risk. Positive examples of how industries protected themselves, how they reduced their losses, and how they were able to get back into operation quickly are needed to stimulate other corporate decision-makers to take similar actions.

The public-sector side of business losses are also important to evaluate. Business and industry depend on local government for many pre- and postdisaster actions. Poorly organized or uncoordinated local emergency management is a threat to business survival. Often the most poorly organized government emergency managers are also the most arrogant and least cooperative. Therefore, it is difficult for industry to determine the real extent of local emergency management's operational strength or competence. Few governmental jurisdictions have welcomed industry representatives (other than representatives from local utility companies) into their disaster planning process. One notable exception is Broward County, Florida, which has organized an Emergency Support Function (ESF) with 19 businesses and industries. Led by the county's Economic Development Council, the results of this initiative should be very interesting to other jurisdictions which have recognized the importance of including business and industry but have not yet developed the institutional linkages to implement it.

NOTE

1   This chapter was prepared prior to George W. Bush's election to the U.S. presidency and his appointment of Joseph Allbaugh to head the Federal Emergency Management Agency.

PART V
APPENDIX

SELECTED INTERNET RESOURCES ON NATURAL HAZARDS AND DISASTERS

David L. Butler

This list of hazard/disaster World Wide Web sites and other Internet resources is taken from the Web site of the Natural Hazards Research and Applications Information Center:

http://www.colorado.edu/hazards

Specifically:

http://www.colorado.edu/hazards/sites/sites.html

This index was compiled in the spring of 2001. Because information on the World Wide Web can be volatile, the interested reader is encouraged to consult the Web itself to determine up-to-date sources. Any hard copy list such as this inevitably is doomed to error as formerly healthy sites change locations or disappear entirely due to the vagaries and mysteries of the Internet. The Hazards Center page listed above is regularly updated and thus should provide a good starting point for any Internet search. Moreover, the sites listed below have been selected in part because of their relative stability and the depth of information provided. In most cases, they too represent good points within the World Wide Web to begin looking for information on a given hazard/disaster topic.

ALL HAZARDS

General hazards information

http://www.fema.gov

The Federal Emergency Management Agency (FEMA) Web site now contains thousands of pages of hazards/disaster information—text, graphics, photos, tables, maps—about the agency itself and its ongoing programs; current disaster situations; and disaster preparedness, response, recovery, and mitigation generally. The home page provides immediate access to news about recent and ongoing disasters, general news about policy and program developments, and much background information about both the agency and natural disasters. Among its *numerous* documents and services, the site provides "Fact Sheets"—including preparedness tips—concerning all kinds of hazards via **http://www.fema.gov/pte/prep.htm**. Indeed, there are hundreds of useful documents available from the FEMA library at **http://www.fema.gov/library/**.

FEMA has also posted a map of the United States that lists federally declared disasters for each state in a given year at **http://www.fema.gov/disasters**. That page links to descriptions of FEMA response and other information about these disasters.

The site's continually updated mitigation section (**http://www.fema.gov/mit/**) offers hundreds of pages on what individuals, families, and businesses can do to lessen disaster impacts. It includes current mitigation news, the latest reports from FEMA's Project Impact, links to mitigation documents available from FEMA, the complete text of the National Mitigation Strategy, and information about FEMA's HAZUS disaster loss estimation software. It also offers extensive sections on mitigation for homeowners, building professionals, communities, businesses, and school and childcare facilities (http://www.fema.gov/mit/reduce.htm), as well as FEMA's Mitigation "How To Series"—specific instructions for protecting property from wildfire, flooding, and earthquakes.

In addition, FEMA hosts such services as a Tropical Storm Watch Page (http://www.fema.gov/fema/trop.htm) with archived information about recent storms and, during the hurricane season, current weather photographs, forecasts, advisories, and situation reports.

The agency also now offers emergency management training materials and courses on-line (see http://www.fema.gov/emi/training.htm), as well as an emergency news distribution service via the Net. Information about both services is available from the Web site. FEMA's Emergency Management Institute (EMI) section also provides much of the information developed by the institute's "Higher Education Project" (http://www.fema.gov/emi/edu)—an effort to enhance and increase emergency management education among colleges and universities in the U.S. The project offers on-line indices of programs offered in the U.S., as well as complete materials for the many courses developed by educators around the U.S. for the project.

Finally, the site provides dozens of hypertext links to other Internet resources via its Global Emergency Management Service (GEMS) Page: http://www.fema.gov/gems/.

http://www.colorado.edu/hazards

The Web site of the Natural Hazards Research and Applications Information Center, University of Colorado, is designed to link academic, government, private, and nonprofit members of the hazards management community. Besides complete descriptions of the center's many services, it offers current and back issues of the

Natural Hazards Observer, Natural Hazards Informer, and *Disaster Research*—the center's print newsletter, topic-specific compendium, and e-mail newsletter—along with information on how to subscribe. It enables users to search HazLit—the Hazards Center on-line library database—and offers both an index of center publications as well as dozens of full-text, on-line papers and monographs. It also provides lists and indices of hazards/disaster-related organizations and institutions, publications, grants, conferences, and other sources of disaster information.

http://www.redcross.org

The American Red Cross provides extensive information on disaster mitigation, management, and recovery. In particular, a large collection of individual and community disaster preparedness, response, and recovery information is available from **http://www.redcross.org/services/disaster**. The material is available in many foreign languages.

http://www.usgs.gov

The U.S. Geological Survey (USGS) maintains many Web sites with much useful information on geologic hazards, including a Hazards Theme Page (**http://www.usgs.gov/themes/hazard.html**) with sections on earthquakes, landslides, volcanoes, floods, and coastal storms (some of which are listed below); as well as maps showing the distribution of hazards.

The USGS extensively monitors and evaluates threats from many natural hazards. Its resources include a global seismic network, a national streamflow monitoring program, regional volcano observatories, and long-standing interagency partnerships in disaster mitigation and response. To help synthesize the vast amount of information available, the USGS has created the Center for Integration of Natural Disaster Information (CINDI). The CINDI Web site—**http://cindi.usgs.gov**—provides background information about the center and serves as "a gateway to information about natural hazards and disasters."

The Survey now provides World Wide Web access to its Publications Database via http://usgs-georef.cos.com. The database includes comprehensive bibliographic information on USGS reports and maps published from 1880 to the present as well as references to non-USGS publications by USGS authors published from 1983 to date. The text of the Survey documents is not included; however, access to on-line publications of the USGS is provided. The database is completely searchable.

http://www.doi.gov/nathaz/index.html

The U.S. Department of the Interior has devoted one portion of its Web site entirely to natural hazards, with sections on wildfires, volcanoes, earthquakes, floods, landslides, wildlife diseases, geomagnetism, storms and tsunamis, and other hazards. For each topic, the site offers selected links—primarily to USGS Web pages—as well as a link to a "Fact Sheet" on the given subject.

http://www.ngdc.noaa.gov/seg/hazard/hazards.html

The National Geophysical Data Center (NGDC) Natural Hazards Data Web site contains databases, slide sets, and publications available from NGDC on geophysical hazards such as earthquakes, tsunamis, and volcanoes, and includes the *Natural Hazards Data Resources Directory* (http://www.ngdc.noaa.gov/seg/hazard/resource) published jointly with the Natural Hazards Research and Applications Information Center.

http://www.nesdis.noaa.gov
http://ns.noaa.gov/NESDIS/NESDIS_Home.html

NOAA's National Environmental Satellite, Data, and Information Service (NESDIS) maintains many tools and assets for observing and analyzing hazards; conducts several programs for detecting, monitoring, responding to, and mitigating hazards; and offers numerous resources (primarily via the Web) to educate the general public about hazards. In addition, NESDIS manages extensive databases concerning historical and current disaster events. The URL above provides an entrée to this great resource of information.

http://www.haznet.org

In 1998, the National Sea Grant network created HazNet, a Web site devoted to coastal hazard awareness and mitigation. The HazNet site gathers information and resources from Sea Grant programs, the National Oceanographic and Atmospheric Administration, and other public- and private-sector sources to help people meet the challenges presented by such natural hazards as hurricanes, flooding, storm surge, coastal erosion, earthquakes, tsunamis, tornadoes, and even volcanoes. Besides providing information about all these hazards, the site lists resources under the headings of Mitigation and the Built Environment, Mitigation Policy/Planning, Hazards Communication/Education, and Hazards Bibliography. The site also supports a bulletin board and discussion group and archives much other data and educational information.

http://www.weather.com/safeside

Project Safeside is a joint effort of The Weather Channel and the American Red Cross intended to educate individuals and families about meteorological hazards and to increase their recognition of the importance of preparing for natural disasters. The Safeside Web site includes information about extreme heat, flooding, hurricanes, lightning, and tornadoes, plus a guide to the creation of a family disaster plan. Besides the Web site, the project offers other information and tools to help educators incorporate weather safety and family preparedness into existing weather curricula.

http://www.sustainable.doe.gov

The Department of Energy's Center of Excellence for Sustainable Development redesigned Web site includes an extensive section on disaster planning that contains segments on key principles, case studies, codes/ordinances, articles/publications (lots of good ones), educational materials, and other resources. These pages offer information on how long-term community sustainability can be incorporated into disaster preparedness, mitigation, and recovery.

http://coe.tamc.amedd.army.mil/

The Center of Excellence in Disaster Management and Humanitarian Assistance provides education, training, and research regarding civil-military operations, particularly those involving international disaster management and humanitarian assistance. COE is a partnership of the United States Pacific Command (USPACOM), the Pacific Health Services Support Area (HSSA) of Tripler Army Medical Center, and the University of Hawaii. Focusing on the Asia-Pacific region, the center conducts needs assessments, program development and evaluation, curriculum development, conferences, training programs, and research. COE is also collaborating with the University of Hawaii to offer Master's-level classes and eventually a degree in humanitarian assistance. The COE Web site features selected links to other disaster-related Internet sites, disaster-related news and weather reports, historical data, Pacific Rim disaster-related information, information on current disasters, electronic journals and newsletters, discussion and e-mail groups, fully searchable full-text publications including country-specific disaster management handbooks and plans, as well as considerable information about the center itself. The center's e-mail discussion lists cover such topics as disaster medicine, disaster communications, disaster management, and disaster terrorism. Information on subscribing to these groups is available from the Web site.

National associations

http://www.iaem.com

The Web site of the International Association of Emergency Managers (IAEM—formerly the National Coordinating Council on Emergency Management) includes information about the association, its mission, and its Certified Emergency Manager program; details about IAEM conferences; a "Topic of the Month" section; lists of IAEM partners and experts; news regarding current issues in emergency management; and copious links to other emergency-management-related sites. The site now also provides on-line registration for the IAEM e-mail list dedicated to the discussion of all hazards and emergency management topics (iaem-list@asmii.com).

http://www.nemaweb.org

The National Emergency Management Association site includes information on NEMA's history, publications, committees, and membership, as well as lists of upcoming conferences, information on regional communications, updates on current legislation and other federal issues, and state contact information. Of particular interest is NEMA's extensive Hazard Mitigation Grant Program (HMGP) database which tracks all federally funded Hazard Mitigation Grant Program projects. This database is intended to help state hazard mitigation officers share ideas and design practicable hazard mitigation projects.

A few state resources

http://www.ag.uiuc.edu/~disaster/disaster.html

The University of Illinois Cooperative Extension Service (CES) Disaster Services offers a broad spectrum of information—on current disasters, disaster preparedness and recovery, other agencies and networks dealing with disasters, and disasters generally.

http://www.ext.vt.edu/pubs/disaster/disaster.html

The Virginia Cooperative Extension's Home Page offers this "After a Disaster Series of Publications"—a wide range of useful material on postdisaster recovery—30 publications in all, grouped into seven categories: safety, food and water, coping with stress, cleaning, insurance and contracts, landscape and agriculture, and roof repairs.

http://www.ceres.ca.gov

The California Environmental Resources Evaluation System (CERES) is an information system developed by the California Resources Agency to facilitate access to a variety of electronic data describing the region's rich and diverse environments. The goal of CERES is to improve environmental analysis and planning by integrating natural and cultural resource information from multiple contributors and by making it available and useful to a wide variety of users. The site provides much information that is useful beyond California; it includes excellent sections on earthquakes (**http://www.ceres.ca.gov/theme/earthquakes.html**), the El Niño phenomenon (**http://www.ceres.ca.gov/elnino**), wildfire (**http://www.ceres.ca.gov/theme/fire.html**), and floods (**http://www.ceres.ca.gov/flood**).

Some international and overseas sites

http://www.disasterrelief.org

The Disaster Relief home page—a joint effort of the American Red Cross, the IBM corporation, and CNN—offers much background information about disasters, disaster relief, and disaster preparedness, as well as news about ongoing and recent events. Moreover, it provides a means for locating worldwide disaster relief organizations and either soliciting or providing aid for specific disasters. During emergencies, it can provide referrals to means for reaching friends and family at risk, as well as referrals to sources of recovery assistance and support. As the organizers state, "Our mission is to help disaster victims and the disaster relief community worldwide by facilitating the exchange of information on the Internet"—and this includes services during actual events. The site also includes an on-line "Forum" for discussing relief issues and a library of disaster facts, figures, and other information.

http://www.reliefweb.int

ReliefWeb is a site maintained by the United Nations Office for the Coordination of Humanitarian Affairs (OCHA) intended to aid national agencies and nongovernmental organizations involved in emergency and disaster relief worldwide. The site addresses prevention, preparedness, and response, and includes country and emergency profiles, a bulletin section with daily updates, a "What's New" feature that directs the reader to recently added information, and various maps of countries and regions where emergency operations are currently underway.

http://www.md.ucl.ac.be/cred/

The Centre for Research on Epidemiology of Disasters (CRED) at the School of Public Health, Catholic University of Louvain, Brussels, Belgium, maintains one of the Web's more comprehensive databases of disaster information. As the creators of this resource state, "In recent years, natural and man-made disasters have been affecting increasing numbers of people throughout the world. Budgets for emergency and humanitarian aid have sky-rocketed. Efforts to establish better preparedness for and prevention of disasters have been a priority concern of donor agencies, implementing agencies and affected countries. For this reason, demand for complete and verified data on disasters and their human and economic impact, by country and type of disaster has been growing. Planners, policy makers, field agencies engaged in preparedness have all expressed need for data for their work. The CRED/OFDA initiative responds to this need by making available a specialized, validated database on disasters that facilitates preparedness, thereby reducing vulnerability to disasters and improving disaster management." The site includes background information; a "what's new" update section; the searchable database covering over 10,000 disasters; "disaster profiles" (now including data on epidemics) presented in three sub-sets: "top 10," "chronological table," and "raw data," and grouped according to country, region, world, and disaster type; summary data; maps; a bibliographic database; and links to other useful sites. Additionally, a country-by-country database compiled by the U.S. Office of Foreign Disaster Assist and CRED is available via the United Nations ReliefWeb site: **http://www.reliefweb.int** [see above].

http://www.worldbank.org/dmf/
http://www.proventionconsortium.org/

Because of its increasing awareness of the effects of disasters on development (and development on disasters), in July 1998 the World Bank established a new Disaster Management Facility (DMF) to ensure that disaster prevention and mitigation are integral parts of development programs. The DMF Web site includes much information on World Bank policy and projects in disaster management, good practices in disaster risk management, market incentives for mitigation investment, recent disasters, World Bank Publications, current disaster news, key readings, and useful links.

As part of its disaster management efforts, the bank has created the ProVention Consortium, whose mission is "to help developing countries build sustainable and successful economies and to reduce the human suffering that too often results from natural and technological catastrophes." More information about specific goals and projects is available from the Web site above.

The bank's Web site offers other pages addressing hazards. For example, see **http://wbln0018.worldbank.org/disastmgmtteam/ disastmgmtteamopenar.nsf/Homepage/Homepage/ OpenDocument**—the section on "Disaster Management and

Mitigation in Latin America and the Caribbean"—with information on all the World Bank disaster recovery projects in that region, as well as news about recent and ongoing disasters.

http://www.unisdr.org/

To continue the efforts initiated during the International Decade for Natural Disaster Reduction (IDNDR—1991-2000) the United Nations has established an International Strategy for Disaster Reduction (ISDR), managed by a small U.N. Secretariat in Geneva. This new Web site provides extensive background information on the IDNDR/ISDR; a list of ISDR and ISDR-related events; on-line versions of the *ISDR Highlights* newsletter; descriptions of various ISDR initiatives; and numerous reports, tools, brochures, and U.N. documents.

http://hoshi.cic.sfu.ca/~anderson/

The Emergency Preparedness Information Exchange (EPIX) was one of the original Internet disaster sites and remains one of the most comprehensive. Established by the Centre for Public Policy Research on Science and Technology, Simon Fraser University, British Columbia, Canada, EPIX contains extensive information about both current situations and disaster management generally.

http://165.158.1.110/english/ped/pedhome.htm

In the fall of 1998, the Pan American Health Organization's Emergency Preparedness and Disaster Relief Coordination Program completely redesigned its Web site. It now provides an overview of the program; a section on "Dealing With Disasters" that includes discussion groups, donation guidelines, an index of contact persons in Latin America and the Caribbean, and situation reports; a catalog of publications including full-text documents and newsletters from PAHO; a link to the Regional Disaster Center in San Jose, Costa Rica (see below) and its extensive database (11,000 documents) of hazards/disaster literature; information on PAHO's Humanitarian Supply Management System (SUMA); a section entitled "Technical Guidelines" that answers many questions about the provision of public health in emergencies (this section includes contact information for experts in various public health areas); and extensive links to other disaster management resources on the Internet. There is also a special section on El Niño; links to World Health Organization Collaborating Centers; the latest issue of the Disaster Section's excellent newsletter, *Disasters: Preparedness and Mitigation in the Americas*; and the findings from several PAHO/WHO conferences on disasters. All information is provided in English *y en español* (**http://165.158.1.110/spanish/ped/pedhome.htm**).

In the fall of 1999, PAHO announced another new Web enhancement—its Virtual Disaster Library (VDL) also at **http://165.158.1.110/english/ped/pedhome.htm**. The VDL is an on-line collection of hundreds of publications in English and Spanish on disasters along with a powerful yet simple search engine that helps one quickly locate information.

http://www.netsalud.sa.cr/crid

The Regional Disaster Information Center (Centro Regional de Informacion Sobre Desastres—CRID) for Latin America and the Caribbean is a multiorganizational project, housed in San Jose, Costa Rica, and supported by several international agencies. The main objective of the center is to facilitate access to technical and scientific information to improve disaster management in all countries of Latin America and the Caribbean. Specific objectives are to improve and broaden the collection, processing, and dissemination of disaster reduction information in the region; strengthen regional, national, and local capability to establish and maintain disaster information and documentation centers; promote communication through the Internet and develop electronic information services; and contribute to the development of a Regional Disaster Reduction Information System. CRID provides bibliographic searches, either via the Internet, by CD-ROM, or through direct contact with the center; publishes bibliographic material; offers direct access via the Internet to an extensive collection of technical documents in full text (in both Spanish and English); publishes original documents and training materials; and offers many other services.

http://www.epc-pcc.gc.ca

Emergency Preparedness Canada (EPC) (which, in 2001, is being reconstituted as the Canadian Office of Critical Infrastructure Protection and Emergency Preparedness) offers emergency management and preparedness information in English or *en francais*. The site also provides access to the SAFE GUARD Project Web site: **http://www.safeguard.ca**; SAFE GUARD is Canada's public/private cooperative effort to increase public awareness of emergency preparedness. Finally, one can also access the Canadian Emergency Preparedness Association's Web site—**http://www.cepa-acpc.ca**—from the EPC site.

http://www.ema.gov.au

In 2001, Emergency Management Australia jazzed up its Web site with a new look and new information. The site includes a section describing the agency's programs and structure as well as pages covering current EMA activities, EMA media releases, and emergency management generally. It offers a virtual library, community information, a summary of available education and training, and a list of conferences. It now also provides an extensive section on "Disaster Education for Schools," with pages for teachers, students, and school communities, as well as a news section and an index of school disaster education resources—from Web sites to books and videos. The site is linked to the Australian Natural Hazards Research Center (**http://www.es.mq.edu.au/nhrc/index1.html**) and its Natural Hazards Research and Researcher Database (**http://www.es.mq.edu.au/nhrc/index1.html**)—a fully searchable on-line index that includes synopses, complete contact information, and other details regarding various hazards research projects in the Australia-South Pacific region.

http://www.mem.govt.nz/MEMwebsite.nsf

The New Zealand Office of Civil Defense deals with natural and technological hazards and emergencies—providing national coordination and a range of support for local government and other emergency services. The office's Web site describes the agency, its programs, and available training; offers tips on personal preparedness, as well as an overview of New Zealand hazards and disasters (particularly volcanoes); provides updates on ongoing emergencies; and furnishes an on-line version of the office's excellent periodical, *Tephra Magazine* (http://www.mem.govt.nz/MEMwebsite.nsf/URL/Publications-TephraMagazine).

http://www.adpc.ait.ac.th/Default.html

The Asian Disaster Preparedness Center (ADPC), in Bangkok, is a regional center committed to the protection of life, property, and the environment in Asia and the Pacific. It assists local, regional, and national governments in developing their capabilities and policies through training, information provision, and technical assistance to mitigate the impact of disasters. The center Web site includes pages about the ADPC, its information and research programs, learning and professional development programs, the Asian Urban Disaster Mitigation Program (AUDMP), its international consultancies and alumni coordination efforts, and its disaster network. The section on "Information, Research and Network Support" offers pages providing disaster information resources (categorized as hazard specific information, country specific information, disaster organizations, and reference resources), the center's newsletter—*Asian Disaster Management News*—several ADPC on-line documents, a description of ADPC library services, and now a searchable, annotated database of the ADPC library holdings.

INDICES OF WEB SITES

http://www.colorado.edu/hazards/sites/sites.html

This listing from the Natural Hazards Center is the basis for this appendix. It is regularly updated and thus should provide a good starting point for any Internet search.

http://ltpwww.gsfc.nasa.gov/ndrd
http://ltpwww.gsfc.nasa.gov/ndrd/disaster/

The Natural Disaster Reference Database Home Page, assembled by the Earth Sciences Directorate of the NASA Goddard Space Flight Center in Greenbelt, Maryland, contains extensive bibliographic information on research results and programs related to the use of satellite remote sensing for disaster mitigation of all kinds. The Disaster Finder, at the second URL above, is "a complete index to the best disaster Web sites on the Internet." It covers hundreds of disaster information sites, and, using a keyword/concept search facility or category/type menu buttons, users can quickly identify specific sites that provide the information they need. The sites found by using the search engine are prioritized according to their probable suitability, and the Disaster Finder even provides short previews of the selections so that individuals can see what kind of information is available.

http://www.disasterlinks.net

This site is just what its name implies—dozens, if not hundreds, of links to disaster Web sites arranged in approximately 30 categories (from "Satellite Images" to "Icebergs")—brought to you by CBS News.

http://www.yahoo.com http://www.google.com etc.

Yahoo and the many other search facilities now available on the World Wide Web are invaluable resources for discovering hazards information. Yahoo includes a section (**http://www.yahoo.com/ Environment_and_Nature/Disasters**) devoted to disasters that interested persons can consult to determine the latest Net resources.

EARTHQUAKES

http://www.eqnet.org

EQNet is a collaborative effort of many of the institutions providing earthquake information in the U.S. It is a free, one-stop source for locating Internet information related to earthquakes and earthquake hazards.

http://earthquake.usgs.gov

The U.S. Geological Survey provides much earthquake information via dozens of different Internet avenues (a few of which are listed below). The amount of information is daunting, and, in the past, that abundance, along with the multiplicity of USGS sites offering information, has sometimes made it difficult to locate and sort out information. To remedy that problem, the Survey has launched this Earthquake Hazards Program site—"Earthquake Hazards on the Web"—as an entry point for all USGS earthquake information. It provides information on both global and regional earthquake hazards and on earthquake activity past, present, and future; earthquake education for children, grownups, and teachers; earthquake products such as maps, publications, fact sheets, videos, etc.; earthquake research; the USGS Regional Centers and regional Web sites; seismic networks; and frequently asked questions about quakes.

http://www.usgs.gov/themes/earthqk.html

The Survey's Theme Page on earthquakes provides information about the many USGS programs and products dealing with seismic hazards as well as a "Hazard-Related Fact Sheet" on earthquakes.

http://quake.wr.usgs.gov/ http://geohazards.cr.usgs.gov

The USGS Regional Centers each offer Web pages on seismic hazards. The Western and Central Regions at the URLs above are good examples. The Western Region Earthquake Hazards Information home page is an excellent place to begin any search for seismic information. It includes pages on the latest seismic events, earthquake hazard preparedness, and all other aspects of earthquakes. It also has an entire section devoted to the 1906 San Francisco earthquake and an extensive annotated list of other Web quake sites.

The Central Region's Geologic Hazards Page covers earthquakes, landslides, and geomagnetism. The earthquake section (**http://geohazards.cr.usgs.gov/eq**) offers numerous products related to the USGS national seismic hazard mapping program. For example, users can look up the seismic hazard in any part of the continental U.S. by zip code, and the section also includes a custom mapping feature through which the user can specify latitude and longitude bounds and produce customized hazard maps of the selected area. Additionally, large versions (24"x36") of the national and western U.S. seismic hazard maps can be ordered using forms available from the Web site.

http://wwwneic.cr.usgs.gov

The USGS's National Earthquake Information Center Web Site comprises pages and pages, maps and maps of seismicity information from around the world. It offers general information about the center and its services, current quake information, general quake information, and access to other earthquake information sources. In addition, users can now search the National Earthquake Information Services (NEIS) historical database to identify historical seismic events (2100 B.C. to the present) for any location, using several user-defined parameters. The site also provides a means for users to report a quake, and a "Products and Services" section that offers an earthquake e-mail notification service, several USGS publications on quakes, and numerous NEIC maps portraying local, national, and global seismicity.

http://peer.berkeley.edu

The Pacific Earthquake Engineering Research (PEER) Center brings together the premier earthquake engineering research universities in the western U.S. with a common objective of developing technologies and strategies to reduce the human and economic risks due to major earthquakes. Center researchers have expertise in diverse areas including earthquake hazards, analysis, design, risk and reliability, and economics and policy planning. The center was established with primary funding from the National Science Foundation and matching funds from the participating states and industry partners. The PEER Web site includes sections describing PEER, its various research programs and studies, PEER's extensive education programs, its "Business & Industry Partners" program, the PEER newsletter, and a link to the NISEE Web site described below, which provides information services for PEER.

http://www.eerc.berkeley.edu/

The Earthquake Engineering Research Center (EERC)/National Information Service for Earthquake Engineering (NISEE) Web site includes links to all EERC/NISEE databases searchable on the Web: *Earthquake Engineering Abstracts*; *Engineering Reports*; computer software for earthquake engineering; protective systems; training resources (demo database). The site also offers numerous papers, reports, and visual images of recent major events and historical quakes of the twentieth century.

http://mceer.buffalo.edu
http://mceer.buffalo.edu/infoService/default.asp

The Multidisciplinary Center for Earthquake Engineering Research (formerly the National Center for Earthquake Engineering Research)

Home Page offers a wide variety of resources, as well as links to other Internet information on earthquake engineering and natural hazards mitigation. The main menu features an interactive connection to MCEER's Quakeline database, background information on MCEER, a list of MCEER technical reports with ordering information, a comprehensive list of upcoming conferences, and access to many of the other information resources and reference services available from the center. In 1998 MCEER began offering "Express News" (*ENews*— **http://mceer.buffalo.edu/infoService/enews/default.asp**) via this site. *ENews* is a customized electronic service that, based on a reader's self-defined interest profile, alerts readers via e-mail to the publication of selected earthquake/hazards information in the most recent issue of *MCEER Information Service News*.

http://mae.ce.uiuc.edu

The Mid-America Earthquake (MAE) Center is a multidisciplinary center of expertise whose aim is to reduce losses in future earthquakes that may strike the central and eastern United States. It supports coordinated research programs on essential facilities and transportation networks in order to acquire the data needed to determine possible earthquake risks and to develop improved strategies for seismic retrofit of constructed facilities. Center research is directed at developing effective and economical mitigation measures considering the unique physical, technical, social, and economic features of the earthquake problem in mid-America. Center programs in education, outreach, and implementation provide the necessary links to extend research results to the public and professional communities.

With PEER and MCEER (described above), the MAE Center is one of three national earthquake engineering research centers established by the National Science Foundation and its partner institutions. The MAE Center consists of a consortium of seven core institutions and is funded by NSF and each core university as well as through joint collaborative projects with industry and other organizations. Center projects fall under four general types: (1) research, (2) implementation of research results, (3) education, and (4) outreach. The center's Web site offers more information about the organization, its goals and products, each of its core programs (coordinated research, essential facilities, transportation networks, hazards evaluation, outreach, and education), as well as recent news from the center.

The common Web site of the three National Science Foundation–sponsored Engineering Research Centers (ERCs) is **http://www.erc-assoc.org**.

http://www.eeri.org

The Earthquake Engineering Research Institute (EERI) is a national, nonprofit, technical society of engineers, geoscientists, architects, planners, public officials, and social scientists—including researchers, practicing professionals, educators, government officials, and building code regulators—all of whom are concerned with mitigating earthquake hazards. The EERI Web site provides an introduction to the institute; a list of upcoming EERI meetings and other events; descriptions of EERI services; a catalog of publications, slides, and videos available from the institute; and other information and news about earthquake hazard mitigation. Additionally, EERI now has several full-text earthquake reconnaissance reports on-line, as well as other information about current events. EERI has established an e-mail listserve to apprise members and other persons interested in earthquake hazard mitigation of special EERI announcements. To subscribe, send an e-mail to **subscribe@eeri.org**, and in the body write, "subscribe eeri-announce." Persons can also subscribe through the EERI Web site.

http://www.scec.org

The Southern California Earthquake Center (SCEC) is a Science and Technology Center of the National Science Foundation that brings scientists together for joint research to reduce vulnerability to earthquake hazards in southern California. The formal mission of the center is to promote earthquake hazard reduction by estimating when and where future damaging earthquakes will occur, calculating their expected ground motion, and disseminating that information to the public. The SCEC home page contains background information about SCEC and links to its many member academic institutions. It includes sections on the center's research programs (with links to databases, publications, and other information); communication, education, and outreach (including maps and other products and publications); and access to *SCEC InstaNET News*, the center's Web and e-mail news service.

http://www.cusec.org/

The Central United States Earthquake Consortium (CUSEC) Home Page provides an overview of the agency and the state agencies and geologists participating in the consortium. It also includes a list of events and programs, describes CUSEC products and services, and provides a catalog of publications (including all Federal Emergency Management Agency publications related to the National Earthquake Hazards Reduction Program). It also provides background information about the New Madrid Seismic Zone earthquake hazard, safety information, useful Internet links, and an on-line version of the consortium's *CUSEC Journal*.

http://www.geohaz.org http://www.geohaz.org/radius.html

One of the major initiatives of the recently completed United Nations International Decade for Natural Disaster Reduction (IDNDR) addressed the issue of reducing seismic risk in large cities of the developing world. The RADIUS (Risk Assessment Tools for Diagnosis of Urban Areas Against Seismic Disaster) Project has resulted in much helpful information and some useful tools for evaluating urban earthquake hazards and planning for their mitigation. GeoHazards International, a consulting firm centrally involved in RADIUS, has provided extensive information about the project via its Web site. Included are a description and outline of the project, case studies from around the world, guidelines for developing local RADIUS-type risk management projects, a description of a "Tool for Earthquake Damage Estimation" developed for RADIUS and available on CD-ROM, a comparative study of the RADIUS initiative around the world, and a project evaluation. The GeoHazards site provides information about subsequent international seismic hazard mitigation efforts such as the Global Earthquake Safety Initiative.

http://incede.iis.u-tokyo.ac.jp/Incede.html

The International Center for Disaster-Mitigation Engineering (INCEDE) site offers general information about INCEDE; abstracts of all INCEDE reports and back issues of INCEDE newsletters; summaries of recent research, including an extensive report on the Kobe earthquake; and other sections on recent disasters, hazard mitigation, severe weather, hydrology, remote sensing, etc.

http://www.abag.ca.gov/bayarea/bayarea.html

The Association of Bay Area Governments (ABAG) maintains a state-of-the-art Web site that includes a series of colorful maps depicting potential earthquake effects in the Bay Area. A person can choose not only a specific locale in the area but also a specific earthquake source (i.e., a specific fault) and then view the consequences of the given scenario. The site includes additional information about the maps, about earthquake hazards in Northern California, and about seismic hazard mitigation generally.

http://www.ce.washington.edu/~liquefaction/html/main.html

This Soil Liquefaction Web site was developed to provide general information for interested lay persons, and more detailed information for engineers, on this seismic phenomenon. Visitors who are not familiar with soil liquefaction can find answers to such typical questions as: What is soil liquefaction?; When has soil liquefaction occurred in the past?; Where and why does soil liquefaction commonly occur?; How can soil liquefaction hazards be reduced? For each question, more detailed information is provided separately for earthquake and engineering professionals. The site is well illustrated with photographs and animated graphics and includes links to much additional information on liquefaction and earthquakes in general.

TSUNAMIS

http://www.geophys.washington.edu/tsunami/welcome.html

This site, entitled "Tsunami!—An On-Line Interactive Resource of Tsunami Information," contains information about the mechanisms of tsunami generation and propagation, great tsunamis in history, the impact of tsunamis on humankind, tsunami warning systems, and tsunami hazard mitigation. The site also includes more detailed material about recent tsunami events and ongoing studies that will be of interest to tsunami and interdisciplinary researchers.

http://www.pmel.noaa.gov/tsunami/
http://www.pmel.noaa.gov/tsunami-hazard

The Pacific Marine Environmental Laboratory (PMEL) Tsunami Program was created to mitigate tsunami hazards affecting the Pacific Coast, Alaska, and Hawaii. The PMEL Web site includes sections on field observations, modeling and forecasting, tsunami events and data, inundation mapping, and (at the second URL above) the National Tsunami Hazard Mitigation Program, a joint effort of a consortium of state and federal agencies. That section outlines the five goals of the program:

- Develop State/NOAA Coordination and Technical Support
- Deploy Tsunami Detection Buoys
- Produce Inundation Maps

- Develop Hazard Mitigation Programs
- Improve Seismic Networks

It also reviews work to date, provides the Tsunami Hazard Mitigation Plan and a brochure describing the Tsunami Hazard Mitigation Program, displays real-time tsunami information, offers information for children, and lists numerous links to other sites with tsunami information.

http://observe.ivv.nasa.gov/nasa/exhibits/tsunami/tsun_start.html

The National Aeronautics and Space Administration (NASA) offers this on-line guide to tsunamis with information about the basic science of tsunamis, tsunami warning, what to do when a tsunami strikes, and great historical tsunamis.

http://walrus.wr.usgs.gov/tsunami

The U.S. Geological Survey Western Region Web site offers information on tsunami research at the USGS, as well as basic background information and on-line tsunami animations.

http://www.ngdc.noaa.gov/seg/hazard/tsu.shtml

This page, "Tsunami Data at NGDC," provides access to the considerable information on tsunamis available from the National Geophysical Data Center—including NGDC's Tsunami Database, which encompasses about 6,600 records of tsunami events from 49 B.C. to the present, tide gage records for over 3,000 tsunamis in the last 150 years, tsunami slide sets, and tsunami publications.

http://wcatwc.gov http://www.tsunami.gov

The home page of the West Coast and Alaska Tsunami Warning Center provides the most recent press releases, advisories, watches, and warnings from the center; messages from and links to other warning centers around the Pacific; and links to earthquake catalogs, tsunami catalogs, and recent historical data on tsunamis. It also describes the center and provides background information on the physics of tsunamis, tsunami safety, and the Great Alaskan Earthquake and Tsunami of 1964.

http://www.nws.noaa.gov/pr/ptwc

Via the Web, the Pacific Tsunami Warning Center offers current bulletins (including estimated times of arrival and maps), as well as historical information concerning tsunamis of the region.

http://www.shoa.cl/oceano/itic/frontpage.html
http://www.shoa.cl/oceano/itic/itsu.html

The home page of the International Tsunami Information Center (ITIC) describes the work of the center and offers both current and historical information about tsunamis, as well as an on-line Post Tsunami Survey Field Guide. The related home page of the International Coordination Group for the Tsunami Warning System in the Pacific (ICG/ITSU) presents background information on tsunamis and tsunami warnings and a description of the Pacific Tsunami Warning System.

http://www.nerc-bas.ac.uk/tsunami-risks/

The Tsunami Risks Project, based in the U.K., was launched to introduce the British insurance industry to tsunamis and the risks they pose and to quantify tsunami hazards by developing frequency-magnitude distributions and evaluations of direct and indirect insurance risks. The project is examining subjects ranging from how tsunamis are generated and how they propagate across the oceans; to the mechanisms by which they cause damage when they make landfall; to the means by which disaster planning can reduce the economic losses that result; and to the sources of postdisaster information and mapping that can be consulted to validate tsunami-related insurance claims. The project's Web site provides details about this initiative, as well as an interactive map with accompanying articles about historic tsunami disasters of the world; a "Risk Atlas"—another interactive map showing tsunami risk around the world; several complete publications; a bibliography; and an index of related Web sites.

LANDSLIDES

http://landslides.usgs.gov/
http://landslides.usgs.gov/html_files/nlicsun.shtml
http://landslides.usgs.gov/html_files/landslides/nationalmap/national.html
http://www.usgs.gov/themes/landslid.html

The landslide Web pages of the U.S. Geological Survey and the Web site for the National Landslide Information Center (NLIC) offer an avalanche of landslide information, as well as indexes to landslide publications available both in hard copy and on-line. The first URL describes the National Landslide Hazards Program, lists landslide program publications and current projects, and describes recent landslide events. The NLIC site provides "real-time" monitoring of an active landslide in California, San Francisco Bay area landslide maps, links to landslide information for each state, landslide images, other useful links, a virtual fieldtrip of a Colorado landslide, and access to a new on-line bibliographic database. A particularly nifty feature, at the third URL above, is an interactive map of landslide hazards across the United States which can be viewed on-line or downloaded in more detail. The USGS's "theme page" on landslides provides more information about USGS landslide programs, as well as a link to a landslide "Fact Sheet" that lists Survey publications in this area.

http://ilrg.gndci.pg.cnr.it/

The International Landslide Research Group (ILRG) is an informal group of individuals concerned with mass earth movement and interested in sharing information on landslide research. The ILRG

Web site currently provides all back issues of the group's newsletter with information about landslide programs, new initiatives, meetings and publications, the experiences of people engaged in landslide research, and "any other information about landslide research that 'normal' journals will not accept." The site also hosts several bibliographies and reference lists.

VOLCANOES

http://volcanoes.usgs.gov/

The USGS Volcano Hazards Program Page addresses the whole range of U.S. Geological Survey programs to understand volcanoes and mitigate volcanic hazards. The site includes sections on volcano hazards, active U.S. volcanoes, reducing volcano risks, resources (including several complete texts, as well as catalogs of many other on-line, print, and other media resources), and USGS work overseas. It also provides links to the USGS regional observatories:

http://www.avo.alaska.edu

The Alaska Volcano Observatory (AVO) site offers information about aviation and volcanic ash hazards and other volcanic risks with sections covering highlights of past eruptions, current events, videos available from the center, and lots of photos and hypertext links to other volcano Web sites.

http://vulcan.wr.usgs.gov/home.html

The Cascades Volcano Observatory (CVO) site includes sections covering news and current events, including a Mount St. Helens update; information about CVO programs and research; background information about the volcanoes of the Pacific Northwest; information on the associated hazards, with reports, maps, graphics, charts, and tables; a photo archive; a list of publications available on-line; information and materials to support educational programs; and much other volcano information.

http://hvo.wr.usgs.gov/

The Hawaiian Volcano Observatory (HVO) pages are entitled "What's New at Kilauea Volcano?," "USGS Publications about Hawai'i or Monitoring of Hawaiian Volcanoes," "Volcano Watch and Archives," "Recent Earthquakes," and, "About Hawaiian Volcanoes."

http://quake.wr.usgs.gov/VOLCANOES/LongValley

The Long Valley Volcano Observatory site presents background information on the USGS effort to monitor volcanic and seismic activity in that area of California. It includes a map of the region, a summary of current conditions, a long-term outlook, on-line monitoring data, the Long Valley Caldera Response Plan, a brief geologic history of the caldera, and selected references and fact sheets about this volcanic hazard.

http://www.usgs.gov/themes/volcano.html

The U.S. Geological Survey also provides this Theme Page on volcanoes, with links to numerous other USGS Web resources, as well as a link to a "Fact Sheet" on volcanoes.

http://www.geo.mtu.edu/volcanoes/

Michigan Tech's volcanoes page offers a worldwide volcano reference map and information on recent events, volcanic hazards mitigation, and remote sensing.

http://www.dartmouth.edu/~volcano/

The Electronic Volcano page bills itself as "a window into the world of information on active volcanoes." The site is a source of many materials on active volcanoes worldwide—such as maps, photographs, and full texts of dissertations. It offers introductory material in Chinese, German, Spanish, Italian, French, and Russian, and then provides guides to catalogs of active volcanoes and volcano literature. The Electronic Volcano also includes lists of volcanic observatories and institutions, abstracts and excerpts from theses related to active volcanism, descriptions of volcanic hazards, a section on current events and research, and a volcano name and country index.

http://volcano.und.nodak.edu

The University of North Dakota's Volcano World site is an encyclopedic compendium of volcano information for everyone from the first grade aa-phile to the veteran volcanologist.

http://www.iavcei.org

The International Association of Volcanology and Chemistry of the Earth's Interior (IAVCEI) Web Site presents information about the association's structure, purpose, and programs and about many of the association's members. It also offers a publication list, a list of safety recommendations for volcanologists and the general public, a list of upcoming conferences, and numerous links to other volcanology sites on the Web.

http://www.volcano.si.edu/gvp/

The Smithsonian Institution's Global Volcanism Program (GVP) is dedicated to better understanding of all volcanoes through documentation of eruptions, large and small, during the past 10,000 years. The program integrates observations of contemporary activity with historical and geological records in order to promote wise preparation for the future. This Web site includes a database of volcanoes of the world, an on-line version of the program's *Bulletin of the Global Volcanism Network* (reports of ongoing eruptions from local observers around the world), a catalog of other GVP products, information about the program staff, and a well-organized set of links to other volcano Web sites.

CLIMATE CHANGE, DROUGHT, AND EL NIÑO

Climate change

http://www.ncdc.noaa.gov

The Web site of the National Climate Data Center is the climate/weather researcher's Shangri-la. It includes data from thousands of weather stations around the world, as well as hundreds of images, numerous technical reports on extreme weather events, and lots of other climate/weather data.

http://www.cip.ogp.noaa.gov

NOAA's Office of Global Programs supports a Climate Information Project (CIP) Web site that offers near-daily and weekly summaries of reported climatological impacts from around the globe. To be added to the project's electronic mailing list, make requests or comments, or for further information, contact the Office of Global Programs, NOAA 1100 Wayne Avenue, Suite 1210, Silver Spring, MD 20910; (301) 427-2089, ext. 194; fax: (301) 427-2073.

http://www.cpc.ncep.noaa.gov/
http://www.cpc.ncep.noaa.gov/products/predictions/threats/threats.html

The mission of the National Oceanic and Atmospheric Administration's Climate Prediction Center is to watch, diagnose, and predict climate fluctuations in order to assist agencies both inside and outside the federal government in coping with such climate-related issues as food supply, energy allocation, water resources, and emergency management. The center's Web site offers much information and many products in support of this mission, including a "U.S. Threats Assessment" page at the second URL. That page "is intended to provide emergency managers, planners, forecasters, and the public with advance notice of potential threats related to climate, weather, and hydrologic events." It integrates existing National Weather Service official medium—(3-5 day), extended—(6-10 day), and long—(monthly and seasonal) range forecasts, and hydrologic analyses and forecasts. The page includes North America maps showing projected temperature/wind, precipitation, and soil/wildfire anomalies and other data such as a table of rivers currently at or above flood stage.

http://www.esig.ucar.edu
http://www.esig.ucar.edu/socasp/index.html
http://www.esig.ucar.edu/pubs.html

The ever-evolving National Center for Atmospheric Research (NCAR) Environmental and Societal Impacts Group (ESIG) Web site offers much information on weather and climate impacts—particularly through its "Societal Aspects of Weather" section at the second URL above. That site includes pages on such climate/weather phenomena as floods, El Niño, tornadoes, extreme temperature, lightning, hurricanes and tropical cyclones, and winter weather. Additionally, it offers pages covering impact statistics, emergency management, insurance, and weather policy. It also serves up ESIG's *Network Newsletter* (**http://www.esig.ucar.edu/newshp**) for people involved in climate impact assessment, and *WeatherZine* (**http://www.esig.ucar.edu/socasp/zine/index.html**), an informal newsletter serving scholars and practitioners interested in the relation between society and weather. To subscribe to the e-mail version of *WeatherZine*, send an e-mail message to **thunder@ucar.edu** with the subject line "Subscribe Zine." Include your name and the e-mail address to which you would like the newsletter sent.

http://www.ipcc.ch/
http://www.meto.gov.uk/sec5/CR_div/ipcc/wg1/ (IPCC Group I)
http://www.usgcrp.gov/ipcc/ (IPCC Group II)
http://www.rivm.nl/env/int/ipcc/ (IPCC Group III)

Recognizing the problems posed by potential global climate change, in 1988 the World Meteorological Organization (WMO) and the United Nations Environment Program (UNEP) established the Intergovernmental Panel on Climate Change (IPCC) to assess the scientific, technical, and socioeconomic information available for understanding the risk posed by human-induced climate change. The panel has not carried out new research; it has based its assessment mainly on published and peer-reviewed scientific technical literature. The IPCC encompasses three working groups and a task force:

- Working Group I, which assessed the scientific aspects of the climate system and climate change;
- Working Group II, which addressed the vulnerability of socioeconomic and natural systems to climate change, negative and positive consequences of climate change, and options for adapting to it; and
- Working Group III, which examined options for limiting greenhouse gas emissions and otherwise mitigating climate change.
- The Task Force that oversees the National Greenhouse Gas Inventories Program.

In early 2001, all three working groups released their final reports summarizing more than two years of work for the IPCC's Third Assessment, and those documents are available from the Web sites above. Working Group I's contribution to the IPCC Third Assessment Report is entitled *Climate Change 2001: The Scientific Basis*; Group II's is *Climate Change 2001: Impacts, Adaptation and Vulnerability*; and Group III's is *Climate Change 2001: Mitigation*.

Each working group report is extensive; however, brief summaries intended to provide basic information to policy makers are available from each group's Web site.

In addition to the reports of the working groups, the IPCC Web site also offers numerous on-line special reports, including: *The Regional Impacts of Climate Change: An Assessment of Vulnerability*; *Aviation and the Global Atmosphere*; *Methodological and Technological Issues in Technology Transfer*; *Emissions Scenarios*; and, *Land Use, Land Use Change and Forestry*.

http://www.nacc.usgcrp.gov/

The U.S. National Assessment of the Potential Consequences of Climate Variability and Change, which contributed to the global assessment mentioned above, was established to provide a detailed understanding of the consequences of climate change for the nation and to examine possible coping mechanisms for adapting to such change. Conducted under the auspices of the U.S. Global Change Research Program (USGCRP), the assessment has examined both geographic/regional and sectoral (economic, environmental, societal) issues in order to come up with a broad synthesis for the nation. The national assessment focuses on what is known about the potential consequences of climate variability and change for the United States over the next 25-30 years (roughly one generation) and also over the next 100 years. The final report of the National Assessment Synthesis Team, *Climate Change Impacts on the United States: The Potential Consequences of Climate Variability and Change*, is now available from this Web site which also provides the more detailed reports from several of the studies focusing on specific regions and issues.

http://www.unep.ch/iuc/submenu/infokit/factcont.htm

The United Nations Environment Program's Information Unit for Conventions offers a "Climate Change Information Kit" via this site. The kit includes fact sheets covering nearly all aspects of climate change, including the implications for natural disasters.

http://www.pacinst.org/ccresource.html

The Pacific Institute for Studies in Development, Environment and Security has developed an index to aid researchers and students dealing with climate change. "A Selective List of Climate Change Resources on the Internet" is updated weekly and contains hundreds of links to climate change science and policy information on the Internet.

Drought

http://enso.unl.edu/ndmc

The National Drought Mitigation Center was created to help people and institutions develop and implement measures to reduce societal vulnerability to drought—focusing on prevention and risk management rather than crisis management. The NDMC site includes sections that describe the center; explain how and why to plan for drought; provide information about current forecasts, monitoring, and impacts—both for the U.S. and worldwide; present historic climate data; discuss the "Enigma of Drought" in depth and measures that have worked to alleviate drought impacts; offer directions for preparing a "Drought Planner's Handbook"; provide a directory of drought planners; and, of course, offer a list of other useful Internet sites. The site's "Drought Monitor"—**http://enso.unl.edu/monitor/monitor.html**—is a comprehensive analysis of the current situation. A joint effort of the USDA, NOAA/CPC, and the NDMC, released each Thursday, the monitor reviews conditions across the country and provides an outlook for the coming weeks.

http://www.drought.noaa.gov/

The National Oceanic and Atmospheric Administration's "Drought Information Center" is "a roundup of the various NOAA Web sites and information on drought and climate conditions." It provides breaking news, including current drought assessments of various kinds; monthly roundups; and considerable background information including sections entitled "All About Drought," "Normal Precipitation for U.S. Stations," "Billion Dollar Weather Disasters," "All About Heat Waves," and "Fire Potential"; and other links to Web sites with information about drought.

http://www.cdc.noaa.gov/
http://www.cdc.noaa.gov/Drought/
http://www.cdc.noaa.gov/ENSO/

The mission of the NOAA-CIRES Climate Diagnostics Center (CDC) is to identify the nature and causes of climate variations on time scales ranging from a month to centuries and thus to predict climate variations on these time scales. The CDC provides several resources including its Map Room Weather Products (**http://www.cdc.noaa.gov/~map/maproom/text/weather_products.shtml**), Map Room Climate Products (**http://www.cdc.noaa.gov/~map/maproom/text/climate_products.shtml**), and Display and Analysis Web Pages for CDC Climate Data (**http://www.cdc.noaa.gov/PublicData/web_tools.html**).

At the second URL above, the CDC offers a page entitled "Current and Anticipated Precipitation Anomalies over the U.S." that provides information on current and emerging drought situations in the U.S. The page includes maps and graphics showing where problems are occurring and other information and forecasts regarding developing precipitation anomalies.

At the third address, the site provides "El Niño/Southern Oscillation (ENSO) Information" covering such questions as What happens during an El Niño/La Niña cycle? What are the effects of El Niño/La Niña on climate and individual weather systems? And, What is the current state of El Niño/La Niña? Also included are FAQs (Frequently Asked Questions), a glossary, other links and publications, forecasts and advisories, and educational resources.

http://www.fsa.usda.gov/drought

The National Drought Policy Commission (NDPC) was established by Congress to provide advice and recommendations on the creation of an integrated, coordinated federal policy that prepares for and responds to serious drought emergencies. Congress charged the NDPC with making recommendations on: How to better integrate federal drought laws and programs with ongoing state, local, and tribal programs; How to improve public awareness of the need for drought mitigation, prevention, and response; and Whether all federal drought preparation and response programs should be consolidated under one existing federal agency, and if so, which agency. The commission report is now available on-line.

http://www.undp.org/seed/unso/ http://www.undp.org/seed/unso/news.htm

The United Nations Development Program (UNDP) Office to Combat Desertification and Drought (UNSO) Dryland Web site provides information about the office as well as an overview of the major tools it uses; a newsletter and list of publications available from UNSO; information about other successful projects, practices, and lessons learned; and a news service that once a week distributes links to articles related to drought and desertification. News service items are available from **http://www.undp.org/seed/unso/news.htm**. To subscribe to the service, send an e-mail message to **webmaster.unso@undp.org** with "Subscribe to news updates" in the subject line.

El Niño

http://www.coaps.fsu.edu/lib/biblio/enso-bib-intro.html

The Center for Ocean-Atmospheric Prediction Studies (COAPS) Library has posted a comprehensive El Niño bibliography on the

Internet at the address above. The bibliography is searchable by author's name and can also be browsed page by page.

http://www.fema.gov/nwz97/elnino.htm

The Federal Emergency Management Agency "El Niño Loss Reduction Center" Web site includes much information about mitigating El Niño hazards, as well as illustrations of the phenomenon itself, news releases, and many links to additional information on El Niño available through the World Wide Web.

http://www.esig.ucar.edu/signal

The National Center for Atmospheric Research (NCAR) Environmental and Societal Impacts Group (ESIG), mentioned above, regularly publishes *ENSO Signal*, a newsletter intended for those interested in the El Niño-Southern Oscillation (ENSO) cycle and its impacts on ecosystems and societies.

http://www.esig.ucar.edu/un/ http://www.esig.ucar.edu/un/enFinal.pdf

In the autumn of 2000, the United Nations released an international study, *Lessons from the 1997-98 El Nino: Once Burned, Twice Shy?*, which asserts that thousands of human casualties and tens of billions of dollars in economic damage will continue to befall the world's developing countries every two to seven years until an investment is made to improve forecasting and preparedness against El Niño.

The creation of regional organizations to prepare collective responses to El Niño is one of the key recommendations in this study undertaken by teams of researchers working in 16 countries in Latin America, Asia, and Africa. The 19-month project was undertaken with the collaboration of four United Nations organizations—the U.N. Environment Programme, the U.N.

University, the World Meteorological Organization (WMO), and the International Strategy for Disaster Reduction (ISDR)—together with the U.S.-based National Center for Atmospheric Research (NCAR). It examined societal impacts of the 1997-98 El Niño in the 16 countries in order to identify "what worked and what didn't with regard to societal responses to the forecasts and impacts of the 1997-98 El Niño event."

The study's findings highlight the need to undertake systematic long-term risk reduction activities, including better understanding of climate-related vulnerability through education and training. Consequently, the U.N. agencies are partnering with NCAR to develop a comprehensive program of "educating educators" in developing countries. The program particularly addresses science, policy, and ethics related to climate change, variability, and extremes.

http://www.ogp.noaa.gov/enso/

The National Oceanic and Atmospheric Administration El Niño-Southern Oscillation (ENSO) home page, produced by the NOAA Office of Global Programs, is self-described as "your one-stop source for the latest on El Niño and the Southern Oscillation." It addresses the questions: How large is the current El Niño?; What is the El Niño forecast?; How will El Niño affect the U.S.?; How will El Niño affect the world?; What is El Niño, and where can I learn more?; What are we doing to learn more about ENSO?

http://elnino.noaa.gov

NOAA also provides this El Niño page entitled "NOAA: El Niño Forecasts, Observations and Research." It provides both El Niño forecasts and status reports regarding current conditions, a "threats assessment" for the entire U.S. and individual states, as well as information on El Niño preparedness. It includes sections entitled, "About El Niño," "What is El Niño?," "Frequently Asked

Questions," "Glossary of Terms," "The Atmosphere During El Niño," and "NOAA's Role." It also covers El Niño impacts regionally, nationally, and globally; provides profuse links to other research institutions and publications; and summarizes NOAA research on this phenomenon.

http://www.pmel.noaa.gov/toga-tao/el-nino/

Similarly, NOAA's Pacific Marine Environmental Laboratory El Niño Theme Page provides access to extensive distributed information related to El Niño. It covers current conditions and recent news releases and also includes sections addressing: What is El Niño?; What are the impacts of El Niño?; What are the current El Niño forecasts?; What is the latest El Niño data?; What are some frequently asked questions?; and, Where can I find more El Niño data and information? This site also provides numerous links to other El Niño information on the Web.

http://www.esig.ucar.edu/la_nina_home/

If it's not one thing, it's another . . . . With the waning of El Niño comes the onset of La Niña—the cooling of eastern Pacific waters off the coast of South American—and with it, global meteorological consequences of many kinds. The Environmental and Societal Impacts group at NCAR launched this La Niña Web page that consolidates and serves as an entry point to many of the other Internet sources of information on this phenomenon.

HURRICANES AND COASTAL HAZARDS

http://www.nhc.noaa.gov

The National Hurricane Center Web site offers much information about hurricanes and tropical cyclones.

http://hurricanes.noaa.gov/
http://www.nws.noaa.gov/om/hurricane/index.shtml

In the spring of 1999, the National Oceanic and Atmospheric Administration (NOAA) and the National Weather Service's Office of Meteorology created two hurricane awareness Web sites offering information on hurricane awareness activities around the country. The NOAA site, entitled "Hurricanes: The Greatest Storms on Earth," presents the latest hurricane news and extensive background information, as well as links to numerous sources of hurricane information, including local sites. The Office of Meteorology site provides several on-line preparedness guides in both Spanish and English, current watches and warnings, on-line tracking charts, and links to many of the national agencies dealing with hurricane preparedness, response, and recovery.

http://www.fema.gov/fema/trop.html
http://www.fema.gov/fema/hurricaf.html

The Federal Emergency Management Agency's Tropical Storm Watch Page provides news and situation reports on current storms as well as archived information on past hurricanes monitored by FEMA. FEMA's "Fact Sheet on Hurricanes" includes information about what to do before, during, and after a hurricane and steps to take to effect long-term mitigation.

http://www.gopbi.com/weather/storm/

Lowe's Home Improvement Stores, in cooperation with the Federal Emergency Management Agency, maintains this site with information about storms of the current hurricane season. It includes the latest weather reports, satellite and radar information, warnings, marine reports, preparedness information, background information about hurricanes and El Niño, an on-line bookstore, a chat room, and other resources. From this site, one can also sign up for an e-mail list that sends out updated information about developing and existing storms.

http://www.usatoday.com/weather/whur0.htm

The *USA Today* "Hurricane Information Guide" offers a wealth of information including the outlook for the coming season, warnings and advisories, graphics, satellite images, maps, *USA Today* stories about current and recent tropical storms, historical information, and other articles covering virtually all aspects of hurricanes—from meteorology to sociology. The site offers the latest storm information from various meteorological centers around the world and includes sections on the physical science of hurricanes, hurricane safety, hurricane terms, climate change and hurricanes, and hurricane science and technology, as well as access to the newspaper's vast library on the science and technology used to predict the development and paths of hurricanes. Users can also use the site to submit questions to a hurricane expert.

http://www.aoml.noaa.gov/hrd/tcfaq/

This Frequently Asked Questions (FAQs) page, assembled and maintained by Chris Landsea of NOAA's Hurricane Research Division, is an excellent resource for general background information on hurricanes.

http://www.usgs.gov/themes/coast.html

The U.S. Geological Survey offers this "Theme Page" on coastal hazards with links to numerous USGS Web resources on various coastal hazards.

http://state-of-coast.noaa.gov

In 1996, Vice President Al Gore challenged federal agencies to develop a "report card" on the state of the nation's environment.

In response, the National Oceanic and Atmospheric Administration created the *State of the Coast Report* provided at this Web site. The foundation of the report is a series of essays on important coastal issues; two of these essays are entitled "Population at Risk from Natural Hazards," and "Reducing the Impacts of Coastal Hazards." These thorough articles include overviews of the problem on a national scale, regional analyses, specific case studies, interviews with experts, suggested readings and references, and glossaries.

http://www.fema.gov/nwz00/erosion.htm
http://www.heinzcenter.org

In 2000, the Federal Emergency Management released a major study on coastal erosion hazards, *Evaluation of Erosion Hazards*, prepared for FEMA by the Heinz Center for Science, Economics and the Environment. The study provides a comprehensive assessment of coastal erosion and its impact on people and property. According to the report, approximately 25 percent of homes and other structures within 500 feet of the U.S. coastline and the shorelines of the Great Lakes will fall victim to the effects of erosion within the next 60 years. For details and links to the Executive Summary and the full report, see either the FEMA or Heinz Center Web site listed above.

TORNADOES, THUNDERSTORMS, HIGH WIND, LIGHTNING, AND OTHER SEVERE WEATHER

Severe weather generally

http://www.fema.gov/pte/prep.htm

The Federal Emergency Management Agency Web site provides "Fact Sheets"—including preparedness tips—concerning many weather related hazards, including: http://www.fema.gov/library/

tornadof.htm—"Fact Sheet: Tornadoes" http://www.fema.gov/library/heatf.htm—"Fact Sheet: Extreme Heat" http://www.fema.gov/library/stormsf.htm—"Fact Sheet: Winter Storms" http://www.fema.gov/library/wntsft.htm—"Winter Safety Tips" http://www.fema.gov/library/thunderf.htm—"Fact Sheet: Thunderstorms and Lightning"

http://www.nssl.noaa.gov

The National Severe Storms Laboratory is one of the Environmental Research Laboratories of the National Oceanic and Atmospheric Administration. Headquartered in Norman, Oklahoma, the NSSL, in partnership with the National Weather Service (see below), is dedicated to improving the nation's severe weather warnings and forecasts in order to save lives and reduce property damage. The NSSL Web site includes general information about the lab and the research it undertakes, as well as a list of lab publications.

http://www.spc.noaa.gov

The Web site of the Storm Prediction Center (SPC) in Norman, Oklahoma, part of the National Centers for Environmental Prediction, includes numerous "Forecast Products" (weather outlooks, storm/tornado watches and warnings, etc.), including "Fire Weather Forecasts"—a section that provides national-scale fire weather guidance for NWS offices and other interested federal agencies dealing with wildland fire management. The site also serves up severe storm statistics and archived weather data, offers a list of SPC publications, and generally describes the work of the center.

http://www.nws.noaa.gov/pa

The National Weather Service Office of Public Affairs Web page provides links to numerous NWS pages addressing various aspects of weather hazards. The links are organized according to NWS background information; data, statistics, indices, reports, research;

severe weather and safety; climate; publications; photos and Webcams; and other resources. The site also lists recent information and press releases from the NWS.

http://www.nws.noaa.gov/om
http://www.nws.noaa.gov/om/tpsterm.htm
http://www.nws.noaa.gov/om/torn.htm
http://www.nws.noaa.gov/om/severeweather/index.shtml
   etc., etc., etc . . . .

The National Weather Service (NWS) Office of Meteorology (OM) Web site is a trove of useful meteorological hazards information. It not only includes information on current conditions but also provides on-line versions of the NWS newsletters *Aware* and *AwareNow*; information on flood, hurricane, tornado, winter weather, and other severe weather preparedness and response; natural hazard statistics and a map of the current year's disasters; as well as drought, climate change, and El Nino information. It links to IWIN—the Interactive Weather Information Network—which shows currently active severe weather warnings, and EMWIN—the Emergency Managers Weather Information Network (see below). It also includes information about the office, complete "Service Assessments" (evaluations of NWS performance before, during, and following disasters), a list of meetings and conferences, a link to the hurricane watch office, maps and information about historical disasters, and a list of the many NWS publications and hazard awareness materials available on-line.

As an example of the many resources available, the second URL above is a page on "Advance Short Term Warnings and Forecasts" that includes information on the short-term prediction process; links to sites providing current short-term watches, warnings, advisories, etc.; information on how to deal with short-term warnings; and, of particular note, individual sections entitled "All About . . ." thunderstorms and lightning, tornadoes, floods/flash floods, droughts, winter storms, hurricanes, heat waves, and wild fires. Each provides background information on the specific phenomenon, in most cases on-line information or a brochure on personal preparedness and response, and many links to other useful information on each hazard.

http://iwin.nws.noaa.gov
http://iwin.nws.noaa.gov/emwin/index.htm

These are the Web sites of the National Weather Service's Integrated Weather Information Network (IWIN)—the service's Internet data source—and the Emergency Managers Weather Information Network (EMWIN)—a cooperative effort of the NWS, FEMA, and other public and private organizations. EMWIN is intended to provide the emergency management community with access to basic NWS warnings, watches, forecasts, and other products and involves a suite of methods—including radio, Internet, and satellite—to provide this basic weather information.

http://www.nws.noaa.gov/stormready/index.htm

The National Weather Service has created a program to help cities, counties, and other local governments implement procedures to reduce the potential for weather-related disasters. By participating in "StormReady," local agencies can earn recognition for their jurisdiction by meeting voluntary criteria established by the NWS in partnership with federal, state, and local emergency management professionals. The StormReady Web site describes these various criteria; offers background information about the program; and provides a list of useful publications, as well as safety information about winter weather, hurricanes, extreme heat, and other severe weather.

http://www.ncdc.noaa.gov
http://www.ncdc.noaa.gov/ol/climate/severeweather/
  extremes.html
http://www5.ncdc.noaa.gov/pubs/publications.html

As mentioned above, the Web Site of the National Climate Data Center includes data from thousands of weather stations around the world, as well as hundreds of images, numerous technical reports on extreme weather events, and lots of other climate/weather information. The second URL above focuses on climate extremes and severe weather related to climate change. It includes sections on U.S. hurricanes, rainfall, temperature extremes, and tornadoes; billion dollar weather disasters; global climate change; El Niño/La Niña; recent weather extremes; historical global extremes; annual climate profiles; U.S. local storm reports; climatic data; satellite images and radar composites. The third URL offers Storm Data Publications—including *Storm Data,* which contains a chronological listing, by state, of hurricanes, tornadoes, thunderstorms, hail, floods, drought conditions, lightning, high winds, snow, temperature extremes, and other weather phenomena. The reports are provided by the National Weather Service and contain statistics on personal injuries and damage estimates.

http://www.esig.ucar.edu
http://www.esig.ucar.edu/socasp/index.html

Also previously mentioned, the Web site of the National Center for Atmospheric Research (NCAR) Environmental and Societal Impacts Group (ESIG) provides much information and interesting statistics on the societal aspects of weather, covering floods, El Nino, tornadoes, extreme temperature, lightning, hurricanes and tropical cyclones, and winter weather. Much other information is available, including a section on "Weather Policy" and the ESIG *WeatherZine* (see **http://www.esig.ucar.edu/pubs.html**).

http://www.weather.com

Not surprisingly, the Weather Channel hosts an extensive Web site that includes news, maps, photographs, and other information about currently developing tornadoes, hurricanes, thunderstorms, and other severe weather, as well as background information about these phenomena and tips on preparing for and surviving them.

http://ww2010.atmos.uiuc.edu/(Gh)/home.rxml
http://ww2010.atmos.uiuc.edu/(Gh)/guides/mtr/hurr/home.rxml

The Weather World 2010 Web site, created by the Department of Atmospheric Sciences at the University of Illinois, Urbana-Champaign, hosts a wealth of information on weather. It integrates real-time and archived data with instructional resources, using new, interactive technologies. The site includes a multimedia *Online Meteorology Guide*, with modules on such specific phenomena as clouds and precipitation, forces and winds, air masses and fronts, weather forecasting, severe storms, El Niño, and, at the second URL above, a new module on hurricanes.

Tornadoes

http://www.tornadoproject.com

The Tornado Project Online site offers basic tornado information, tornado data, tornado myths and FAQs (frequently asked questions), tornado oddities, personal accounts of tornado experiences, information on tornado safety, tornado history, as well as information about recent tornadoes. It also lists the many tornado videos and books available from the Tornado Project and information on other disasters.

http://whyfiles.news.wisc.edu/013tornado/index.html

The mission of the Why Files is to "explore the issues of science, math, and technology that lurk behind the headlines of the day." Founded in 1996 as a project of the National Institute for Science Education with support from the National Science Foundation and produced by the Graduate School of the University of Wisconsin, the files include this section, which provides background information on tornadoes, including an explanation of these phenomena and information on prediction, personal protection, and tornado effects on the natural landscape.

High wind

http://www.aawe.org

The American Association for Wind Engineering Web site is now maintained at the State University of New York at Buffalo where it has been extensively updated. The site now includes "Wind Net"— a communication network intended to facilitate communication to and among wind engineering centers and to the political establishment that controls wind engineering funding. It also provides general information about the group, the association newsletter, a list of publications, and much information and data about wind hazards.

http://www.wind.ttu.edu/

Texas Tech University is the home of the Wind Engineering Research Center, which conducts interdisciplinary research on the effects of wind on civil engineering infrastructure, including buildings, bridges, transmission-line towers, and various other structures. This Web site provides a complete description of the center, its programs, and projects; lists center publications; explains research and other projects that have been undertaken; offers research data and findings, and provides links to other useful Internet resources. Additionally, the holdings of Texas Tech's Wind Engineering Library have recently been put on-line. This resource covers more than 4500 articles and provides a simple system for searching the catalog.

Lightning

http://www.lightningsafety.com

The National Lightning Safety Institute Web Page provides information about lightning losses in the United States and offers lightning safety tips for protecting both human beings and human structures.

http://www.glatmos.com/lightinfo/recommendations.html

An ad hoc "Lightning Safety Group" convened at the 1998 American Meteorological Society Annual Conference to outline recommended appropriate actions for individuals to take under various circumstances when lightning threatens. Their guidelines are presented at this URL.

http://www.nwstc.noaa.gov/d.MET/Lightning/Ltng_home.htm

The National Weather Service Training Center (NWSTC) is responsible for training weather forecasters in the use of the new technology that forms the backbone of the modernized weather service. To support that training, this section of the NWSTC Web site offers extensive information about lighting—from fundamental physics, to detection, to safety, to climatology. The site includes several case studies of severe lightning outbreaks as well as references and a glossary of lightning terms.

FLOODS

http://www.fema.gov/library/flood.htm
http://www.fema.gov/library/floodf.htm

FEMA's background pages on floods and its "Flood Fact Sheet" provide basic, but essential, information about flood hazards and flood damage and injury prevention.

http://www.fema.gov/nfip
http://www.fema.gov/nfip/crs.htm
http://www.fema.gov/nfip/manual.htm

The Federal Insurance Administration (FIA—the Federal Emergency Management Agency department in charge of the National Flood Insurance Program [NFIP]) maintains an NFIP Web site. The NFIP information is intended for both the general public and the many organizations and agencies participating in the program. It includes much information about the NFIP and other flood disaster assistance available from the federal government and access to the newly revised NFIP booklet *Answers to Questions about the National Flood Insurance Program.*

The Community Rating System (CRS) is a program of the NFIP that provides incentives, in the form of reduced premiums, to communities that undertake various kinds flood mitigation activities in excess of NFIP minimum standards. To support the program, FEMA/NFIP has made many CRS publications and guidelines available on the World Wide Web at the second address above. Other background information is available there, as well. Additional supporting materials are available from the on-line FEMA Mitigation Library—**http://www.fema.gov/library/lib06.htm**.

In May and October of each year, the FIA publishes changes to NFIP policies and procedures in an updated *NFIP Flood Insurance Manual.* Subscriptions to the manual are available from the FEMA Map Service

Center, P.O. Box 1038, Jessup, MD 20794-1038; 1-800-358-9616. The complete manual is also available on-line at the third URL above.

http://www.fema.gov/mit/tsd

The Flood Hazard Mapping page from FEMA's Technical Services Division provides an overview of the National Flood Insurance Program (NFIP) and FEMA's map modernization program. It covers everything that homeowners, business owners, lenders, insurers, planners, engineers, surveyors, floodplain managers, and community officials need to know about the NFIP floodplain mapping program and offers on-line hazard maps and tutorials.

http://www.fema.gov/msc http://web1.msc.fema.gov

... speaking of flood maps. Now you can order all of your National Flood Insurance Program mapping products on-line through FEMA's National Flood Insurance Program (NFIP) Map Service Center (MSC) Flood Map Store. Products available include: flood maps issued by the NFIP, NFIP manuals, Flood Insurance Studies, community status books, Letters of Map Change (LOMCs), Flood Map Status Information System (FMSIS), Q3 digital flood data, and coastal barrier resource area data. This new MSC site is just one of a suite of on-line services planned to expedite the dissemination of FEMA's flood map and insurance products.

http://www.usgs.gov/themes/flood.html

The U.S. Geological Survey "Theme Page" on floods provides links to other USGS Web resources on floods, as well as a link to a "Fact Sheet" on this hazard.

http://ks.water.usgs.gov/Kansas/pubs/fact-sheets/fs.024-00.html

The U.S. Geological Survey (USGS) has issued a fact sheet entitled *Significant Floods in the United States During the 20th Century—*

USGS Measures a Century of Floods, by Charles A. Perry. According to Perry, during the twentieth century, floods were the number one disaster in the United States, both in terms of lives lost and property damage. Since 1900 flooding has killed more than 10,000 people, and property damage from flooding now totals over $1 billion a year. The fact sheet discusses 32 significant floods that occurred during the twentieth century, broken down into six types: large regional floods, flash floods, storm surge, ice-jam floods, dam and levee failure, and mudflow. The fact sheet names the flood brought on by the September 1900 Galveston Hurricane as the worst of the century. Besides examining these disasters, this Web site also describes the survey's efforts to measure floods and lists additional sources of flood information on the Internet.

http://water.usgs.gov
http://water.usgs.gov/public/realtime.html

The USGS's "Water Resources of the United States" page offers current U.S. water news; extensive current (including real-time) and historic water data; numerous fact sheets and other publications; various technical resources; descriptions of ongoing Survey water programs; local water information; and connections to other sources of water information. From the USGS real-time streamflow page interested persons can monitor stream levels around the nation and watch as floods evolve.

http://www.usace.army.mil/inet/functions/cw/cecwp/ nfpc.htm

The U.S. Army Corps of Engineers National Flood Proofing Committee (NFPC) has published numerous documents on floodproofing that are available from this Web site.

http://www.floods.org

The Association of State Floodplain Managers is an organization of professionals involved in floodplain management; flood hazard mitigation; the National Flood Insurance Program; and flood preparedness, warning, and recovery. The group has become a respected and influential voice in floodplain management practice and policy in the United States because it represents flood hazard specialists from across jurisdictions and disciplines. ASFPM supports comprehensive nonstructural and structural management of the nation's floodplains and related water resources and believes that, through coordinated, well-informed efforts, the public and private sectors can reduce loss of human life and property damage resulting from flooding, preserve the natural and cultural values of floodplains, and avoid actions that exacerbate flooding. To help reach these goals, ASFPM fosters communication among those responsible for flood hazard activities; provides technical advice to governments and other entities about proposed actions or policies that will affect flood hazards; and encourages flood hazard research, education, and training. The ASFPM Web site includes information on how to become a member, the organization's constitution and bylaws, directories of officers and committees, a publications list, information on upcoming conferences, a history of the association, the publication *Mitigation Success Stories in the United States*, and other useful information and Internet links.

http://www.floodplain.org

The Floodplain Management Web site was established by the Floodplain Management Association (FMA) to serve the entire floodplain management community. It includes sections containing full-text articles, a calendar of upcoming events, a list of positions available in the discipline, an index of publications available free or at nominal cost, a list of associations, a list of firms and consultants in floodplain management, an index of newsletters dealing with flood issues (with hypertext links, if available), a section on the basics of floodplain management, a list of frequently asked questions

(FAQs) about the Web site, and, of course, a copious catalog of Web links.

http://www.ag.ndsu.nodak.edu/flood/home.htm

The North Dakota State University Extension Service offers this thorough Web section, entitled "Coping with Floods," which covers resources for homeowners and family members and discusses both how to prepare for flooding and steps to take after a flood. It includes detailed information on everything from assessing damaged electrical systems and appliances to dealing with financial concerns.

http://www.louisianafloods.org

The Louisiana State University Agricultural Center's Cooperative Extension Service maintains a Web page with much information about various types of flooding—from hurricane storm surge to flash floods. The site has sections on flood conditions, safety and recovery, emergency protection, and flood damage prevention. It provides numerous downloadable publications on flood mitigation and floodproofing, including a publication prepared by the Extension Service entitled *Beyond the Basics* which covers flood risk and flood protection.

http://www.extension.unr.edu/Flood/Flood.html

The University of Nevada Reno's extension office has prepared this "Flood Facts" Web site with much information about flood preparedness, the history of flooding in the region, flood dynamics, and long-term mitigation and solutions to flood problems. It also includes regional flood maps and survival tips. Besides being a good resource in itself, this site provides a model for similar resources that could be developed elsewhere to address local or regional flood hazards.

http://www.nws.noaa.gov/oh
http://www.nws.noaa.gov/oh/hic/
http://www.nws.noaa.gov/oh/hic/nho/index.shtml

As one might suspect, the National Weather Service's Office of Hydrology (OH) and its Hydrological Information Center (HIC) offer much information on floods and other aquatic disasters. Besides information about the various components of the office, the OH site offers current and historical data including an archive of past flood summaries, information on current hydrologic conditions, water supply outlooks, as well as an *Automated Local Flood Warning Systems Handbook*, *Natural Disaster Survey Reports*, and other scientific publications on hydrology and flooding. The site also provides information and order forms for the office's video on the dangers of *Low Water Crossing*.

The Hydrological Information Center subsection describes the mission of the center, provides much additional information on flood impacts, and offers extensive flood impact data on deaths and economic losses. The center also serves up a regularly updated map (at the third URL above) showing flood potential across the nation, along with explanatory text. The information does not provide specific forecasts of flood location and severity, but it does identify areas that warrant careful monitoring. The site also provides access to more detailed information on local conditions provided by NWS field offices.

http://www.dartmouth.edu/artsci/geog/floods/

The Dartmouth College "Flood Remote Sensing Page" is the home page of the Global Flood Monitoring and Analysis Project—the work of a team of geographers at Dartmouth who are using satellite technology and other means to develop up-to-the-minute flood maps to support flood management and relief efforts and to further flood prediction around the world. This site includes an evolving database of extreme floods and a collection of satellite images of several major events—in short, a wealth of data on recent major floods around the world for response agencies and researchers alike.

http://www.cira.colostate.edu/fflab/

In part because several devastating flash floods have occurred along the foothills of northern Colorado in the last quarter century, Colorado State University, located in Fort Collins, now hosts a Flash Flood Laboratory. The lab's Web site describes the mission and current projects of the institution, recent news and a list of recent flash floods, information on how to prepare for and survive a flash flood, links to other flash flood related sites, and more information and research concerning the 1997 flash flood in Fort Collins—including lessons learned by the university (whose library and several other facilities were severely damaged) concerning flood recovery.

http://www.fhrc.mdx.ac.uk/choice.html

The Flood Hazard Research Centre (FHRC) is an interdisciplinary center based in the Research and Post Graduate Studies Institute of the School of Social Science at Middlesex University in the U.K. The research center specializes in the interaction between people and the environment, together with the analysis and appraisal of environmental policies. A large part of its research refers to the management of the water environment. The FHRC Web pages were created for the dissemination of scientific knowledge and include sections on the purposes, people, and publications of the center; new information available from the center; research themes; and useful links.

WILDFIRE

http://online.anu.edu.au/Forestry/fire/firenet.html

FireNet, which includes a node at the Australian National University (http://online.anu.edu.au/Forestry/fire/firenet.html) and Charles

Stuart University (**http://www.csu.edu.au/firenet/firenet.html**), supports an international network of landscape fire information.

http://www.firewise.org/

Firewise is a Web site created for people who live, vacation, or own structures in wildfire-prone areas. It offers online wildfire protection information and checklists, as well as listings of other publications, videos, and conferences. The interactive home page allows users to ask questions of fire protection experts and to register and receive further information as it becomes available.

http://www.usgs.gov/themes/wildfire.html

The U.S. Geological Survey "Theme Page" on wildfire provides links to other Web resources, as well as a link to a "Fact Sheet" on this hazard.

http://www.iawfonline.org/

The International Association of Wildland Fire/*Wildfire Magazine* Web site offers all kinds of information about the incendiary hazard, including excerpts and previews from the magazine; recent wildfire news; a wildfire calendar of meetings, training, and other events; lists and catalogs of publications and journals available from the International Association of Wildland Fire; databases, directories, and bibliographies; access to the association's library services; tips on wildfire safety; and much more.

http://www.nifc.gov

The National Interagency Fire Center in Boise, Idaho, is the nation's primary logistical support center for wildland fire suppression. The center is home to federal wildland fire experts in fields as diverse as fire ecology, fire behavior, technology, aviation, and weather. Working together and in concert with state and local agencies, NIFC's role is to provide national response to wildfire and other emergencies and to serve as a focal point for wildland fire information and technology. The NIFC Web site provides current fire information (including daily incident management situation reports, National Weather Service fire weather forecasts, and national fire news), information about and links to cooperating agencies, and other information about NIFC projects.

http://www.uni-freiburg.de/fireglobe

For most types of wildfire, research of the last decade provides sufficient knowledge to support decision making regarding fire policy and management. However, in many developing countries this knowledge is not known or not accessible, either for developing appropriate fire policies and management strategies or for responding to large fire emergencies. To help solve this problem, in 1998 the Global Fire Monitoring Center (GFMC) was established at the Max Planck Institute for Chemistry in Germany. The GFMC provides real-time or near-real-time information related to fire. The GFMC Web site provides emergency contacts; fire data and information on fire monitoring; current news; information on global fire models; details about other programs and projects; and indices to meetings and jobs, literature, databases, and links to other information on the Internet.

http://www.fs.fed.us/fire

Persons concerned about the current wildfire situation in the U.S. might want to consult this U.S. Forest Service Fire and Aviation Web site, which provides up-to-date reports and news regarding current and recent wildfires, as well as wildfire potential, across North America. The library section offers numerous downloadable publications.

http://www.spc.noaa.gov/fire/

The Web site of the Storm Prediction Center (SPC) in Norman,

Oklahoma, part of the National Centers for Environmental Prediction, includes numerous "Forecast Products," including these "Fire Weather Forecasts/Information Pages" that provide national-scale fire weather guidance for NWS offices and other interested federal agencies dealing with wildland fire management.

http://www.fema.gov/library/wildlanf.htm

The Federal Emergency Management Agency (FEMA) provides this "Fact Sheet: Wildland Fire" via the URL above.

http://www.nofc.forestry.ca/fire/

The mission of the Canadian Forest Service's Fire Management Network is to "increase our understanding and ability to manage wildland fires within the context of sustainable development of Canada's forests." The FMN Web site offers information about the network and its various science programs; a directory; a publications list; a glossary of fire acronyms; information about the Canadian Forest Fire Danger Rating System (CFFDRS), Forest Fire Weather Index System (FWI), Forest Fire Behavior Prediction System (FBP); and other links and indices of information.

SNOW AVALANCHE

http://www.csac.org

The Cyberspace Snow and Avalanche Center covers current snow conditions, avalanche education and research (including bibliographies and other publications), archived avalanche bulletins, summaries of avalanche incidents, professional resources (including lists of research centers, weather resources, conferences, and additional avalanche Web sites), and other snow and avalanche information such as statistics and personal accounts of disasters.

http://www.avalanche.org
http://www.avalanche.org/~aaap/

The Westwide Avalanche Network Home Page, created in affiliation with the American Association of Avalanche Professionals (AAAP), describes the network's products and services, offers accounts of recent accidents, provides a means for data information exchange among network members, lists recent avalanche research, provides a library of downloadable software and a library of avalanche photographs, lists North American avalanche centers and provides their most recent reports, lists avalanche educational resources, provides a weather page, offers information on the use of avalanche search dogs, lists upcoming avalanche conferences, and provides numerous links to other public and privates sources of avalanche information—including the AAAP and its *Avalanche Review* newsletter at the second URL above.

EXTRATERRESTRIAL/SPACE HAZARDS

http://impact.arc.nasa.gov/index.html

This "Asteroid and Comet Impact Hazards" Web site, offered by the Ames Space Science Division of NASA, provides background information about these space hazards, recent and archived news stories, reports and position papers from NASA, transcripts of recent congressional testimony on this threat, a list of currently known near-earth objects (NEOs) and projected future close encounters, an overview and reports regarding the NASA NEO Project, photographs and artist renderings of asteroid/comet encounters, and a bibliography and fact sheet on the NEO hazard.

http://www.nearearthobjects.co.uk

In January 2000 the U.K. Minister for Science announced the establishment of a Task Force on Potentially Hazardous Near Earth

Objects (NEOs—asteroids and comets whose orbits bring them close to the earth). He invited the task force to recommend how the U.K. could best contribute to international efforts to determine and mitigate the risk of collisions with NEOs. This Web site offers the final report of the task force, including 14 recommendations. The task force notes that the risk due to objects whose orbits are known can be predicted with temporal and spatial accuracy, whereas the risk due to uncharted objects can only be computed using statistical averages and thus cannot include predictions of time or place of occurrence. Thus, a principle focus of the recommendations is the establishment of facilities (i.e., telescopes) and programs to identify and chart the orbits of NEOs. They also address the British role in a greater international effort, the assessment of risks, and measures to mitigate future impacts. The report also provides a comprehensive review of current knowledge about this hazard.

http://geomag.usgs.gov/

This is the U.S. Geological Survey "Themes Page" on geomagnetism and geomagnetic storms created by the National Geomagnetic Information Center. It includes background information about the center, as well as numerous charts, publications, and other information and tools for studying and monitoring this phenomenon.

MENTAL HEALTH AND DISASTER MEDICINE

http://apollo.m.ehime-u.ac.jp/GHDNet/
http://www1.pitt.edu/~ghdnet/GHDNet/

The Global Health Disaster Network was created to make disaster health and medical information available globally. It includes nodes in both Japan and the United States that offer an extensive bibliography of journal articles on disaster medicine, on-line publications, and links to other sites with disaster medicine information.

http://www.cdc.gov/nceh/publications.htm

The National Centers for Disease Control's Center for Environmental Health (NCEH) has issued several on-line *Prevention Guides to Promote Your Personal Health and Safety Before, During, and After Emergencies and Disasters* in both English and Spanish. The guides cover earthquakes, extreme cold, extreme heat, floods, hurricanes, and tornadoes. The earthquake guide, to take one example, covers general information about earthquakes, how to prepare for a quake, inspection of a home for possible hazards, what to do during and after an earthquake, and issues concerning people with special needs. It includes several checklists, including suggested first aid and survival kits for the home, automobile, and workplace.

http://www.mentalhealth.org/cmhs/EmergencyServices/index.htm

The Emergency Services and Disaster Relief Branch (ESDRB) of the Center for Mental Health Services (CMHS) (part of the U.S. Department of Health and Human Services), in partnership with the Federal Emergency Management Agency, administers the Crisis Counseling Assistance and Training Program, overseeing national efforts to provide emergency mental health services to survivors of presidentially declared disasters. The branch's activities are divided into three areas:

- Services to individuals and communities affected by disasters
- Services to state and local mental health administrators
- Services to other groups

The ESDRB Web site provides information about each of these areas as well as a half dozen documents and manuals on the provision of mental health services following disaster, including *Psychosocial Issues for Older Adults in Disasters* and a *Best Practices* document that describes exemplary disaster crisis counseling programs implemented across the country.

http://165.158.1.110/english/ped/pedhome.htm

The Pan American Health Organization's Emergency Preparedness and Disaster Relief Coordination Program Web site provides *extensive* information about disasters and disaster management generally, as well as specific information and publications about disaster preparedness by hospitals and other health sectors, disaster medicine, public health management following disasters, and disaster mental health. The "Technical Guidelines" section answers many questions about the provision of public health in emergencies and includes contact information for experts in various public health areas. PAHO's newsletters and an on-line virtual disaster library are also available.

http://www.HINAP.org

The World Health Organization has launched its "Health Information Network for Advanced Planning" (HINAP) on the World Wide Web. HINAP consolidates baseline health information by country, identifies health issues of primary concern, and makes this information available for program planning. Up-to-date information is provided during an emergency, permitting program adjustment due to changing circumstances, thereby minimizing mortality and morbidity from preventable causes. HINAP currently includes health indices, profiles, and analyses, plus outbreak verification, for nine countries: Albania, Angola, Colombia, Kosovo, Macedonia, Indonesia, Nigeria, Tajikistan, and Uganda.

http://www.ph.ucla.edu/cphdr/

The UCLA Center for Public Health and Disaster Relief has defined its mission as developing "a curricular focus area and research agenda that examines how natural and human-generated disasters relate to the public's health." The curriculum is being designed to prepare public health professionals for the interdisciplinary roles they play in preparing communities prior to disaster and during the recovery period following a mass population emergency. This Web site provides information about this new center, its goals and programs, people and organizations involved in its development, and its current offerings.

http://www.usd.edu/dmhi/

The Disaster Mental Health Institute (DMHI) is a State of South Dakota Board of Regents Center of Excellence offering an undergraduate minor in disaster response and a doctoral specialty track in clinical/ disaster psychology at the University of South Dakota. The institute also hosts an annual "Conference on Innovations in Disaster Mental Health." The DMHI Web site provides in-depth information about the institute and conference, a list of available publications, and several on-line booklets on coping with the aftermath of disasters.

http://www.trauma-pages.com/

The Trauma Information Pages offer information about emotional trauma, post traumatic stress disorder (PTSD), and disaster mental health.

LISTS/NEWSLETTERS/DISCUSSION GROUPS

Disaster Research (DR), published by the Natural Hazards Center, covers not only disaster research but also recent policy developments, new institutions and information sources, training, and conferences. It also provides a means for subscribers to post queries to the entire readership of over 2,500. Subscribe by sending an e-mail message to **listproc@lists.colorado.edu** with the single command in the body of the message: "subscribe hazards [your name]."

Networks in Emergency Management (Nets) focuses on computer networks and networking in emergency management. Subscribe by sending an e-mail message to **majordomo@sfu.ca** with the single command in the body of the message: "subscribe nets-em [your e-mail address]."

FEMA E-Mail News Service. Subscribe by sending an e-mail message to **majordomo@fema.gov** with the single command in the body of the message: "subscribe news."

Natural-Hazards-Disasters is an e-mail discussion list for anyone interested in natural hazards and disasters; it is hosted by the Benfield Greig Hazard Research Centre, University College London. Subscribe by sending an e-mail message to **mailbase@mailbase.ac.uk** with the command in the body of the message: "join natural-hazards-disasters [your first name, your last name]."

Disaster Grads is a discussion list intended for informal discussion and information sharing among students—both graduate and undergraduate—who conduct research in the area of hazards and disasters. Subscribe by sending e-mail to **listproc@lists.colorado.edu**, and in the body of the message write "subscribe disaster_grads [your first name] [your last name]." For more information about this service, contact Lori Peek, Natural Hazards Information Center, Campus Box 482, University of Colorado, Boulder, CO 80309-0482; e-mail: **peek@sobek.colorado.edu**.

WSSPC-L is a list managed by the Western States Seismic Policy Council. For subscription information, see **http://www.wsspc.org**.

EMERGENCY-MANAGEMENT is an e-mail discussion list for anyone interested in the protection of local communities from hazards and resulting emergencies. Subscribe by sending an e-mail message to **listserv@zipcode1.office.aol.com** with the single command in the body of the message: "subscribe emergency-management [your real name]."

Emergency Management is a list maintained by the International Association of Emergency Managers (IAEM) for discussion of issues of relevance to local emergency managers. Subscribe by sending e-mail to **majordomo@asmii.com** with the message "subscribe iaem-list [your e-mail address]."

Emergency Management is yet another list for all emergency management professionals. Subscribe by sending e-mail to **emergency-management-subscribe@yahoogroups.com**, or see **http://groups.yahoo.com/group/Emergency-Management**.

Cal-EPI is a forum for discussion of disaster information technology issues, but also a community in which emergency managers, journalists, and others can discuss the larger challenges of alerting, informing, and reassuring the public during emergencies. Subscribe by sending e-mail to **Cal-EPI-request@incident.com** with the word "subscribe" in the body of your message.

HAZMIT is a list for the global hazard mitigation community, covering both natural and anthropogenic disasters. Subscribe by sending and e-mail message to **hazmit-request@mitigation.com** with the command "subscribe" in either the subject or body of the message. Alternatively, subscribe via the Web page **http://www.mitigation.com/listserv.htm**.

EMERG-UNIV is intended to support university and college emergency planning. Send messages to **emerg-univ@julian.uwo.ca**. The list is manually managed; to subscribe, send an e-mail message to **emerg-univ-request@julian.uwo.ca**, with the subject: "subscribe" and a short note in the body of the message stating your full name, interests, affiliation, and signature information. Complete details about this campus emergency management list are available from http://www.uwo.ca/emerg/list.html.

RISKANAL is the e-mail discussion list of the Society for Risk Analysis. Subscribe by sending an e-mail message to **listserv@listserv.pnl.gov** with the single command in the body of the message: "subscribe riskanal [your name]."

Risk-com is a list promoting discussion about risk communication. Subscribe by sending an e-mail message to **risk-com-request@umich.edu** with the word "SUBSCRIBE" as the subject of the message.

RISKWeb is a Web-based discussion forum that supports global conversation of risk and insurance issues. See **http://www.riskweb.com**.

OFDA-L is the distribution list for the U.S. Agency for International Development Office of Foreign Disaster Assistance, reporting on disasters in which OFDA is involved. Subscribe by sending an e-mail message to **listproc@info.usaid.gov** with the single command in the body of the message: "subscribe OFDA-L [your name]."

EMERG-L is a list for emergency services personnel. Subscribe by sending an e-mail message to **listserv@bitnic.educom.edu** with the single command in the body of the message: "subscribe EMERG-L [your name]."

Floodsystem List is a forum for professionals involved in flood warning. To subscribe, see **http://www.egroups.com/group/Floodsystems**.

DisastMH is an e-mail discussion forum for disaster mental health professionals. To subscribe, send an e-mail message to **listserv@maelstrom.stjohns.edu** with "subscribe DisastMH [your first name, your last name]" in the body of the message.

HAZUSNET-USA is an Internet forum for persons interested in HAZUS, the Federal Emergency Management Agency's hazard evaluation software program. Subscribe by sending an e-mail message to **listserv@listserv.buffalo.edu**; leave the subject line blank and send only the message: "sub hazusnet-usa-list [First Name Last Name]."

The Lightning Protection Group is a forum for discussing the protection of humans, infrastructure, and personal property from lightning hazards. See: **http://groups.yahoo.com/group/ LightningProtection**

DisasterInfo provides disaster information relevant to the Americas from the Pan American Health Organization. Subscribe by sending an e-mail message to **DisasterInfo@paho.org** containing your postal and e-mail addresses.

The U.S. Geological Survey has established several e-mail lists that will automatically provide subscribers with the latest news releases, bulletins, and other information about USGS activities as issued by the Survey's Office of Outreach. The listserves are organized by topic:

- *water-pr*
- *geologic-hazards-pr*
- *biological-pr*
- *mapping-pr*
- *products-pr*
- *lecture-pr*

To subscribe to any of these lists, send an e-mail message to **listproc@listserver.usgs.gov** and in the body of the message type: "subscribe [name of list] [your name]."

There are many other lists available.

SATELLITES, REMOTE SENSING, AND GIS

http://ltpwww.gsfc.nasa.gov/ndrd

The Natural Disaster Reference Database Home Page, assembled by the Earth Sciences Directorate of the NASA Goddard Space Flight Center in Greenbelt, Maryland, contains extensive bibliographic information on research results and programs related to the use of satellite remote sensing for disaster mitigation of all kinds.

http://www.ceos.noaa.gov

The Committee on Earth Observation Satellites (CEOS) "Disaster Management Support Project" was created to strengthen natural and technological disaster management on a worldwide basis by fostering improved use of existing and planned earth observation satellite data. Supporting the committee's work, the National Oceanic and Atmospheric Administration has launched this Web site to provide ongoing specific satellite data on volcanic ash, floods, tropical storms, wildfire, El Niño, earthquakes, drought, and oil spills. It also provides special coverage of emerging crises (for example, the 1998 fires in Indonesia), as well as background information about the entire project.

http://www.cla.sc.edu/GEOG/hrl/

The Hazards Research Lab (HRL) in the Department of Geography at the University of South Carolina was established in 1995 to conduct research and graduate training on the use of geographic information processing techniques in environmental hazards analysis and management. The HRL Home Page provides information on the laboratory mission, staff, current projects, and publications. This site includes an index of individuals and groups using, or studying the use of, Geographical Information System (GIS) technology in hazards/disaster management and research: **http://www.cla.sc.edu/GEOG/hrl/gislist.html**.

DATA ON DISASTERS AND DISASTER COSTS

United States

http://www.esig.ucar.edu/socasp/toc_img.html
http://www.esig.ucar.edu/socasp/stats.html
http://www.esig.ucar.edu/sourcebook/index.html

The Web site of the Environmental and Societal Impacts Group, National Center for Atmospheric Research provides considerable data on natural disasters, primarily focusing on meteorological events in the U.S. The last URL above provides the *Extreme Weather Sourcebook*, an encyclopedic compilation of statistics regarding "Economic and Other Societal Impacts Related to Hurricanes, Floods, Tornadoes, Lightning, and Other U.S. Weather Phenomena."

http://www.nws.noaa.gov/pa
http://www.nws.noaa.gov/om/hazstats.htm

The National Weather Service Office of Public Affairs Web page provides links to several NWS pages with weather and weather disaster data. Specifically, at the second URL, the Weather Service offers a "Summary of Natural Hazards Statistics," which covers fatalities due to all hazards, severe weather, lightning, tornadoes, tropical storms, heat, floods/flash floods, cold, winter storms, and high wind.

http://books.nap.edu/catalog/5782.html

Disasters by Design: A Reassessment of Natural Hazards in the United States, a recent volume by sociologist Dennis Mileti, examines the status of hazards/disaster knowledge and research in the U.S. at the end of the twentieth century. Chapter 3 covers "Losses, Costs, and Impacts." The entire text is available on-line; just click the "Read" button on the left.

http://www.cbo.gov
http://www.cbo.gov/byclasscat.cfm?class=0&cat=7

The Congressional Budget Office (CBO) has examined the issue of the rising cost of natural disasters and, at this address, published two reports: *Emergency Spending Under the Budget Enforcement Act* and *Emergency Spending Under the Budget Enforcement Act: An Update*, which provide useful data on federal expenditures for natural disasters.

http://www.nws.noaa.gov/oh/hic/flood_stats/index.html

This Weather Service site provides data on both flood fatalities and flood damage.

http://www.ncdc.noaa.gov/
http://www.ncdc.noaa.gov/ol/climate/severeweather/
 extremes.html
http://www.ncdc.noaa.gov/ol/reports/billionz.html
http://www5.ncdc.noaa.gov/pubs/publications.html

The National Climatic Data Center (NCDC) maintains much information about extreme weather and climate events, including lists of billion dollar weather disasters in the U.S., 1980-1999. At the final URL above, the NCDC provides information on how to obtain *Storm Data*, the center's ongoing compendium of—you guessed it—storm data. Monthly issues contain a chronological listing, by states, of occurrences of storms and unusual weather phenomena. The reports include information on storm paths, deaths, injuries, and property damage.

http://www.ibhs.org
http://www.ibhs.org/ibhs2/html/ibhs_projects/
 projects_paidloss.htm

The Institute of Business and Home Safety (IBHS) Web site maintains insurance disaster loss data for the U.S. The second URL above provides information about the IBHS "Paid Loss Database," which the institute describes as the nation's only database of actual insurance payments from natural catastrophes.

http://www.disasterrelief.org/Library/WorldDis/index.html

The DisasterRelief Web site provides aggregate statistics on all sorts of natural hazards—primarily for the U.S., but also for the entire globe—including hurricanes (deadliest, most expensive, most intense); earthquakes (deadliest, largest, distribution in the U.S. by region); forest fires (10 largest in the U.S.); volcanoes; billion-dollar U.S. disasters; etc.

http://www.disastercenter.com/disaster/TOP100P.html
http://www.disastercenter.com/disaster/TOP100T.html
http://www.disastercenter.com/disaster/TOP100K.html
http://www.disastercenter.com/disaster/TOP100C.html

The Disaster Center Web site offers lists of what it considers to be:

- The Most Deadly Technological Disasters of the 20th Century
- The 100 Most Expensive Technological Disasters of the 20th Century
- The 100 Most Deadly Natural Disasters of the 20th Century
- The 100 Most Expensive Natural Disasters of the 20th Century

http://www.disastercenter.com/tornado.htm

The Disaster Center Web site also includes 50 pages dealing with data from a NOAA tornado statistics database. The site developers have compiled cost, injury, and frequency data, and then conducted comparisons across states as well as calculations based on population. The information includes data for individual states.

http://www.nssl.noaa.gov/~holle/techmemo-sr193.html

This 1997 report provides data on *Lightning Fatalities, Injuries, and Damage Reports in the United States from 1959-1994*. The data are analyzed along many dimensions and broken down from national to regional numbers.

http://www.consumerfed.org/disasinspr.pdf

In 1998, the Consumer Federation of America issued a report on skyrocketing federal expenditures for disasters and the uneven distribution of federal disaster relief among states. A summary of that report is available from this federation Web site.

Worldwide

http://www.md.ucl.ac.be/cred/

The Centre for Research on the Epidemiology of Disasters (CRED) Database (EM-DAT) documents natural and technological disasters around the world. EM-DAT contains essential core data on the occurrence and effects of thousands of mass disasters in the world from 1900 to the present. The database is compiled from various sources, including U.N. agencies, nongovernmental organizations, insurance companies, research institutes, and press agencies. At present, the EM-DAT databases are in two forms: standard and enhanced. The standard database consists of the data taken from an existing disaster database, checked for duplications, but not having any additional information included. As the databases are enhanced (i.e., additional information added) they are added to the Web site. Data are downloadable in several formats. Information in the databases includes:

- The date when the disaster occurred
- Country
- Region

- Continent
- Dead
- Affected Population
- Injured
- Homeless
- Primary source of disaster information
- A unique disaster number for each disaster
- Information on whether OFDA responded to a disaster

The site also provides "Disaster Profiles" derived from data on natural disasters occurring during the last century. The profiles cover three areas: Top 10, Chronological Table, and Raw Data; and there are profiles covering individual countries, regions, the entire globe, and various disaster types. The site also offers other summary data and information on disaster trends in the form of graphs, charts, and maps.

http://www.munichre.com/ http://www.swissre.com/

The Munich and Swiss Reinsurance companies have both issued numerous publications documenting disasters around the globe. In particular, Munich Re publishes an annual report that catalogs such events and their consequences for the insurance industry. A recent report, *Topics 2000: Natural Catastrophes—The Current Position*, examines disasters throughout the second millennium and offers numerous charts and graphs indicating numbers of events, economic losses, fatalities, and insured losses. Much of this information is available from the company's Web site.

http://www.ifrc.org/publicat/

The *World Disasters Report*, published annually by the International Federation of Red Cross and Red Crescent Societies (IFRC), offers an accounting of disasters world-wide.

http://www.epc-pcc.gc.ca/research/epcdatab.html

This Emergency Preparedness Canada (EPC) site provides a downloadable 32-page report that describes EPC's effort to consolidate data on Canadian disasters.

http://www.nhc.noaa.gov/pastdeadlya1.html

The U.S. National Hurricane Center offers much information on hurricanes, past and present, including data on the deadliest, most expensive, and most intense storms. Its Web site also lists U.S. strikes by decade and U.S. strikes by state. The data on the deadliest hurricanes are derived from the monograph *The Deadliest Atlantic Tropical Cyclones, 1492—Present*, by Edward N. Rappaport and Jose Fernandez-Partagas, the complete text of which is available.

http://www.ngdc.noaa.gov/seg/hazard/tsu.shtml

This NOAA site provides a tsunami database that can be searched along any of several parameters, including location, date, magnitude, and number of deaths.

PHOTOGRAPHS AND OTHER VISUAL IMAGES OF HAZARDS AND DISASTERS

For index of Web sites providing photographs and other images of hazards and disasters, see: **http://www.colorado.edu/hazards/sites/photos.html**.

BIBLIOGRAPHY

Acker, J. (1991) "Hierarchies, jobs, bodies: a theory of gendered organizations," in J. Lorber and S. Farrell (eds.), *The Social Construction of Gender*, Newbury Park CA: Sage.
—— (1992) "From sex roles to gendered institutions," *Contemporary Sociology* 21: 565-568.
Adams, T. (1918) "The planning of the new Halifax," *Contract Record and Engineering Review* 32: 682.
Aday, L. (1996) *Designing and Conducting Health Surveys: A Comprehensive Guide*, San Francisco CA: Jossey-Bass.
Adnan, S. (1991) *Floods, People and the Environment*, Dhaka: Research and Advisory Group.
Aktion Katastrophenvorbeugung (1992) *Organisation und Aufgaben des Deutschen IDNDR Komittees*, Bonn: IDNDR.
Aldrich, J. (1985) "Dr. Charles E. Puttner: pharmacist extraordinaire," *Canadian Pharmaceutical Journal* 118: 574-576.
Allport, G. (1953) "The trend in motivational theory," *American Journal of Orthopsychiatry* 23: 107-119.
Anderson, M. (1991) "Sources strategies and the communication of environmental affairs," *British Journal of Sociology* 4: 455-483.
Anderson, W. (1969) *Local Civil Defense in Natural Disaster: From Office to Organization*, Columbus OH: Disaster Research Center, The Ohio State University.
—— (1978) "Social science disaster research in the United States," paper presented at the World Congress of Sociology, Uppsala, Sweden.
—— (1990) "Nurturing the next generation of hazards researchers," *Natural Hazards Observer* 15: 1-2.

Andorka, R. (1990) "The use of time series in international comparison," in E. Øyen (ed.), *Comparative Methodology*, London: Sage.

Andrews, P. (1983) "A system for predicting the behavior of forest and range fires," in J. Carroll (ed.) *Computer Simulation in Emergency Planning*, Simulation Series Vol. 11, No. 2, San Diego CA: Society for Computer Simulation.

Anon. (1955) "The Great London blackout, a report to the Committee on Disaster Studies, National Research Council, Washington; of an enquiry into responses to unfamiliar stimuli," London: unpublished report, Institute of Community Studies.

Armer, M. (1973) "Methodological problems and possibilities in comparative research," in M. Armer and A. Grimshaw (eds.), *Comparative Social Research: Methodological Problems and Strategies*, New York: John Wiley and Sons.

Armer, M. and Grimshaw, A. (eds.) (1973) *Comparative Social Research: Methodological Problems and* Strategies, New York: John Wiley and Sons.

Association of Bay Area Governments (1982) *Using Earthquake Hazard Maps to Analyze the Vulnerability of Lifeline System Locations*, Working Paper No. 18, Berkeley CA: Association of Bay Area Governments.

Baddeley, A. (1936) "3000 Tons of T.N.T. blew up," *Evening Standard* (December 6): 22, National Archives of Canada RG 24 Vol. 1825 G.A.Q. S-55.

Baker, G. and Chapman, D. (eds.) (1962) *Man and Society in Disaster*, New York: Basic Books.

Baker, M. (1986) *Property Values and Potentially Hazardous Production Facilities: A Case Study of the Kanawha Valley, West Virginia*, Morgantown WV: unpublished Ph.D. dissertation, University of West Virginia.

Balassa, B. (1964) "The purchasing power parity doctrine: a reappraisal," *Journal of Political Economy* 72: 584-596.

Barinaga, M. (1992) "Who controls a researcher's files?" *Science* 256: 1620-1621.

Barton, A. (1962) "The emergency social system," in G. Baker and D. Chapman (eds.), *Man and Society in Disaster*, New York: Basic Books.

—— (1969) *Communities in Disaster: A Sociological Analysis of Collective Stress Situations*, Garden City NY: Doubleday and Company, Inc.

Bates, F. (ed.) (1982) *Recovery, Change, and Development: A Longitudinal Study of the Guatemala Earthquake*, Athens GA: International Laboratory for Socio-Political Ecology, University of Georgia.

Bates, F. and Peacock, W. (1987) "Disasters and social change," in R. Dynes, B. De Marchi, and C. Pelanda (eds.), *Sociology of Disasters*, Milano, Italy: Franco Angeli.

—— (1989) "Long-term recovery," *International Journal of Mass Emergencies and Disasters* 7: 349-365.

—— (1993) *Living Conditions, Disasters, and Development: An Approach to Cross-Cultural Comparisons*, Athens GA: University of Georgia Press.

Bates, F. and C. Pelanda. (1994) "An ecological approach to disasters," in R. Dynes and K. Tierney (eds.), *Disasters, Collective Behavior, and Social Organization,* Newark DE: University of Delaware Press.

Becker, H. (1967) "Whose side are we on?" *Social Problems* 14: 239-247.

Belcher, J. (1972) "A cross-cultural household level of living scale," *Rural Sociology* 37: 208-220.

Bell, F. (1918) *A Romance of the Halifax Disaster*, Halifax: H. H. Marshall.

Beniger, J. (1986) *The Control Revolution: Technological and Economic Origins of the Information Society*, Cambridge MA and London: Harvard University Press.

Berke, P. and Ruch C. (1985) "Application of a computer system for hurricane emergency response and land use planning," *Journal of Environment Management* 21: 117-134.

Bernknopf, R., Brookshire, D., and Thayer, M. (1990) "Earthquake and volcano hazard notices: an economic evaluation of changes in risk perceptions," *Journal of Environmental Economics and Management* 18: 35-49.

Bernoulli, D. (1896) *Die Grundlage der modernen Wertlehre*, Leipzig.

Bertaux, D. (1990) "Oral history approaches to an international social movement," in E. Øyen (ed.), *Comparative Methodology*, London: Sage.

Berting, J. (1979) "What is the use of international comparative research," in J. Berting, F. Geyer, and R. Jurkovich (eds.), *Problems in International Comparative Research in the Social Sciences*, New York: Pergamon Press.

Biderman, A. (1966) "Anticipatory studies and stand-by research capabilities," in R. Bauer (ed.), *Social Indicators*, Cambridge MA and London: The M.I.T. Press.

Bird, M. (1962) *The Town That Died*, Toronto: McGraw-Hill Ryerson.

Blaikie, P., Cannon, T., Davis, I., and Wisner, B. (1994) *At Risk: Natural Hazards, People's Vulnerability, and Disasters*, London: Routledge.

Blalock, H. (1969) *Theory Construction: From Verbal to Mathematical Formulations*, Englewood Cliffs NJ: Prentice-Hall.

Bligh, C. (1920) "New public health programme for Halifax, Nova Scotia and Dartmouth, Nova Scotia, Canada," *Canadian Nurse* 16 (February).

Blinn, L. and Harrist, A. (1991) "Combining native instant photography and photo-elicitation," *Visual Anthropology* 4: 75-192.

Block, F. (1990) *Postindustrial Possibilities: A Critique of Economic Discourse*, Berkeley: University of California Press.

Blumer, H. (1966) "Sociological implications of the thought of George Herbert Mead," *American Journal of Sociology* 71: 535-544.

Bogdan, R. and Biklen, S. (1992) *Qualitative Research for Education*, Second Edition, Needham Heights MA: Simon and Schuster.

Bolin, R. and Bolton, P. (1978) "Modes of family recovery following disaster: a cross-national study," in E. Quarantelli (ed.), *Disasters: Theory and Research*, Beverly Hills CA: Sage.

—— (1986) *Race, Religion, and Ethnicity in Disaster Recovery*, Boulder CO: Monograph No. 42, Institute of Behavioral Science, University of Colorado.

Borton, J., and York, S. (1987) "Experiences of collection and use of micro-level data in disaster preparedness and managing emergency operations," Disasters 2: 173-181.

Bourque, L. and Becerra, R. (1993) *Report of the Ad Hoc Committee to Review the Survey Research Center*, Los Angeles CA: Institute for Social Science Research, University of California, Los Angeles.

Bourque, L. and Clark, V. (1992) *Processing Data: The Survey Example*, Newbury Park CA: Sage.

Bourque, L. and Fielder, E. (1995) *How to Conduct Self-Administered and Mail Surveys*, Thousand Oaks CA: Sage.

Bourque, L. and Russell L. (1994) *Experiences During and Responses to the Loma Prieta Earthquake*, Oakland CA: Governor's Office of Emergency Services, Earthquake Program.

Bourque, L., Russell, L., and Goltz, J. (1993) "Human behavior during and immediately after the earthquake," in P. Bolton (ed.), *The Loma Prieta, California Earthquake of October 17, 1989—Public Response*, Washington DC: U.S. Geological Survey.

Braedt, J., and Jungwirth, F. (1988) "Überblick beim Umweltschutz. Satellitenbilder verbessern Einblick in die Umwelt," *Siemens-Magazin COM* 23: 23-25.

Bremer, S. (1987) *The Globus Model: Computer Simulation of Worldwide Political and Economic Developments*, Frankfurt am Main: Campus, and Boulder CO: Westview Press.

Bremer, S. and Gruhn, W. (1988) *Micro GLOBUS: A Computer Model of Long-Term Global Political and Economic Processes*, Berlin: Edition Sigma (includes 5 diskettes).

Britton, N. (1986) "Developing an understanding of disaster," *Australian and New Zealand Journal of Science* 22: 254-271.

Britton, N. and Clapham, K. (1991) *Annotated Bibliography of Australian Hazards and Disaster Literature, 1969-(1989)* (Volume 1: Author Indexed; Volume 2: System Level Indexed), Armidale, New South Wales, Australia: Centre for Disaster Management, University of New England.

Bronson, H. (1918) "Some notes on the Halifax explosion," in *Transactions of the Royal Society of Canada*, Volume XII, Section III.

Brooks, H. (1976) Letter of May 18, 1976 to C. Fritz. (Copy in the Disaster Research Center archives of National Academy of Sciences archival material).

Brookshire, D., Thayer, M., Tschirhart, J., and Schulze, W. (1985)

"A test of the expected utility model: evidence from earthquake risks," *Journal of Political Economy* 93: 368-389.

Brown, C. (1917) Hand written statement promising film, once developed, would be taken to the censors. RG 24 Vol. 4548.

Brown, J. (1972) *The Economic Effects of Floods*, New York: Springer-Verlag.

Bucher, R., Fritz, C. and Quarantelli, E. (1956a) "Tape recorded interviews in social research," *American Sociological Review* 21: 359-364.

—————— (1956b) "Tape recorded research: some field and data processing problems," *Public Opinion Quarterly* 20: 427-439.

Butler, D. (1997) "Selected Internet sites on natural hazards and disasters," *International Journal of Mass Emergencies and Disasters* 15: 197-215.

Butler, J. and Doessel, D. (1981) "Efficiency and equity in natural disaster relief," *Public Finance* 36: 193-213.

Calderon, F., and Piscitelli, A. (1990) "Paradigm crisis and social movements: a Latin American perspective," in E. Øyen (ed.), *Comparative Methodology*, London: Sage Publications.

California Universities for Research in Earthquake Engineering (1995) *Directory of Northridge Earthquake Research*, Richmond CA: California Universities for Research in Earthquake Engineering.

—————— (1997) *Proceedings of the Northridge Earthquake Research Conference,* Richmond CA: California Universities for Research in Earthquake Engineering.

Canan, P. and Pring, G. (1988) "Strategic lawsuits against public participation," *Social Problems* 35: 506-519.

Canan, P., Satterfield, G., Larson, L., and Kretzmann, M. (1990) "Political claims, legal derailment, and the context of disputes," *Law and Society Review* 24: 923-952.

Cannell, W. and Otway, H. (1988) "Audience perspective in the communication of technological risks," *Futures* 5: 519-531.

Carroll, J., and Jackson, L. (1985) "Simulation of human estimation," in J. Carroll (ed.) *Emergency Planning*, Proceedings of the Conference on Emergency Planning, 24-26 January 1985, San Diego CA, Simulation Series Vol. 15, No. 1, San Diego CA: Society for Computer Simulation.

Carstens, C. (1917) "From the ashes of Halifax: the relief work for the blind, maimed and the orphans," *The Survey* (December 29): 360-361.

Carter, M., Kendall, S., and Clark, J. (1983) "Household response to warnings," *International Journal of Mass Emergencies and Disasters* 1: 95-104.

CASRO (1982) *On the Definition of Response Rates: A Special Report of the CASRO Task Force on Completion Rates*, New York: Council of American Survey Research Organizations.

Cecil, J. and Boruch, R. (1988) "Compelled disclosure of research data: an early warning and suggestions for psychologists," *Law and Human Behavior* 12: 181-189.

Cernea, M. (1990) *From Unused Social Knowledge to Policy Creation: The Case of Population Resettlement*, Cambridge MA: Harvard Institute for International Development, Harvard University.

—— (1994) *Bridging the Research Divide: Studying Refugees and Development Oustees*, Washington DC: Environment Department, The World Bank.

Chambers, B. (1927) *Salt Junk Naval Reminiscences*, London: Constable and Co. Ltd.

Chapman, D. (1954) "Introduction," *The Journal of Social Issues* 10: 2-4.

Chartland, R. and Punaro, T. (1995) *Information Technology Utilization in Emergency Management*, Washington DC: Report No. 85-74. Congressional Research Services, Library of Congress.

Chen, L. (ed.) (1973) *Disaster in Bangladesh*, New York: Oxford University Press.

Cheshire, P., Furtdo A., and Magrini, S. (1996) "Quantitative comparisons of European cities and regions," in L. Hantrias and S. Mangen (eds.), *Cross-National Research Methods in the Social Sciences*, London: Pinter.

Chetkovich, C. (1997) *Real Heat: Gender and Race in the Urban Fire Service*, New Brunswick NJ: Rutgers University Press.

Chowdhury, A. (1989) *Let Grassroots Speak*, Dhaka: University Press Limited.

—— (1992) "Flood action plan: one-sided approach?" in *Five Views of the Flood Action Plan for Bangladesh*, Boulder CO: University of Colorado: Natural Hazards Research.

Cisin, I. and Clark, W. (1962) "The methodological challenge of disaster research," in G. Baker and D. Chapman (eds.), *Man and Society in Disaster*, New York: Basic Books.

Clarke, E. (1976) "The Hydrostone neighbourhood," in Nova Scotia Association of Architects (eds.), *Exploring Halifax*, Toronto: Grey de Pencier.

Clarke, L. (1995) "An unethical ethics code?" *The American Sociologist* 26: 12-21.

Clifford, R. (1956) *The Rio Grande Flood: A Comparative Study of Border Communities in Disaster*, Washington DC: Disaster Study No. 7, Committee on Disaster Studies, National Academy of Sciences-National Research Council.

Cobler, S. (1980) "Herold gegen alle. Gespräch mit dem Präsidenten des Bundeskriminalamts," *Transatlantik* No. 11.

—— (1981) "DAZUSY, PSI und MOPPS. Computer auf den Spuren von Risikopersonen," *Kursbuch* 66: 7-18.

Cochrane, H. (1974) "Predicting the economic impact of earthquakes," in H. Cochrane, J. Haas, M. Bowden, and R. Kates (eds.), *Social Science Perspectives on the Coming San Francisco Earthquake*, Boulder, CO: Natural Hazards Research Paper No. 25, University of Colorado.

Cohn, R. and Wallace, W. (1992) "The role of emotion in organizational response to disaster: an ethnographic analysis of videotapes of the Exxon Valdez accident," Boulder CO: Institute of Behavioral Science, Working Paper Series #74.

Collier, J. and Collier, M. (1986) *Visual Anthropology: Photography as a Research Method*, Albuquerque NM: University of New Mexico Press.

Colombo, M. (1995) *Convivere con i rischi ambientali. Il caso Acna-Valle Bormida*, Milano: Franco Angeli.

Congressional Budget Office (1992) "A descriptive analysis of federal relief, insurance, and loss reduction programs for natural hazards," Washington DC: Report to the Subcommittee on

Policy Research and Insurance of the Committee of Banking, Finance, and Urban Affairs.

Connelly, W. (1987) "The Halifax Nova Scotia explosion of 1917: an epilogue," *Journal of the Royal Society of Medicine* (December): 774-775.

Cordes, J. and Yezer, M. (1998) "In harm's way: does federal spending on beach enhancement and protection induce excessive development in coastal areas?" *Land Economics* 74: 128-145.

Corner, J., Richardson, K., and Fenton, N. (1990) "Textualizing risk: TV discourse and the issue of nuclear energy," *Media, Culture and Society* 12: 105-124.

Creighton, H. (1961) "The Halifax explosion," *Maritime Folk Songs*. Toronto: Ryerson.

Cullen, J. (1976) *Comparison of Lifeline System Vulnerability in Two Large Regional Disasters: The Wyoming Valley Flood and the Projected Puget Sound Earthquake*, Washington DC: U.S. Department of Housing and Urban Development, Federal Disaster Assistance Administration.

Cummerwie, H. (1989) "Maßstäbe für den Umweltschutz. SICAD verknüpft Planungsdaten zu thematischen Karten," *Siemens-Magazin COM* 24: 21-23.

Cummins, J. and Geman, H. (1995) "Pricing Catastrophe Insurance Futures and Call Spreads: An Arbitrage Approach," *Journal of Fixed Income* 5: 46-57.

Currey, B. (1978) *Mapping of Areas Liable to Famine in Bangladesh* (mimeo).

—— (1980) "The famine syndrome: its definition for relief and rehabilitation in Bangladesh," in J. Robson (ed.), *Famine: Its Causes, Effects and Management*, New York: Gordon.

Curry, T. and Clarke, C. (1977) *Introducing Visual Sociology*, Dubuque IA: Kendall/Hunt.

Dacy, D. and Kunreuther, H. (1969) *The Economics of Natural Disasters: Implications for Federal Policy*, New York: The Free Press.

Darnstaedt, T. (1983) *Gefahrenabwehr und Gefahrenvorsorge. Eine Untersuchung über Struktur und Bedeutung der Prognosebestände im Recht der öffentlichen Sicherheit und Ordnung.*

Dash, N. (1997) "The use of geographic information systems in disaster research," *International Journal of Mass Emergencies and Disasters* 15: 135-146.

Deacon, J. (1917) "When the city burns," *The Survey* (December 15): 302-310.

De Alessi, L. (1975) "Toward an analysis of postdisaster cooperation," *American Economic Review* 65: 127-138.

DeLiberty, T. (1995) *Class Notes from Seminar in Geographic Information Systems*, Newark DE: University of Delaware.

De Marchi, B. (1991) "The Seveso directive: an Italian pilot study in enabling communications," *Risk Analysis* 2: 207-215.

Denzin, N. and Lincoln, Y. (eds.) (1994) *Handbook of Qualitative Research*, Thousand Oaks CA: Sage.

Department of Regional Development and Environment (1991) *Primer on Natural Hazard Management in Integrated Regional Development Planning*, Washington DC: Department of Regional Development and Environment, Executive Secretariat for Economic and Social Affairs Organization of American States.

Derogatis, L. and Spencer P. (1982) *Administration and Procedures: BSI Manual I*, Clinical Psychometric Research. di Bella, A. (1987) "I telegiornali e l'alluvione in Valtellina," *Problemi dell'informazione* 4: 479-488.

Dombrowsky, W. (1983a) "Katastrophensoziologische Erkenntnisse zur gegenwärtigen Notwendigkeit von Kriegen," in K. Betz and A. Kaiser (eds.) *Wissenschaft zwischen Krieg und Frieden*, Berlin: Elefanten Press. ("Findings of sociology of disaster regarding the present necessity of wars.")

—— (1983b) "Solidarity during snow-disasters," *International Journal of Mass Emergencies and Disasters* 1: 189-206.

—— (1987) "Critical theory in sociological disaster research," in R. Dynes, B. de Marchi, and C. Pelanda (eds.), *Sociology of Disasters: Contribution of Sociology to Disaster Research*, Milano: Franco Angeli.

—— (1989) "Digitale Katastrophen und Digitalisierter Katastrophenschutz," *Zivil-verteidigung* 4: 24-33.

—— (1991) "Computereinsatz im Katastrophenschutz. Möglichkeiten und Grenzen," *Zivilschutzforschung,* Neue Folge Band 4, Schriftenreihe der Schutzkommission beim Bundesminister des Innern, hrsg. vom Bundesamt für Zivilschutz, Bonn: BZS.

—— (1995) "Zum Teufel mit dem Bindestrich. Zur Begründung der Katastrophen-Soziologie in Deutschland durch Lars Clausen," in W. Dombrowsky and U. Pasero (eds.) *Wissenschaft, Literatur, Katastrophe. Festschrift zum sechzigsten Geburtstag von Lars Clausen,* Wiesbaden: Westdeutscher Verlag.

Donohue, G., Olien, C., and Tichenor, P. (1987) "Media access and knowledge gaps," *Critical Studies in Mass Communication* 4: 87-92.

Drabek, T. (1965) *Laboratory Simulation of a Police Communication System Under Stress,* Columbus OH: doctoral dissertation, Department of Sociology, The Ohio State University.

—— (1968) *Disaster in Aisle 13,* Columbus OH: College of Administrative Science, The Ohio State University.

—— (1969a) *Laboratory Simulation of a Police Communication System Under Stress,* Columbus OH: College of Administrative Science, The Ohio State University.

—— (1969b) "Social processes in disaster: family evacuation," *Social Problems* 16: 336-349.

—— (1970) "Methodology of studying disasters: past patterns and future possibilities," *American Behavioral Scientist* 13: 331-343.

—— (1983a) "Shall we leave? a study on family reactions when disaster strikes," *Emergency Management Review* 1: 25-29.

—— (1983b) "Sociology research needs," in S. Changnon et al. (eds.), *A Comprehensive Assessment of Research Needs. on Floods and Their Mitigation,* Champaign IL: Illinois State Water Survey.

—— (1985) "Managing the emergency response," *Public Administration Review* 45: 85-92.

—— (1986) *Human System Responses to Disaster: An Inventory of Sociological Findings,* New York: Springer-Verlag.

—— (1987a) "Emergent structures," in R. Dynes, B. de Marchi, and C. Pelanda (eds.), *Sociology of Disasters: Contribution of Sociology to Disaster Research*, Milano: Franco Angeli.

——— (1987b) *The Professional Emergency Manager: Structures and Strategies for Success*, Boulder CO: Institute of Behavioral Science, University of Colorado.

——— (1989) "Taxonomy and disaster: theoretical and applied issues," in G. Kreps (ed.), *Social Structure and Disaster*, Newark DE: University of Delaware Press.

——— (1990) *Emergency Management: Strategies for Maintaining Organizational Integrity*, New York: Springer-Verlag.

——— (1991) *Microcomputers in Emergency Management: Implementation of Computer Technology*, Boulder CO: Program on Environment and Behavior, Monograph No. 51, Institute of Behavioral Science, The University of Colorado.

——— (1994a) "Away from home," in *Bridges: A Special Edition of the NCCEM Bulletin* 21-24.

——— (1994b) *Disaster Evacuation and the Tourist Industry*, Boulder CO: Institute of Behavioral Science, University of Colorado.

——— (1994c) "Disaster in aisle 13 revisited," in R. Dynes and K. Tierney (eds.), *Disasters, Collective Behavior, and Social Organization*, Newark DE: University of Delaware Press.

——— (1995) "Disaster planning and response by tourist business executives," *Cornell Hotel and Restaurant Administration Quarterly* 36: 86-96.

——— (1996a) *Disaster Evacuation Behavior: Tourists and Other Transients*, Boulder CO: Institute of Behavioral Science, University of Colorado.

——— (1996b) *The Social Dimensions of Disaster: Instructor Guide*, Emmitsburg MD: Emergency Management Institute, Federal Emergency Management Agency.

——— (1996c) *Sociology of Disaster: Instructor Guide*, Emmitsburg MD: Emergency Management Institute, Federal Emergency Management Agency.

Drabek, T. and Boggs, K. (1968) "Families in disaster: reactions and relatives," *Journal of Marriage and the Family* 30: 443-451.

Drabek, T. and Haas, J. (1967) "Laboratory simulation: myth or method?" *Social Forces* 45: 337-346.

―――― (1969) "Laboratory simulation of organizational stress," *American Sociological Review* 34: 223-238.

―――― (1974) *Understanding Complex Organizations*, Dubuque IA: Wm. C. Brown Company.

Drabek, T. and Hoetmer, G. (eds.) (1991) *Emergency Management: Principles and Practice for Local Government*, Washington DC: International City Management Association.

Drabek, T. and Key, W. (1976) "The impact of disaster on primary group linkages," *Mass Emergencies* 1: 89-105.

―――― (1984) *Conquering Disaster: Family Recovery and Long-Term Consequences*, New York: Irvington Publishers.

Drabek, T., Mushkatel, A., and Kilijanek, T. (1983) *Earthquake Mitigation Policy: The Experience of Two States*, Boulder CO: Institute of Behavioral Science, University of Colorado.

Drabek, T. and Quarantelli, E. (1967) "Scapegoats, villains, and disasters," *Transaction* 4: 12-17.

Drabek, T. and Stephenson, J. (1971) "When disaster strikes," *Journal of Applied Social Psychology* 1: 187-203.

Drabek, T., Tamminga, H., Kilijanek, T., and Adams, C. (1981) *Managing Multiorganizational Emergency Responses: Emergent Search and Rescue Networks in Natural Disasters and Remote Area Settings*, Boulder CO: Institute of Behavioral Science, University of Colorado.

Dynes, R. (1970) *Organized Behavior in Disaster*, Lexington MA: Heath Lexington Books.

―――― (1983) "Problems in emergency planning," *Energy* 8: 653-660.

―――― (1988) "Cross-cultural international research: sociology and disaster," *International Journal of Mass Emergencies and Disasters* 6: 101-129.

―――― (1994) "Community emergency planning: false assumptions and inappropriate analogies," *International Journal of Mass Emergencies and Disasters* 12: 141-158.

―――― (1998) *The Lisbon Earthquake in 1755: Shifts in Meaning in the First Modern Disaster*. Paper presented on RC 39 Session 11 at the World Congress of Sociology, Montréal.

——— (2000) "The dialogue between Voltaire and Rousseau on the Lisbon earthquake: the emergence of a social science view," *International Journal of Mass Emergencies and Disasters* 18: 97-115.

Dynes, R., De Marchi, B., and Pelanda, C. (eds.) (1987) *Sociology of Disaster*, Milan: Franco Angelia Libre.

Dynes, R., Haas, J., and Quarantelli, E. (1967) "Administrative, methodological and theoretical problems of disaster research," *Indian Sociological Bulletin* 4: 215-227.

Eco, U. and Fabbri, P. (1978) "Progetto di ricerca sull'utilizzazione dell'informazione ambientale," *Problemi dell'informazione* 4: 555-597.

Edwards, W. and von Winterfeldt, D. (1984) "Patterns of conflict about risk technologies," *Risk Analysis* 1: 55-68.

Elias, N. 1987. *Involvement and Detachment*. Translated by Edmund Jephcott. Oxford, U.K., and New York: Basil Blackwell.

Ellison, R., Milliman, J., and Roberts, R. (1984) "Measuring the regional economic effects of earthquakes and earthquake predictions," *Journal of Regional Science* 24: 559-579.

Emergency Communications Research Unit (1987) *The Gander Air Crash, December (1985)*, Ottawa: Emergency Preparedness Canada.

Enarson, E. and Morrow, B. (eds.) (1998) *The Gendered Terrain of Disaster: Through the Eyes of Women*, Westport CT: Praeger.

Erickson, B., Mostacci, L., Nosanchuk, T., and Dalyrymple, C. (1978) "The flow of crisis information as a probe of work relations," *Canadian Journal of Sociology* 3: 71-87.

Erikson, K. (1976) *Everything in its Path*, New York: Simon & Schuster.

——— (1994) *A New Species of Trouble*, New York: W.W. Norton and Company.

——— (1995) "Commentary," *The American Sociologist* 26: 4-11.

Erlandson, D., Harris, E., Skipper, B., and Allen, S. (1993) *Doing Naturalistic Inquiry: A Guide To Methods*, Thousand Oaks CA: Sage.

ESRI, Inc. (1993) *Understanding GIS: The ARC/INFO Method, Rev. 6 for Workstations*, New York: John Wiley and Sons.
—— (1995) "Florida Department of Environmental Protection wins major distinction for marine spill GIS," *ARC/NEWS*, Spring: http://www.esri.com/headlines/arcnews/spring95aricles/florida.html.
Eysenck, H. (1977) *Kriminalität und Persönlichkeit*, Wien: Europaverlag.
Farley, J., Barlow, H., Finkelstein, M., and Riley, L. (1993) "Earthquake hysteria, before and after: a survey and follow-up on public response to the Browning forecast," *International Journal of Mass Emergencies and Disasters* 11: 305-22.
Fergusson, C. (1976) "The Halifax explosion," *Journal of Education* 17: 25-31.
Fischer, H. and Krause, M. (1988) "Vegetations—und Bodensimulation mit dem Rechner," *Chip Plus* 4: 12-16.
Fonow, M., and Cook, J. (1991) *Beyond Methodology: Feminist Scholarship as Lived Research*, Bloomington IN: Indiana University Press.
Form, W., Nosow, S., Stone, G., and Westie, C. (n.d.) "Preliminary Progress Report of the Flint-Beecher Tornado," East Lansing MI: unpublished report, Continuing Education Service, Social Research Service, Michigan State University.
—— (1954) "Final Report on the Flint-Beecher Tornado," East Lansing MI: unpublished report, Social Research Service, Continuing Education Service, Michigan State University.
Forrest, B. and Nishenko, S. (1996) "Losses due to natural hazards," *Natural Hazards Observer* 21: 16-17.
Foster, T. (1917-1918) "The Halifax disaster," *Journal of the Maine Medical Association*, pp. 199-203.
Fothergill, A. (1996) "Gender, risk and disaster," *International Journal of Mass Emergencies and Disasters* 14: 33-56.
Frame, D. (1998) "Housing, natural hazards, and insurance," *Journal of Urban Economics* 44: 93-109.
Friesema, H., Caporaso J., Goldstein G., Lineberry R., and McCleary, R. (1979) *Aftermath: Communities After Natural Disasters*, Beverly Hills CA: Sage.

Fritz, C. and Mathewson, J (1957) *Convergence Behavior in Disasters: A Problem of Social Control*, Washington DC: Committee on Disaster Studies, National Academy of Sciences-National Research Council.

Fürst, D. (1990) "Umweltqualitätsstandards im System der Regionalplanung," *Landschaftstadt + Stadt* 22: 73-77.

Gamson, W. and Modigliani, A. (1989) "Media discourse and public opinion on nuclear power: a constructionist approach," *American Journal of Sociology* 95: 1-37.

Gatz, M. (1996) *Multi-Generational Predictors of Earthquake Impact an Preparedness: Final Report to National Science Foundation*, Los Angeles: Andrus Gerontology Center, University of Southern California.

Gearhart, W. and Pierce, J. (1989) "Fire control and land management in the Chaparral," *UMAP* (The Journal of Undergraduate Mathematics and its Application) 10: 47-80.

Geertz, C. (1973) "Thick description: toward an interpretive theory of culture," in C. Geertz (ed.), *The Interpretation of Cultures: Selected Essays*, New York: Basic Books.

General Officer Commanding Military District No. 6 (1917) Telegram to Corporal Frank Rickets December 17. National Archives of Canada RG 24 Vol. 4547 78-1-1.

Gerth, H. and Mills, C. W. (eds.) (1958 [1946]) *From Max Weber: Essays in Sociology*, New York: Oxford University Press.

Ghiglione, A. (1992) "Immaginario catastrofico e informazione nel dopo Chernobyl," *Comunicazioni Sociali* 1: 49-69.

Gillespie, D., Colignon, R., Banerjee, M., Murty, S., and Rogge, M. (1993) *Partnerships for Community Preparedness*, Boulder CO: Institute of Behavioral Science, University of Colorado.

Glaser, B. (1978) *Theoretical Sensitivity*, Mill Valley CA: The Sociology Press.

—— (1992) *Basics of Grounded Theory Analysis: Emergence vs. Forcing*, Mill Valley CA: The Sociology Press.

Glaser, B. and Strauss, A. (1965) *Awareness of Dying*, Chicago: Aldine.

—— (1967) *The Discovery of Grounded Theory: Strategies for Qualitative Research*, Chicago: Aldine.

Glueck, S. and Glueck, E. (1959) *Predicting Delinquency and Crime,* Cambridge MA:
—— (eds.) (1972) *Identification of Predelinquents,* New York:
Goffman, E. (1959) *The Presentation of Self in Everyday Life,* New York: Anchor.
Goltz, J., Russell, L., and Bourque, L. (1992) "Initial behavioral response to a rapid onset disaster: a case study of the October 1, 1987, Whittier Narrows earthquake," *International Journal of Mass Emergencies and Disasters* 10: 43-69.
Greenberg, D., Murty, T., and Ruffman, A. (1993) "A numerical model for the Halifax Harbour tsunami due to the 1917 explosion," *Marine Geodesy* 16: 153-167.
Griersmith, D. and Kingwell, J. (1988) *Planet Under Scrutiny: An Australian Remote Sensing Glossary,* Canberra, Australia: Bureau of Meteorology, Department of Administrative Services.
Grimshaw, A. (1973) "Comparative sociology: in what ways different from other sociologies," in M. Armer and A. Grimshaw (eds.), *Comparative Social Research: Methodological Problems and Strategies,* New York: John Wiley and Sons.
Groves, R. and Kahn, R. (1979) *Surveys by Telephone: A National Comparison with Personal Interviews,* New York: Academic Press.
Gruntfest, E. and Weber, M. (1998) "Internet and emergency management: prospects for the future," *International Journal of Mass Emergencies and Disasters* 16: 55-72.
Guba, E. (1981) "Criteria for assessing the trustworthiness of naturalistic inquiries," *Educational Communication and Technology Journal* 29: 75-92.
Guetzkow, H. (1962) "Joining field and laboratory work in disaster research," in G. Baker and D. Chapman (eds.), *Man and Society in Disaster,* New York: Basic Books.
Guimaraes, P., Hefner, F., and Woordward, D. (1993) "Wealth and income effects of natural disasters: an econometric analysis of Hurricane Hugo," *The Review of Regional Studies* 23: 97-114.
Guilette, E. (1993) "Applying an age-group approach to leadership during hurricane recovery," Miami FL: National Research Conference on Hurricane Hazard Mitigation.

Gurney, J. (1985) "Not one of the guys: the female researcher in a male-dominated setting," *Qualitative Sociology* 8: 42-62.

Gurr, T. (1990) *Polity II: Political Structures and Regime Change, 1800-1986*, Ann Arbor MI: Inter-University Consortium for Political and Social Research, ICPSR 9263.

Gusfield, J. (1984) "On the side: practical action and social constructivism in social problems theory," in J. Schneider and J. Kitsuse (eds.), *Studies in the Sociology of Social Problems*, Norwood NJ: Ablex.

Haas, J. and Drabek, T. (1973) *Complex Organizations: A Sociological Perspective*, New York: The Macmillan Company.

Hammond, P. (ed.) (1964) *Sociologists at Work*, New York: Basic Books, Inc.

Hansen, A. (1991) "The media and the social construction of environment," *Media, Culture and Society* 13: 443-458.

Hantrias, L., and Mangen. S. (eds.) (1996a) *Cross-National Research Methods in the Social Sciences*, London: Pinter.

—— (1996b) "Method and management of cross-national social research," in L. Hantrias and S. Mangen (eds.), *Cross-National Research Methods in the Social Sciences*, London: Pinter.

Harding, S. (1987) *Feminism and Methodology*, Bloomington IN: Indiana University Press.

—— (1991) *Whose Science?, Whose Knowledge? Thinking from Women's Lives*, Ithaca NY: Cornell University Press.

Hatfield, L. (1990) *Sammy the Prince*, Lancelot: Hantsport.

Haynes, V. and Bojcun, M. (1988) *The Chernobyl Disaster: The True Story of a Catastrophe—An Unanswerable Indictment of Nuclear Power*, London: The Hogarth Press.

Hemispheric Congress on Disasters and Sustainable Development (1996) *Proceedings*, Miami FL: International Hurricane Center, Florida International University.

Henry, W. (1905) "Henry S. Colwell," *The Maritime Merchant* 13: 31-194.

—— (1917) Letter to Alex Johnston, Deputy Minister of Marine Public Archives of Canada RG 12 (December 22).

Hewitt, K. (1995) "Excluded perspectives in the social construction of disaster," *International Journal of Mass Emergencies and Disasters* 13: 317-339.

—— (1998) "Excluded perspectives in the social construction of disaster," in E. Quarantelli (ed.), *What Is a Disaster? Perspectives in the Question*, London: Routledge.

Hirschleifer, J. (1987) *Economic Behavior in Adversity*, Chicago, IL: University of Chicago Press.

Hockings, P. (ed.) (1995) *Principles of Visual Anthropology*, Berlin: Mouton de Gruyter.

Hoover, G. (1989) "Intranational inequality: a cross-national dataset," *Social Forces* 67: 1008-1026.

Hoover, G. and Bates, F. (1985) "The impact of a natural disaster on the division of labor in twelve Guatemalan communities: a study of social change in a developing country," *International Journal of Mass Emergencies and Disasters* 3: 9-26.

Hossain, H. et al. (1992) *From Crisis to Development: Coping With Disaster in Bangladesh*, Dhaka: University Press Ltd.

Huber, F. (1958) *Daniel Bernoulli*, Basel, Switzerland.

Hudson, B. (1954) "Anxiety in response to the unfamiliar," *Journal of Social Issues* 10: 53-60.

Hwang, S., Sanderson, W., and Lindell, M. (2001) "State emergency management agencies' hazard analysis information on the Internet," *International Journal of Mass Emergencies and Disasters* 19: 85-106.

Iklé, F. and Kincaid, H. (1956) *Some Social Aspects of Wartime Evacuation of American Cities*, Washington DC: Disaster Study No. 4, Committee on Disaster Studies, National Academy of Sciences-National Research Council.

International Federation of Red Cross and Red Crescent Societies (1995) *World Disasters Report 1996*, New York: Oxford University Press.

Jackson, B. (1987) *Fieldwork*, Urbana IL: University of Illinois Press.

Jaffee, D. and Russel, T. (1997) "Catastrophic insurance, capital markets, and uninsurable risks," *Journal of Risk and Insurance* 64: 205-230.

Jäger, W. (1977) *Katastrophe und Gesellschaft: Grundlegungen und*

Kritik von Modellen der Katastrophensoziologie, Darmstadt und Neuwied: Luchterhand.

Jaggers, K. and Gurr, T. (1996) *Polity III: Political Structures and Regime Change, 1800-1994*, Ann Arbor MI: Inter-University Consortium for Political and Social Research, ICPSR 6695.

Janis, I. (1954) "Problems of theory in the analysis of stress behavior," *Journal of Social Issues* 10: 12-24.

Jefferson, H. (1958) "Day of disaster," *Atlantic Advocate* 48: 5 pp.

Jobb, D. (1991) "Assigning the blame," in *Crime Wave: Con Men, Rogues and Scoundrels from Nova Scotia's Past*, Porter's Lake: Pottersfield Press.

Johnstone, D. (Circa 1919) "The tragedy of Halifax: the greatest American disaster of the war," unpublished manuscript, Public Archives of Nova Scotia.

Kaplan, A. (1964) *The Conduct of Inquiry: Methodology for Behavioral Science*, San Francisco: Chandler.

Karim, A. (1960) "The methodology for a sociology of East Pakistan," in P. Bessaignet (ed.), *Social Research in East Pakistan*, Dacca [Dhaka]: Asiatic Society.

Kasperson, R., et al. (1988) "The social amplification of risk: a conceptual framework," *Risk Analysis* 2: 177-187.

Keane, T. and Wolfe, J. (1990) "Comorbidity in post-traumatic stress disorder: an analysis of community and clinical studies," *Journal of Applied Social Psychology* 20: 1776-88.

Khondker, H. (1984) "Governmental response to famine: a case study of the 1974 famine in Bangladesh," Pittsburgh PA: Ph.D. Thesis, University of Pittsburgh.

────── (1992) "Floods and politics in Bangladesh," *Natural Hazards Observer* 14, No. 4.

Killian, L. (1952) "The significance of multi-group membership in disaster," *American Journal of Sociology* 57: 309-314.

────── (1954) "Some accomplishments and some needs in disaster study," *The Journal of Social Issues* 10: 66-72.

────── (1956) *An Introduction to Methodological Problems of Field Studies in Disasters*, Washington DC: National Academy of Sciences-National Research Council Publication 465.

Killian, L. and Rayner, J. (1953) *An Assessment of Disaster Operations Following the Warner Robins Tornado*, Washington DC: National Academy of Sciences-National Research, Committee on Disaster Studies.

Kish, L. (1965) *Survey Sampling*, New York: John Wiley.

Kitz, J. (1989) *Shattered City: The Halifax Explosion and the Road to Recovery*, Halifax: Nimbus.

—— (1992) *Survivors: Children of the Halifax Explosion*, Halifax: Nimbus.

Kline, B. (1994) "Post cards on the 1917 explosion," in A. Ruffman and C. Howell (eds.), *Ground Zero: A Reassessment of the 1917 Explosion in Halifax Harbour*, Halifax: Nimbus Publishing Limited.

Kline, I. (1995) *Social Class as the Determinant of Private-Sector Accountability*, Los Angeles CA: policy paper, Master's Degree Program in Public Policy, University of Southern California.

Kling, R. (1997) "The culture of cyberspace: the Internet for sociologists," *Contemporary Sociology* 26: 434-444.

Koehler, G. (ed.) (1996) *What Disaster Response Management Can Learn From Chaos Theory: Conference Proceedings, May 18-19, 1995*, Sacramento CA: California Research Bureau, California State Library.

Kohn, M. (1989a) "Cross-national research as an analytical strategy," in M. Kohn (ed.), *Cross-National Research in Sociology*, Newberry Park CA: Sage.

—— (1989b) *Cross-National Research in Sociology*, Newberry Park CA: Sage.

Kravis, I., Heston, A., and Summers, R. (1982) *World Product and Income: International Comparisons of Real Gross Products*, Baltimore MD: Johns Hopkins University Press.

Kreps, G. (1984) "Sociological inquiry and disaster research," *Annual Review of Sociology* 10: 309-330.

—— (1985) "Disasters and the social order," *Sociological Theory* 3: 49-64.

—— (1987) "Classical themes, structural sociology, and disaster research," in R. Dynes, B. de Marchi, and C. Pelanda (eds.),

Sociology of Disasters: Contribution of Sociology to Disaster Research, Milan, Italy: Franco Angeli.

Kreps, G. and Bosworth, S., with Mooney, J., Russell, S., Myers, K. (1994) *Organizing, Role Enactment, and Disaster: A Structural Theory*, Newark DE: University of Delaware Press.

Kreps, G. and Drabek, T. (1996) "Disasters are nonroutine social problems," *International Journal of Mass Emergencies and Disasters* 14: 129-153.

Krimsky, S. and Plough, A. (1988) *Environmental Hazards: Communicating Risks as a Social Process*, Dover MA: Auburn House Publishing Co.

Kroll-Smith, S. and Gunter, V. (1998a) "Legislators, interpreters, and disasters: the importance of how as well as what is a disaster," in E. Quarantelli (ed.), *What Is a Disaster? Perspectives in the Question*, London: Routledge.

―――― (1998b) "A plea for heterodoxy: response to Perry's remarks," in E. Quarantelli (ed.), *What Is a Disaster? Perspectives in the Question*, London: Routledge.

Kunreuther, H. (1978) *Disaster Insurance Protection: Public Policy Lessons*, New York: John Wiley and Sons.

―――― (1996) "Mitigating disaster losses through insurance," *Journal of Risk and Uncertainty* 12: 171-187.

Kunreuther, H. and Kleffner, A. (1992) "Should earthquake mitigation measures be voluntary or required?" *Journal of Regulatory Economics* 4: 321-333.

Kunreuther, H. and Miller, L. (1985) "Insurance versus disaster relief: an analysis of interactive modeling for disaster policy planning," *Public Administration Review* 45: 147-154.

Kunreuther, H. and Roth, R. (eds.) (1998) *Paying The Price: The Status and Role of Insurance Against Natural Disasters in the United States*, Washington DC: Joseph Henry Press.

Lammers, C. (1955) *Studies in Holland Flood Disaster 1953*. Volume II. *Survey of Evacuation Problems and Disaster Experiences*, Amsterdam and Washington DC: Institute for Social Research in the Netherlands and National Academy of Sciences-National Research Council, Committee on Disaster Studies.

Lane, J. (1990) "Data archives as an instrument for comparative research," in E. Øyen (ed.), *Comparative Methodology*, London: Sage.

Leiberson, S. (1985) *Making It Count: The Improvement of Social Research and Theory*, Berkeley: University of California Press.

Lewis, C. and Murdock, K. (1996) "The role of government contracts in discretionary reinsurance markets for natural disasters," *Journal of Risk and Insurance* 62: 567-597.

Lewis, T. and Nickerson, D. (1989) "Self-insurance against natural disasters," *Journal of Environmental Economics and Management* 16: 209-233.

Lincoln, Y. and Guba, E. (1985) *Naturalistic Inquiry*, Newbury Park CA: Sage.

Litjen, R., et al. (1978) *Mathematical Simulation of Chaparral Management Alternatives: Final Report to the U.S. Forest Service Fire Laboratory*, Claremont CA: Claremont Colleges.

Lofland, J. (1971) *Analyzing Social Settings*, Belmont CA: Wadsworth.

Lofland, J. and Lofland, L. (1995) *Analyzing Social Settings: A Guide to Qualitative Observation and Analysis*, Third Edition, Belmont CA: Wadsworth.

Logan, L., Killian, L., and Marrs, W. (1952) *A Study of the Effect of Catastrophe on Social Organization*, Chevy Chase MD: Operations Research Office.

Lombardi, M. (1989) "Gestione dell'informazione nelle emergenze di massa. Note intorno al caso Chernobyl," *Studi di Sociologia* 2: 216-227.

—— (1993) *Tsunami. Crisis management della comunicazione*, Milano: Vita e Pensiero.

—— (1996) *Rischio ambientale e comunicazione*, Milano: Franco Angeli.

Lotz, J. (1981) *The Sixth of December*, Markham: Paperjacks.

MacDonald, D., Murdoch, J., and White, H. (1987) "Hazards and insurance in housing," *Land Economics* 63: 361-371.

MacLennan, H. (1938) "Concussion," *Lower Canada College Magazine* MCMXVIII: 27-30.

———— (1941) *Barometer Rising*, Toronto: Collins.

McNabb, S. (1995) "Social research and litigation: good intentions versus good ethics," *Human Organization* 54: 331-335.

McNeil, R. (1992) *Burden of Desire*, Toronto: Doubleday.

Marquardt, K. (1976) *Computersimulation der Folgen kommunalpolitischer Entscheidungen*, Frankfurt am Main: Haag und Herchen.

Marshall, C. and Rossman, G. (1995) *Designing Qualitative Research*, Newbury Park CA: Sage.

Marshall, E. (1993) "Court orders 'sharing' of data," *Science* 261: 284-286.

Martin, J. (1957) *The Story of Dartmouth*, Privately published.

Mascherpa, B. (1990) *La stampa quotidiana e la catastrofe di Seveso*, Milano: Viata e Pensiero.

Mazur, A. (1981) "Media coverage and public opinion on scientific controversies," *Journal of Communication* 2: 106-115.

Merton, R. (1969) "Foreword," in A. Barton (ed.), *Communities in Disaster: A Sociological Analysis of Collective Stress Situations*, Garden City NY: Doubleday and Company, Inc.

Metson, G. (1978) *The Halifax Explosion December 6, 1917*, Toronto: McGraw-Hill Ryerson.

Miles, M. and Huberman, A. (1994) *Qualitative Data Analysis*, Second Edition, Thousand Oaks CA: Sage.

Mileti, D. (1987) "Sociological methods and disaster research," in R. Dynes, B. de Marchi, and C. Pelanda (eds.), *Sociology of Disasters: Contributions of Sociology to Disaster Research*, Milan, Italy: Franco Angeli.

Mileti, D. (1999) *Disasters by Design: A Reassessment of Natural Hazards in the United States*, Washington DC: Joseph Henry Press.

Mileti, D. and Darlington, J. (1995) "Societal response to revised earthquake probabilities in the San Francisco Bay area," *International Journal of Mass Emergencies and Disasters* 13: 119-145.

Mileti, D., Darlington, J., Passerini, E., Forrest, B., and Myers, M. (1995) "Toward an integration of natural hazards and sustainability," *The Environmental Professional* 17: 117-126.

Mileti, D., Drabek, T., and Haas, J. (1975) *Human Systems in Extreme Environments*, Boulder CO: Institute of Behavioral Science, University of Colorado.

Mileti, D. and O'Brien, P. (1992) "Warnings during disaster: normalizing communicated risk," *Social Problems* 39: 40-57.

Miller, K. and Simile, C. (1992) "They could see stars from their beds.: the plight of the rural poor in the aftermath of Hurricane Hugo," Newark DE: Disaster Research Center Preliminary Paper #175.

Mills, C. (1959) *The Sociological Imagination*, New York: Oxford University Press.

Monnon, M. (1977) *Miracles and Mysteries The Halifax Explosion December 6, 1917*, Hantsport: Lancelot Press Limited.

Moore, D. and Okamoto, T., et al. (1985) "The FEMA earthquake damage and loss estimation system (FEDLOSS)," in J. Carroll (ed.) *Emergency Planning*, Proceedings of the Conference on Emergency Planning, 24-26 January 1985, San Diego CA, Simulation Series Vol. 15, No. 1, San Diego: Society for Computer Simulation.

Moore, H. (1958) *Tornadoes Over Texas: A Study of Waco and San Angelo in Disaster*, Austin TX: University of Texas Press.

Moore, H., Bates, F., Layman, M., and Parenton, V. (1963) *Before the Wind: A Study of Response to Hurricane Carla*, Washington DC: Disaster Study No. 19, National Academy of Sciences-National Research Council.

Morrow, B. and Enarson, E. (1994) "Making the case for gendered disaster research," paper presented at the 13th World Congress of Sociology, Bielefeld, Germany, July 18-23.

——— (1996) "Hurricane Andrew through women's eyes: issues and recommendations," *International Journal of Mass Emergencies and Disasters* 14: 5-22.

Morton, R. (1986) *Behind the Headlines*, Halifax: Nimbus Publishing Limited.

Mowat, F. (1956) *The Grey Seas Under*, Toronto: McClelland and Stewart.

Mukherjee, R. (1944) "Effect of the food crisis of 1943 on the rural population of Noakhali, Bengal," *Science and Culture* 10: 185-91 and 231-238.

—— (1948) "Economic structures of rural Bengal: a survey of six villages," *American Sociological Review* 13: 6.

Murdoch, J. Singh, H., and Thayer, M. (1993) "The impact of natural hazards on housing values: the Loma Prieta earthquake," *Journal of the American Real Estate and Urban Economics Association* 21: 167-184.

Musson, R. (1986) "The use of newspaper data in historical earthquake studies," *Disasters* 10: 217-223.

National Center for Geographic Information and Analysis (1990) *CGIA Core Curriculum in GIS*, Boulder CO: Natural Hazards Observer.

—— (1995) "What cities can take from Kobe's catastrophe," *Natural Hazards Observer* 14: 5.

National Science Foundation (1980) *A Report on Flood Hazard Mitigation*, Washington DC: National Science Foundation.

Neal, D. and Phillips, B. (1995) "Effective emergency management: reconsidering the bureaucratic approach," *Disasters* 19: 327-337.

Nemetz, P. and Dushnisky, K. (1994) "Estimating potential capital losses from large earthquakes," *Urban Studies* 31: 99-121.

Nielsen, J. (ed.) (1990) *Feminist Research Methods*, Boulder CO: Westview Press.

Norris, F. (1990) "Screening for traumatic stress: a scale for use in the general population," *Journal of Applied Social Psychology* 20: 1704-18.

Novak, J. (1994) *Bangladesh: Reflections on the Water*, Dhaka: University Press Limited.

Oliver-Smith, A. (1992) *The Martyred City: Death and Rebirth in the Peruvian Andes*, Prospect Heights IL: Waveland.

—— (1996) "Anthropological research on hazards and disasters," *Annual Review of Anthropology* 25: 303-328.

Omvik, A. (1994) "Onsker opplysinger om sjofolk ombord pa <<IMO>>." *Sandefjord Blad* (August 9): 1.

Øyen, E. (1990a) "The Imperfection of Comparisons," in E. Øyen (ed.) *Comparative Methodology: Theory and Practice in International Social Research*, London: Sage.

―――― (ed.) (1990b) *Comparative Methodology: Theory and Practice in International Social Research*, London: Sage.

Palm, R. (1995) *Earthquake Insurance: A Longitudinal Study of California Homeowners*, Boulder, CO: Westview Press.

Palm, R., Hodgson, M., Blanchard, R., and Lyons, D. (1990) *Earthquake Insurance in California, Environmental Policy and Individual Decision-Making*, Boulder CO: Westview Press.

Payzant, J. and L. Payzant. (1979) *Like A Weaver's Shuttle: A History of the Halifax-Dartmouth Ferries*, Halifax: Nimbus.

Peacock, W., Hoover, G., and Killian, C. (1989) "Divergence and convergence in international development," *American Sociological Review* 53: 838-852.

Peacock, W., Killian, C., and Bates, F. (1987) "The effects of disaster damage and housing aid on household recovery following the 1976 Guatemalan earthquake," *International Journal of Mass Emergencies and Disasters* 5: 63-88.

Peacock, W., Morrow, B., and Gladwin, H. (eds.) (1997) *Hurricane Andrew: Ethnicity, Gender, and the Sociology of Disaster*, London: Routledge.

Pelanda, C. (1982) *Disaster and Order: Theoretical Problems in Disaster Research*, Gorizia, Italy: Institute of International Sociology.

Pennings, J. (1981) "Strategically interdependent organizations," in P. Nystrom and W. Starbuck (eds.), *Handbook of Organizational Design,* Volume 1, *Adapting Organizations to Their Environments*, New York: Oxford University Press.

Perrow, C. (1993) Letter to the author, 6 May.

Perrow, C. and Guillén, M. (1990) *The AIDS Disaster: The Failure of Organizations in New York and the Nation*, New Haven and London: Yale University Press.

Perry, R. and Mushkatel, A. (1986) *Minority Citizens in Disasters*, Athens GA: University of Georgia Press.

Phillips, B., Garza, L., and Neal, D. (1994) "Intergroup relations in disasters: service delivery barriers after Hurricane Andrew," *Journal of Intergroup Relations* 21: 18-27.

Picou, J. (1996a) "Toxins in the environment, damage to the community: sociology and the toxic tort," in P. Jenkins and S.

Kroll-Smith (eds.), *Witnessing for Sociology: Sociologists in Court*, Westport CT: Praeger.

——— (1996b) "Sociology and compelled disclosure: protecting respondent confidentiality," revised version of presidential address presented at the annual meeting of the Mid-South Sociological Association, October, (1995), Mobile AL: Department of Sociology and Anthropology, University of South Alabama.

Ponting, J., Fitzpatrick, J., and Quarantelli, E. (1975) "Police perceptions of riot participants and dynamics," *International Journal of Group Tensions* 5: 163-170.

Powell, J. (1954) "An introduction to the natural history of disaster," College Park MD: unpublished report, the Psychiatric Research Institute, University of Maryland.

Powell, J., Rayner, J., and Finesinger, J. (1953) "Response to disaster in American cultural groups," in *Symposium on Stress*, Washington DC: Walter Reed Army Medical Center, Army Medical Service Graduate School.

Presser, S. (1994) "Presidential address: informed consent and confidentiality in survey research," *Public Opinion Quarterly* 58: 446-459.

Prince, S. (1920) *Catastrophe and Social Change*, New York: Columbia University.

Pring, G. and Canan, P. (1996) *SLAPPs: Getting Sued for Speaking Out*, Philadelphia PA: Temple University Press.

Przeworski, A. and Teune, H. (1970) *The Logic of Comparative Social Inquiry*, New York: Wiley-Interscience.

Quarantelli, E. (1980) *The Study of Disaster Movies: Research Problems, Findings and Implications*, Newark DE: Preliminary Paper No. 65, Disaster Research Center, University of Delaware.

——— (1983) *Delivery of Emergency Medical Services in Disasters: Assumptions and Realities*, New York: Irvington Publishers, Inc.

——— (1984) *Sociobehavioral Responses to Chemical Hazards: Preparations for and Responses to Acute Chemical Emergencies at the Local Community Level*, Newark DE: Disaster Research Center, University of Delaware.

—— (1987a) "What should we study? questions and suggestions for researchers about the concept of disasters," *International Journal of Mass Emergencies and Disasters* 5: 7-32.

—— (1987b) "Disaster studies: an analysis of the social historical factors affecting the development of research in the area," *International Journal of Mass Emergencies and Disasters* 5: 285-310.

—— (1987c) *Research In The Disaster Area: What Is Being Done and What Should Be Done?*, Newark DE: Disaster Research Center Preliminary Paper No. 118.

—— (1988) "The NORC research on the Arkansas tornado: a fountainhead study," *International Journal of Mass Emergencies and Disasters* 6: 283-310.

—— (1993) "Converting disaster scholarship into effective disaster planning and managing: possibilities and limitations," *International Journal of Mass Emergencies and Disasters* 11: 15-39.

—— (1994) "Disaster studies: the consequences of the historical use of a sociological approach in the development of research," *International Journal of Mass Emergencies and Disasters* 12: 25-49.

—— (1995a) *Draft of a Sociological Disaster Research Agenda for the Future: Theoretical, Methodological and Empirical Issues*, Newark DE: Preliminary Paper No. 228, Disaster Research Center, University of Delaware.

—— (1995b) "What is a disaster?" *International Journal of Mass Emergencies and Disasters* 13: 221-229.

—— (1996) "The future is not the past repeated: projecting disasters in the 21st Century from present trends," *Journal of Contingencies and Crisis Management* 4: 228-240.

—— (ed.) (1998) *What Is a Disaster? Perspectives on the Question*, London and New York: Routledge.

Raddall, T. (1977) *In My Time, A Memoir*, Toronto: McClelland and Stewart Limited.

Ragin, C. (1987) *The Comparative Method*, Berkeley CA: University of California Press.

Renn, O. (1992) "Concept of risk: a classification," in S. Krimsky and D. Golding (eds.), *Social Theories of Risk*, London: Praeger.

Richards, T. and Richards, L. (1994) "Using computers in qualitative research," in N. Denzin and Y. Lincoln (eds.), *Handbook of Qualitative Research*, Thousand Oaks CA: Sage.

Richardson, L. (1990) *Writing Strategies: Reaching Diverse Audiences*, Newbury Park CA: Sage.

—— (1994) "Writing: a method of inquiry," in N. Denzin and Y. Lincoln (eds.), *Handbook of Qualitative Research*, Thousand Oaks CA: Sage.

Richardson, R., Erickson, B., and Nosanchuk, T. (1979) "Community size, network structure and the flow of information," *Canadian Journal of Sociology* 4: 379-392.

Rose, A. and Allison, T. (1989) "On the Plausibility of the Supply-Driven Input-Output Model: Empirical Evidence on Joint Stability," *Journal of Regional Science* 29: 451-458.

Rose, A., Benavides, J., Chang, S., Szczesniak, S., and Lim, D. (1997) "The regional economic impact of an earthquake: direct and indirect effects of electricity lifeline disruptions," *Journal of Regional Science* 37: 437-458.

Ross, G. (1919) "The Halifax disaster and the re-housing," *Construction: A Journal for the Architectural Engineering and Contracting Interests in Canada* 12: 293-294.

Ross, J. (1923) "Attacking infant and maternal mortality in the city: the Halifax experiment," *Public Health Nurse* 16: 125-127.

Rubin, C. (1982) "Managing recovery from a natural disaster," *Management Information Service Representative* 14 : 1-15.

Rubin, D., et al. (1976) *Report of the Public's Right to Know Information Task Force*, Washington DC: The President's Commission on Three Mile Island.

Ruffman, A. and Howell, C. (eds.) (1994) *Ground Zero: A Reassessment of the 1917 Explosion in Halifax Harbour*, Halifax: Nimbus Publishing Limited.

Russell, L., Goltz, J., and Bourque, L. (1995) "Preparedness and hazard mitigation actions before and after two earthquakes," *Environment and Behavior* 27: 744-70.

Saarinen, T. (ed.) (1982) *Perspectives on Increasing Hazard Awareness*, Boulder CO: Program on Environment and Behavior Monograph No. 35, Institute of Behavioral Science, University of Colorado.

Sawatsky, J. (1984) *GOUZENKO The Untold Story*, Toronto: Macmillan of Canada.

Scanlon, J. (1971) "News flow about release of kidnapped diplomat researched. by j-students," *Journalism Educator* (Spring): 35-38.

────── (1972) "A new approach to the study of newspaper accuracy," *Journalism Quarterly* (Autumn): 587-590.

────── (1974a) *The North Bay/Slater Study*, Ottawa: National Emergency Planning Establishment.

────── (1974b) "They didn't know she was listening," *The Carleton Alumneye* 1: 5-6.

────── (1976) "The not so mass media: the role of individuals in mass communications," in G. Adam (ed.), *Journalism, Communications and the Law*, Scarborough: Prentice-Hall of Canada, Ltd.

────── (1977a) "Post-disaster rumor chains: a case study," *Mass Emergencies* 2: 121-126.

────── (1977b) "The Sikhs of Vancouver," in J. Halloran (ed.), *Ethnicity and the Media*, Paris: UNESCO.

────── (1979) "Day one in Darwin: once again the vital role of communications," in J. Reid (ed.), *Planning for People in Natural Disasters*, Townsville: James Cook University.

────── (1985a) "Students practice detection 39 years after defection," *Bulletin*, Ottawa: The Centre for Investigative Journalism, pp. 7-8.

────── (1985b) "Crisis communications in Canada," in B. Singer (ed.), *Communications in Canadian Society*, Toronto: Copp Clark Publishing.

────── (1988) "Disaster's little known pioneer: Canada's Samuel Henry Prince," *International Journal of Mass Emergencies and Disasters* 6: 213-232.

────── (1992a) *Convergence Revisited: A New Perspective on a Little*

Studied Topic, Boulder CO: Natural Hazards Research and Information Applications Center.

——— (1992b) "The man who helped Sammy Prince write," *International Journal of Mass Emergencies and Disasters* 10: 189-206.

——— (1994) "EMS in Halifax after the 6 December 1917 explosion: testing Quarantelli's theories with historical data," in R. Dynes and K. Tierney (eds.), *Disasters, Collective Behavior, and Social Organization*, Newark DE: University of Delaware Press.

——— (1996) "Not on the record: disasters, records and disaster research," *International Journal of Mass Emergencies and Disasters* 14: 265-280.

Scanlon, J. and Hiscott, R. (1982) "Canadian hospital fire reports: an incomplete record," *Fire Journal* (November): 85-88.

Scanlon, J., Jefferson, J., and Sproat, D. (1976) *The Port Alice Slide*, Ottawa: Emergency Planning Canada.

Scanlon, J., Luukko, R., and Morton, G. (1978) "Media coverage of crises: better than reported, worse than necessary," *Journalism Quarterly* 55: 68-72.

Scanlon, J. and Prawzick, A. (1985) *The 1978 San Diego Air Crash Emergency Response to an Urban Disaster*, Ottawa: Emergency Preparedness Canada.

Scanlon, J. and Sylves, R. (1990) "Conflict and co-ordination in responding to aviation disaster: the San Diego and Gander experiences compared," in R. Sylves and W. Waugh (eds.), *Cities and Disaster North American Studies in Emergency Management*, Springfield IL: Charles C. Thomas Publisher.

Scanlon, J. and Taylor, B. (1977) "A stand-by research capacity," *Mass Emergencies* 2: 35-41.

Scarce, R. (1994) "(No) trial (but) tribulations: when courts and ethnography conflict," *Journal of Contemporary Ethnography* 23: 123-149.

Scawthorn, C., Iemura, H., and Yamada, Y. (1982) "The influence of natural hazards on urban housing location," *Journal of Urban Economics* 11: 242-251.

Schatzman, L. and Strauss, A. (1973) *Field Research: Strategies for a Natural Sociology*, Englewood Cliffs NJ: Prentice-Hall.

Scheuch, E. (1990) "The development of comparative research: toward causal explanations," in E. Øyen (ed.), *Comparative Methodology*, London: Sage.

Schrage, E. and Engel, P. (1982) "The decision maker's dilemma: balancing risks on the fine line between cost and compassion," *The Sciences* (August/September): 26-31.

Schwandt, T. and Halpern, E. (1988) *Linking Auditing and Metaevaluation*, Newbury Park CA.: Sage.

Schwartz, H. and Jacobs, J. (1979) *Qualitative Sociology: A Method to the Madness*, New York: Free Press.

Scott, W. (1955) *Public Reactions to a Surprise Civil Defense Alert in Oakland, California*, Ann Arbor MI: unpublished report, Survey Research Center, University of Michigan.

Shamgar-Handelman, L. (1983) "The social status of war widows," *International Journal of Mass Emergencies and Disasters* 1: 153-170.

Shilling, J., Sirmans, C., and Benjamin, J. (1989) "Flood insurance, wealth redistribution, and urban property values," *Journal of Urban Economics* 26: 43-53.

Simile, C. (1995) *Disaster Settings and Mobilization for Contentious Collective Action: Case Studies of Hurricane Hugo and the Loma Prieta Earthquake*, Newark DE: doctoral dissertation, Department of Sociology and Criminal Justice, University of Delaware.

Simon, J., Simon E., and Taeger, J. (1981) "Wer sich umdreht oder lacht. Rasterfahndung: Ein Beitrag zur Gewährleistung der inneren Sicherheit," *Kursbuch* 66: 20-36.

Simon, J. and Taeger, J. (1981) *Rasterfahndung. Entwicklung, Inhalt und Grenzen einer kriminalpolizeilichen Fahndungsmethode*, Baden Baden, Germany: Nomos.

Slovic, P. (1987) "Perception of risk," *Science* 236: 280-285.

Smelser, N. (1998) "The rational and the ambivalent in the social sciences," *American Sociological Review* 63: 1-16.

Smillie, R. and Ayoub, M. (1986) "Accident causation theories: a simulation approach," *Journal of Occupational Accidents* 1: 47-68.

Smith, S. (1918) *Heart Throbs of the Halifax Horror*, Halifax: Gerald E. Weir.

Sorensen, J., Vogt, B., and Mileti, D. (1987) *Evacuation: An Assessment of Planning and Research*, Oak Ridge TN: Oak Ridge National Laboratory.

Spaeth, M. (1990) "CATI Facilities at Academic Survey Research Organizations," *Survey Research* 21: 11-14.

Spradley, J. (1980) *Participant Observation*, Orlando FL: Harcourt, Brace, Jovanovich.

Stake, R. (1995) *The Art of Case Study Research*, Thousand Oaks CA: Sage.

Stallings, R. (1986) *National Science Foundation Field Report: The Miamisburg (Ohio) Train Derailment and Toxic Fire of July 8, 1986*, Los Angeles: School of Public Administration, University of Southern California.

—— (1995) *Promoting Risk: Constructing the Earthquake Threat*, Hawthorne NY: Aldine de Gruyter.

Stallings, R. and Quarantelli, E. (1985) "Emergent citizen groups and emergency management," *Public Administration Review* 45: 93-100.

Stern, P. (1991) "Learning through conflict: a realistic strategy for risk communication," *Policy Sciences* 24: 99-119.

Strauss, A. and Corbin, J. (1992) *Basics of Qualitative Research: Grounded Theory Procedures and Techniques*, Newbury Park CA: Sage.

Streeter, C. (1991) "Disasters and development: disaster preparedness and mitigation as an essential component of development planning," *Social Development Issues* 13: 100-110.

Stretton, A. (1976) *The Furious Days*, Sydney, Australia: William Collins Publishers Pty. Ltd.

Summers, R. and Heston, A. (1988) "A new set of international comparisons of real product an price level estimates for 130 countries, 1950-1985," *Review of Income and Wealth* 34: 1-25.

Swanson, G. (1971) "Frameworks for comparative research: structural anthropology and the theory of action," in I. Vallier (ed.), *Comparative Methods in Sociology*, Berkeley CA: University of California Press.

Sykes, G. and Drabek, T. (1969) *Law and the Lawless: A Reader in Criminology*, New York: Random House.

Sztompka, P. (1988) "Conceptual frameworks in comparative inquiry: divergent or convergent?" *International Sociology* 3: 207-218.

Taylor, C. and Jodice, D. (1983) *World Handbook of Political an Social Indicators*, New Haven CT: Yale University Press.

Taylor, J., Zurcher, L., and Key, W. (1970) *Tornado: A Community Responds to Disaster*, Seattle WA: University of Washington Press.

Taylor, V. (1976) *The Delivery of Mental Health Services in the Xenia Tornado: A Collective Behavior Analysis of an Emergent System Response*, Columbus OH: doctoral dissertation, Department of Sociology, The Ohio State University.

—— (1978) "Future directions for study," in E. Quarantelli (ed.), *Disasters: Theory and Research*, London: Sage Publications Limited.

Taylor, V., Ross, G. and Quarantelli, E. (1976) *Delivery of Mental Health Services in Disasters: The Xenia Tornado and Some Implications*, Newark DE: Disaster Research Center, University of Delaware.

Teune, H. (1990) "Comparing countries: lessons learned," in E. Øyen (ed.), *Comparative Methodology*, London: Sage.

Thompson, J. (1967) *Organizations in Action*, New York: McGraw-Hill.

Tierney, K., Nigg, J., and Dahlhamer, J. (1996) "The impact of the 1993 Midwest floods: business vulnerability and disruption in Des Moines," in R. Sylves and W. Waugh (eds.), *Disaster Management in the U.S. and Canada*, Second Edition, Springfield IL: Charles C. Thomas.

Tierney, K. and Webb, G. (1995) *Managing Organizational Impressions in Crisis Situations: Exxon Corporation and the Exxon Valdez Oil Spill*, Newark DE: Disaster Research Center, Preliminary Paper No. 235.

Tonnessen, J. and Johnsen, A. (1982) *The History of Modern Whaling*, translated by R. Christopherson, London: C. Hurst and Company.

Tooke, F. (1918) "An experience through the Halifax disaster," *The Canadian Medical Association Journal Volume* 8: 308-320.

Tsumi, M., Nomura, N. and Shibuya, T. (1985) "Simulation of post-earthquake restoration for lifeline systems," *International Journal of Mass Emergencies and Disasters* 3: 87-105.

Turner, R. (1990) "Comparative content analysis of biographies," in E. Øyen (ed.), *Comparative Methodology*, London: Sage.

Turner, R., Nigg, J., and Paz, D. (1986) *Waiting for Disaster: Earthquake Watch in California*, Berkeley CA: University of California Press.

Ungerson, C. (1996) "Qualitative methods," in L. Hantrias and S. Mangen (eds.), *Cross-National Research methods in the Social Sciences*, London: Pinter.

U. S. Census (1990) *U.S. Census, 5% PUMS*, Washington DC: U.S. Census Bureau.

Valentini, T. (1992) *Analisi e comunicazione del rischio tecnologico*, Naples, Italy: Liguori.

Van Maanen, J. (1988) *Tales of the Field: On Writing Ethnography*, Chicago IL: University of Chicago Press.

—— (ed.) (1995) *Representation in Ethnography*, Thousand Oaks CA: Sage. van Schendel, W. (1985) "Living and working with villagers in Rangpur: a few remarks," in A. Chowdhury (ed.) *Pains and Pleasures of Fieldwork*, Dhaka: NILG.

Vilma, N. (1997) "Cybergangsters Paradise. Straftaten im Internet," *PC Magazin* 10: 62-64.

Wagner, J. (ed.) (1979) *Images of Information: Still Photography in the Social Sciences*, Beverly Hills CA: Sage.

Wallace, A. (1956a) *Human Behavior in Extreme Situations: A Survey of the Literature and Suggestions for Further Research*, Washington DC: Disaster Study No. 1, Committee on Disaster Studies, National Academy of Sciences-National Research Council.

—— (1956b) *Tornado in Worcester: An Exploratory Study of Individual and Community Behavior in an Extreme Situation*, Washington DC: Disaster Study No. 3, Committee on Disaster Studies, National Academy of Sciences-National Research Council.

Warheit, G. and Quarantelli, E. (1969) *An Analysis of the Los Angeles Fire Department Operations During Watts*, Newark DE: Disaster Research Center, University of Delaware.

Warren, C. (1988) *Gender Issues in Field Research*, Newbury Park CA: Sage.

Webb, E., Campbell, D., Schwartz, R., Sechrest, L., and Grove, J. (1981) *Nonreactive Measures in the Social Sciences*, Second Edition, Boston: Houghton Mifflin.

Weber, M. (1949) *The Methodology of the Social Sciences*, translated by Edward A. Shils and Henry A. Finch, New York: The Free Press.

——— (1958) *The Protestant Ethic and the Spirit of Capitalism*, translated by Talcott Parsons, New York: Charles Scribner's Sons.

Weitzman, E. and Miles, M. (1995) *Computer Programs for Qualitative Data Analysis*, Newbury Park CA: Sage.

Wenger, D. (1989) "Appendix to part one: the role of archives for comparative studies of social structure and disaster," in G. Kreps (ed.), *Social Structure and Disaster*, Newark DE: University of Delaware Press.

West, C. and Lenze, D. (1994) "Modeling the regional impact of natural disaster and recovery: a general framework and application to Hurricane Andrew," *International Regional Science Review* 17: 121-150.

White, G. and Haas, J. (1975) *Assessment of Research on Natural Hazards*, Cambridge MA and London: The MIT Press.

Whyte, W. (1943) *Street Corner Society: The Social Structure of an Italian Slum*, Chicago: University of Chicago Press.

Wilkins, L. and Patterson, P. (1987) "Risk analysis and the construction of news," *Journal of Communication* 3: 80-92.

Williams, H. (1954) "Fewer disasters, better studied," *Journal of Social Issues* 10: 5-11.

——— (1964) "Human factors in warning-response systems," in G. Grosser, H. Wechsler, and M. Greenblatt (eds.), *Threat of Impending Disaster*, Cambridge MA: The MIT Press.

Wolcott, H. (1990) *Writing Up Qualitative Research*, Newbury Park CA: Sage.

——— (1994) *Transforming Qualitative Data: Description, Analysis and Interpretation*, Thousand Oaks CA: Sage.

Wolensky, R. and Wolensky, K. (1990) "Local government's problem with disaster management: a literature review and structural analysis," *Policy Studies Review* 9: 703-725.

World Resources Institute (1990) *World Resources, 1990-91*, New York: Oxford University Press.

Worth, S. and Adair, J. (1972) *Through Navajo Eyes: An Exploration in Film Communication and Anthropology*, Bloomington IN: Indiana University Press.

Wotherspoon, W. and Scanlon, J. (1974) "Crisis communication," *Royal Canadian Mounted Police Gazette* 36: 8-11.

Wright, J., Rossi, P., Wright, S., and Weber-Burdin, E. (1979) *After the Clean-up: Long-Range Effects of Natural Disasters*, Beverly Hills CA: Sage.

Wynne, B. (1987) *Risk Management and Hazardous Waste: Implementation and the Dialectic of Credibility*, Berlin: Springer.

Yezer, A. (1992) "Differential impact of earthquake events," in *The Economic Consequences of a Catastrophic Earthquake*, Washington DC: National Academy of Sciences Press.

Yezer, A. and Rubin, C. (1987) *The Local Economic Effects of Natural Disasters*, Boulder CO: Working Paper No. 61, Institute of Behavioral Science, University of Colorado.

Yin, R. (1984) *Case Study Research: Design and Methods*, Beverly Hills CA: Sage Publications.

—— (1989) *Case Study Research*, Second Edition, Newbury Park CA: Sage.

# INDEX

A

Acadia University (Wolfville, Nova Scotia) 279, 293
Acker, J. 369
Adair, J. 206
Adams, T. 277
Aday, L. 160, 161
Adnan, S. 338, 344
Africville section of Halifax 295
Ahmed, O. 346
AIDS 39
air crash
 Gander, Newfoundland 268
 San Diego, California 267
Alaskan earthquake 97, 105
Aldrich, J. 277
Allison, T. 229
Allport, G. 80
America Online (AOL) 312
American Red Cross 82, 99, 240
American Society of Public Administration 371
American Sociological Association 358
Amherst, Massachusetts 287
Amherst, Nova Scotia 283
analysis of quantitative data 87
Anderson, M. 261
Anderson, W. 139, 201, 207, 272, 300
Andorka, R. 243
Andrews, P. 319
Andrus Center for Gerontology, University of Southern California California 162
Anne of Green Gables 292
Archiv für Sozialwissenschaft 41
Armer, M. 238
Aronmore 292
Association of Bay Area Governments 319
Atkinson, S. 209
Atlantic Advocate 277
Augusta, Maine 279
Australian Journal of Emergency Management, The 146
Awareness of Dying 195

503

B

Baddeley, A. 274
Baker, G. 23, 24, 137
Baker, M. 231
Balassa, B. 245
Baldwin Hills (California) dam break 97
Bangladesh 36, 303, 334, 346
  a field study of two villages in disaster 347
  disaster research in 339, 348
  disasters in 335
Bangladesh Nationalist Party 340
Barinaga, M. 355
Barometer Rising 274
Barton, A. 137, 272
Bates, F. 242, 246, 248, 249
Bay of Bengal 335, 339, 346
Beaton Institute, Sidney, Nova Scotia 279
Becerra, R. 161
Becker, H. 40
Belcher, J. 248
Bell, F. 274, 279
Beniger, J. 317
Benjamin, J. 146
Benson 298
Berger, R. 312
Berke, P. 356
Bernknopf, R. 216
Bernoulli, D. 315, 316
Berting, J. 243

Best, C. 291
Biderman, A. 25
Big Bear Lake (California) earthquakes 141
Big Thompson River Canyon (Colorado) flood 135
Biklen, S. 196, 205
Biman Bangladesh Airlines 335
Bird, M. 275, 300
Blalock, H. 241
Bligh, C. 277
Blinn, L. 206
blizzard, Chicago 359
Block, F. 245
Blumer, H. 134, 200
Board of Education, London (Ontario) 291
Bogdan, R. 196, 205
Boggs, K. 134
Bojcun, M. 318
Bolin, R. 157, 160, 248, 369
Bolton, P. 157, 160, 248, 369
Boruch, R. 359
Boston (Massachusetts) Public Library 280
Boston Express (train) 297, 298
Bourque, L. 29, 144, 155, 159, 160, 161, 162, 163, 174, 190
Brajuha, M. 355
Bremer, S. 319
Bridgewater, Nova Scotia 287
Britton, N. 146, 321
Bronson, H. 277, 291

Brooks, H. 122
Brookshire, D. 216, 218, 222
Broward County, Florida 385
Brown, C. 278
Brown, J. 215, 216
Bucher, R. 120, 366
businesses
  losses in disaster 386
  readiness for disaster 379
Butler, D. 38, 309
Butler, J. 225

C

c-model (conditions, characteristics, careers, and consequences of disasters) onsequences of disasters) 100, 204
Cable News Network 310
Calderon, F. 238
California Earthquake Authority 233
California Universities for Research in Earthquake Engeering
ngeering 363
California Universities for Research in Earthquake Engineering 363
Calonne 280
Canadian Army Medical Corps 274, 279
Canadian Forces Base Esquimalt 285
Canadian Journal of Sociology 269

Canadian Nurse 278
Canadian Pacific Railroad 283, 294
Canadian Pharmaceutical Journal 277
Canadian Police College 285
Canadian Press 285
Canadian Prime Minister Robert Borden 279, 292, 295
Canan, P. 357
Caribbean Disaster Management Project 384
Carleton University, Ottawa 267, 285, 290, 291
Carstens, C. 276
Carter, M. 196
Catastrophe and Social Change 272
CBC Radio 274, 296, 298, 307
Cecil, J. 359
Chambers, B. 280, 285
chaos theory 137
Chapman, A. 288
Chapman, D. 23, 24, 50
Charleston, South Carolina 215
Charlottetown, Prince Edward Island 273, 292
Chen, L. 336
Chernobyl 318
Cheshire, P. 243
Chetkovich, C. 369
Chicago School, fieldwork tradition 194, 195, 200, 349

Chittagong, Bangladesh 339, 345, 346
Choto Chonua, Bangladesh 344, 346
Chowdhury, A. 336, 339
CIA Nation Reports 319
Cisin, I. 23, 142, 158
Clapham, K. 146
Clark, V. 159, 163, 190
Clark, W. 23, 142, 158
Clarke, C. 206
Clarke, E. 277
Clarke, L. 358
Clifford, R. 91
Clinton administration 377, 380
Coastal Barriers Resources Act of 1982 223
Coastal Barriers Resources System 223
Cobler, S. 316
Cochrane, H. 229
Cohn, R. 206
Colchester Historical Society Museum (Truro, Nova Scotia)
otia) 279, 287
Cold War 24, 39
Coleman, C. 294
Collier, J. 206
Collier, M. 206
Colombo, M. 258
Columbia University (New York) 272
Columbus (Ohio) Police Department 133

Colwell Brothers, Inc. 293
Colwell, H. 277
Committee on Disaster Studies, National Research Council
ncil 22, 47, 50, 56, 373
comparative research 31, 35, 47, 146, 155, 245, 248, 249, 340
versus cross-national research 238
Compuserve 312
computer software 33, 208
computer technology background 307
computer-assisted telephone interviewing (see also survey research)
urvey research) 160, 161, 162, 177, 179, 187, 190, 191
confirmatory studies 24, 149
research designs for 29
Congressional Budget Office 226, 233
Connelly, W. 277
ConstructionA Journal for the Architectural Engineering and Contracting Interests in Canada
ing and Contracting Interests in Canada 278
content analysis
of interviews 78
of radio broadcasts 82
convergence behavior 37, 57, 65, 114, 271, 363

Cook, J. 365
Corbin, J. 196, 204
Cordes, J. 224
Cornell University (Ithaca, New York) 358
Corner, J. 262
Council of American Survey Research Organizations 174, 177
Cranwell, J. 292
Creighton, J. 278
Crime WaveCon Men, Rogues and Scoundrels 276
cross-national research (see also comparative research) ch) 244, 250
cross-national research (see also comparative researcross-national databases
cross-national databases 243
cross-national research (see also comparative researthe problem of equivalence in the problem of equivalence in 248
Cummins, J. 226, 234
Curaca 280, 298
Curry, T. 206
cyclone in Bangladesh 346
Cyclone Tracy, Darwin (Australia) 267, 301

D

Dacy, D. 225
Dade County, Florida 325, 326
Dakin, E. 294
Dalhousie University (Halifax, Nova Scotia) 276, 277, 278, 279, 285, 291, 298
dam collapse 64, 97, 255
Darlington, J. 143, 149
Dartmouth (Nova Scotia) Relief Commission 276
Dartmouth, Nova Scotia 289
Dash, N. 35, 38, 40, 241, 250, 303
data collection 86
  projective tests 80
  questionnaires 73
  sociometric measures 81
databases
  cross-hazards 152
  cross-national 147
Davidson, O. 38, 375
De Marchi, B. 257, 272
Deacon, J. 276
DeLiberty, T. 322
Denver, 1965 flood in 140
Denzin, N. 198, 372
Derogatis, L 163
descriptive studies (see also field studies) 23, 24
development
  disaster and 32
  residential 219, 220, 224, 225, 228
  social and economic 147, 220, 239, 240, 248, 342
Dhaka, Bangladesh 334

di Bella, A. 258
Disaster Forum 334
Disaster in Aisle 13 40, 131
Disaster Management 146
disaster phases 51, 62, 143, 146, 150, 151
Disaster Recovery Business Alliance 379
Disaster Research 209
Disaster Research Center 24, 25, 27, 28, 29, 37, 40, 45, 127, 129, 131, 132, 133, 134, 145, 146, 200, 201, 202, 270, 309, 350, 352, 354, 362, 366, 367, 373, 385
  history of 101
disaster research, future needs in 152
Disaster Research Group, National Research Council 373
disaster victims 32, 34, 37, 39, 40, 53, 65, 71, 74, 76, 84, 85, 86, 88, 107, 114, 128, 134, 135, 142, 143, 144, 183, 184, 187, 201, 203, 239, 271, 275, 285, 291, 308, 318, 332, 346, 351, 354
disaster warning 24, 51, 62, 118, 134, 136, 149, 298, 318, 345
Disasters 146, 196, 204, 335
disasters
  as context for research 53

cost assessments of 148
economic effects of 220, 230
economics of 31, 143, 249
effects of on insurance markets 227
historical 33, 34, 155
Disasters Management 204
DisastersTheory and Research 267
Discovery of Grounded Theory, The 195
diversity
  issues of in field research 369
documents
  use of in research 107, 117, 194, 205, 206, 357
Doessel, D. 225
Dombrowsky, W. 34, 35, 38, 196, 303, 317
Dominion Atlantic Railroad 280, 283, 295
Donohue, G. 313
Doubleclick 313
Drabek, R. 137, 153
Drabek, T. 24, 25, 27, 29, 30, 36, 38, 40, 43, 44, 45, 94, 131, 132, 133, 134, 135, 136, 137, 138, 139, 140, 141, 142, 143, 150, 152, 205, 238, 246, 249, 272, 300, 328, 372
Dynes, R. 25, 28, 34, 95, 127, 129, 136, 137, 200, 238, 239, 272, 300, 321

E

Eaman, R. 286
Earthquake Engineering
  Research Center Library,
  University of California,
  Berkeley
  ersity of California, Berkeley
  190, 373
Earthquake Engineering
  Research Institute 193,
  362, 363
earthquake prediction 215,
  231
  study of 161, 162
earthquakes 30, 50, 143, 151,
  191, 204, 214, 215, 233,
  318, 378
Earthwatch 356
Eaton, J. 295
Eco, U. 257
economic impact analysis 213
economic models
  information effects 220
Edwards, W. 255
Elias, N. 41, 42
Ellison, R. 215, 229
Emergency Communications
  Research Unit 268, 269,
  301
emergency management
  professionalization of,
    implications for field
    research 372
  study of programs 139
  study of technology in 140

Emergency Management
  Australia 309
Emergency Management
  Institute 133, 143, 371
Emergency Operations
  Centers 107, 109, 112,
  114, 116, 119, 123, 328,
  352, 353, 360, 370
emergent multiorganizational
  systems 137, 151
Enarson, E. 201, 203, 368
Engel, P. 316
entrée in field research 36, 88,
  89, 119, 350, 361
equilibrium theory of regional
  development 213
Erickson, B. 243, 269
Erikson, K. 152, 359
Erlandson, D. 198, 199, 202,
  203
ESRI 321
Ethnograph (computer
  software) 208
eugenics 316
evaluation research 38, 150,
  348
Evening Standard 274
experimental designs 23, 24,
  25, 52, 57, 84, 133, 135,
  143, 144, 149, 150, 158,
  167, 188, 192, 353
explanatory research 24
exploratory research (see also
  field studies) 22, 45, 54,
  57

Exxon Mobil Corporaton 356, 357
Exxon Valdez 356
 oil spill 206
Eysenck, H. 315, 316

F

Fabbri, P. 257
Farley, J. 162
Federal Emergency Management Agency 96, 99, 102, 133, 143, 149, 159, 183, 202, 309, 320, 325, 327, 329, 361, 370, 371, 377, 379, 381, 386
Federal Response Plan 379
Fergusson, G. 277
field studies 23, 24, 25, 27, 29, 30, 37, 48, 49, 52, 53, 54, 56, 57, 58, 65, 68, 73, 76, 78, 81, 83, 84, 85, 86, 89, 91, 95, 97, 98, 99, 107, 110, 111, 123, 125, 133, 303, 339, 373
 DRC prefield procedures 106
 history of in disaster research 351
 problems in doing research on organizations 362
 problems with convergence of researchers 365
 reconnaissance in 352, 353, 373
 visual data in 209

findings, reporting 90, 120
Finesinger, J. 51
Fischer, H. 201
Fitzpatrick, J. 94
Florida City (Miami, Florida, neighborhood) 325
Florida International University, Miami 201
Florida Joint Underwriting Authority 233
Florida Marine Spill Analysis System 324
Focalink Communications 313
Focus 310
Fonow, M. 365
Form, W. 47, 59, 92
Forrest, B. 148
Foster, T. 277
Fothergill, A. 368
Frame, D. 218, 219, 220, 224
Friesema, H. 217, 242, 246
Fritz, C. 37, 47, 120, 271, 300
From, H. 273, 281, 289
Furtdo, A. 243

G

Gamson, W. 262
Gatz, M. 158, 162, 168, 188, 189, 193
Gearhart, W. 319
Geertz, C. 204
Geman, H. 226, 234

generating hypotheses (see also exploratory research) 29
Geographic Information Systems 35, 38, 40, 192, 241, 303, 326, 382
  definition of 323
  use of during Hurricane Andrew 326
  use of in future disaster research 333
  use of in natural hazards 324
Gerth, H. 42
Ghiglione, A. 258
Gillespie, D. 136, 150
Gladwin, H. 250, 369
Glaser, B. 101, 195, 196
Global Disaster Handbook 319
Global Disaster Information Network 38, 382
Glueck, E. 315, 316
Glueck, S. 315, 316
Goffman, E. 202
Goltz, J. 162
graduate research assistants 28, 37, 98, 101, 102, 103, 104, 105, 106, 108, 111, 112, 114, 115, 118, 120, 121
Graham, G. 280
Graunt, J. 315
Greenberg, D. 277
Greenwell, M. 193
Grey Seas Under, The 292
Grimshaw, A. 235, 238
Ground Zero 277, 289
grounded theory 101, 196
Groves, R. 160
Gruhn, W. 319
Gruntfest, E. 309
Guba, E. 196, 198, 199, 202
Guetzkow, H. 23, 24
Guilette, E. 205
Guillén, M. 39
Gunter, V. 40
Gurney, J. 367
Gurr, T. 240
Gusfield, J. 41

H

Haas, J. 25, 27, 127, 131, 133, 135, 137, 144, 372
Halifax (Nova Scotia) munitions ship explosion 33, 155, 206, 272
Halifax Chronicle-Herald 293
Halifax County Academy 287
Halifax, Nova Scotia 33, 155, 194, 206, 244
Halifax Relief Commission 275, 276, 278, 288
Halifax-Massachusetts Relief Committee 280, 295
Halley, E. 315
Halpern, E. 199
Hammond, P. 128
Hansen, A. 260
Hantrais, L. 237, 243
Harding, S. 365
Harris, F. 278, 291

Harrist, A. 206
Hatfield, L. 276, 290
Havana, Cuba 284
Haynes, V. 318
Hays, T. 193
Hazards Reduction and Recovery Center, Texas A&M University 200
Hearts Throbs of the Halifax Horror 276
Hemispheric Congress on Disasters and Sustainable Development 240
Henry, W. 277, 279
Heston, A. 245
Hewitt, K. 40
Hilford 271, 294, 295, 298
Hiscott, R. 269
historical research
  problems of memory recall in 288
  role of inference in 297
  use of archives and libraries 282
  use of biographies and novels in 278
  use of contacts in 294
  use of newspapers in 274
  use of official records in 286
  validity of materials in 299
History of Modern Whaling, The 289
Hockings, P. 206
Hoetmer, G. 143
Hollerith, H. 317
Hoover, G. 243, 246

Hose, W. 274
Hossain, H. 338
Hovland 281
Howell, C. 277, 289
Huber, F. 316
Huberman, A. 196
Hudson, B. 84
human subjects 37
  in disaster research 351, 355
  institutional review boards 37, 352, 353
Human System Responses to Disaster 138, 300
Hurricane Andrew 35, 141, 149, 201, 203, 222, 226, 233, 320, 322, 324, 325, 326, 327, 373
Hurricane Betsy 112
Hurricane Camille 99
Hurricane Frederic 136
HyperSearch (computer software) 208

I

Iben Browning earthquake prediction 162
Iklé, F. 51
IMO 270, 273, 279, 281, 282, 289, 290, 297, 299, 300
impact zone 51, 65, 66, 67, 69, 77
Independent 334
Indian Statistical Institute 337

Indianapolis (Indiana) Coliseum explosion 97, 131
Industrial and Environmental Crisis Quarterly 146
informants
  interviewing of 68, 70, 92
  selecting for research 71
Institute for Business and Home Safety 384
Institute for Social Science Research, University ofCalifornia, Los Angeles California, Los Angeles 161, 162, 176, 179, 190, 193
Institute of Emergency Administration and Management, University of North Texas 201
Institute of Youth Research, Hamburg (Germany) 312
insurance 88
  earthquake 171, 172, 221, 222
  hazards 31, 72, 148, 213, 214, 219, 220, 226, 227, 233
  problems and issues with 385
Intercolonial Railroad 283, 295, 298
International City Management Association 143
International Committee of the Red Cross 309
International Decade for Natural Disaster Reduction 239, 309, 323
International Federation of Red Cross and Red Crescent Societies 240
International Journal of Mass Emergencies and Disasters 146, 195, 204
Internet (World Wide Web) 34, 35, 38, 125, 209, 211, 303, 319, 360, 382
  background issues 317, 382
  use of in disaster research 311, 319
interviewers 53, 72, 73, 74, 75, 76, 77, 78, 79, 84, 85, 87, 88, 89, 93, 98, 120, 123, 160, 161, 174, 178, 341, 354
  supervising 77, 163
interviewing in field research 52, 64, 69, 72, 73, 75, 77, 78, 79, 91, 93, 102, 103, 104, 107, 108, 112, 113, 117, 121, 123, 129, 134, 139, 160, 161, 175, 176, 179, 191, 201, 203, 205, 270, 341, 350, 353
  DRC guidelines for 111
  problems of recall in interviewing 53, 85, 86, 92, 93
  recording interviews 79

involvement versus detachment in disaster research 44

J

Jackson, B. 205
Jacobs, J. 195
Jaffee, D. 226, 227
Jäger, W. 40
Jaggers, K. 240
Janis, I. 134
Jefferson, H. 277
Jefferson, J. 268
Jobb, D. 276
Jodice, D. 240
Johnsen, A. 289
Johnstone, D. 276, 291, 296
Journal of Contemporary Ethnography 195
Journal of Economic Literature 231
Journal of Education 278

K

Kahn, R. 160
Kanter, J. 383
Kaplan, A. 42
Karim, A. 338
Kasperson, R. 251, 252
Keane, T. 163
Kentville, Nova Scotia 271, 274, 280, 282, 295
Key, W. 134, 135
Khondker, H. 36, 303, 337, 338
Killian, C. 246
Killian, L. 23, 26, 27, 36, 45, 50, 158
Kincaid, H. 51
King Fredric II of Prussia 316
Kings College, London 290, 291
Kish, L. 164, 166, 167, 174
Kitz, J. 275, 276, 300
Kleffner, A. 214, 221, 233
Kline, B. 278
Kline, I. 35
Kling, R. 306, 308
Knights of Columbus 272
Kobe (Japan) earthquake 329, 373
Koehler, G. 137
Kohn, M. 235, 238, 243
Kravis, I. 245
Kreps, G. 138, 145, 151, 152, 238, 249, 321
Kroll-Smith, S. 40
Kunreuther, H. 214, 221, 225, 226, 233, 316

L

LA RED 240
laboratory studies 24, 25, 29, 52, 94, 132, 133, 150
Lammers, C. 86
Lane, J. 235, 241
Lenze, D. 229
Lewis, C. 226
Lewis, T. 225
Lewiston, Maine 273

Liberal Roman Catholic Union Party 277
Lincoln, Y. 196, 198, 199, 336, 372
Lisbon (Portugal) earthquake 34
Lismer, A. 299
litigation
　threat of in field research 359
Litjen, R. 319
Lloyds of London 290
locally-unwantedlanduses 358
Lofland, J. 195, 196, 204
Lofland, L. 196, 204
Logan, L. 26
Loma Prieta (California) earthquake 158, 162, 163, 169, 171, 174, 175, 176, 177, 178, 179, 187, 190, 193, 222, 373
Lombardi, M. 32, 155, 258, 259
London School of Economics 285
London smog incident 59, 91
longitudinal research 205, 207
Longitudinal Study of Generations, University of Southern Caifornia 168
looting 90, 99, 112, 116, 145, 149
Los Angeles Business and Industry Council for Emergency Preparedness and Planning 379

Los Angeles, City of 97, 145, 163, 171, 331, 352, 361, 373, 378, 379
Los Angeles, County of 161, 162, 164, 166, 167, 172, 175, 176, 178, 188, 189, 193
Los Angeles County Social Survey 176, 177
Lotz, J. 275
Lovett, G. 294
Lower Canada College Magazine 275
Lulan 281

M

MacDonald, D. 216
MacIntosh, C. 293
MacLennan, H. 274, 275
MacMeachan, A. 276, 278, 279, 298
Magrini, S. 243
Mammoth, California 216
Mangen, S. 237, 238, 243
Marine Geodesy 278
Maritime Merchant 277
Maritime Museum (Canada) 273, 289, 298
Marrs, W. 26
Marshall, C. 207
Marshall, E. 356
Martin (computer software) 208
Martin, J. 208, 277
Mascherpa, B. 258

mass media and coverage of risk 265
study of 265
Mathewson, J. 37, 271
Mayor Tom Bradley, Los Angeles 378
Mazur, A. 261
McClellan, S. 301
McNabb, S. 356, 357
McNeil, R. 275
Menninger Foundation, Topeka, Kansas 135
Merton, R. 137
Metropolitan Life Insurance Company 383
Metson, G. 275
Mexico City earthquake 373
Michigan State University, East Lansing 47, 59, 91, 92
Miles, M. 196, 208
Mileti, D. 21, 27, 137, 142, 143, 144, 149, 151, 157, 160, 201, 205, 241
Miller, K. 203, 316
Millersville (Pennsylvania) State University 201
Mills, C. 42, 128, 266, 268
Ministry of Relief and Rehabilitation, Bangladesh 337
Miracles and Mysteries 276
Modigliani, A. 262
Moncton, New Brunswick 271, 283, 295, 297
Monnon, M. 276

Mont Blanc 270, 273, 274, 279, 280, 281, 282, 284, 289, 294, 295, 298, 299, 300
Moore, D. 316
Moore, H. 32, 134
Morrow, B. 201, 203, 250, 368, 369
Morton, R. 276
Mount St. Helens volcano eruption 136
Mount St. Vincent 292
Mowat, F. 292
Mukherjee, R. 337
Multidisciplinary Center for Earthquake Engineering Research 362, 373
Munich Ruck Insurance Company 318
Murdock, K. 226
Murray, Lt.-Cdr. 294
Mushkatel, A. 369
Musson, R. 206
Myers, M. 201

N

National Academy of Sciences 50, 122, 349, 362, 373
National Aeronautics and Space Administration 25, 202, 209
National Archives of Canada, Ottawa 274, 279, 281, 293
National Archives, Washington, D.C. 280

National Center for Geographic Information and Analysis 322
National Coordinating Council on Emergency Management 143, 371
National Earthquake Hazards Reduction Program 39, 363
National Emergency Management Association 379
National Flood Insurance Program 221, 222, 223, 224, 233
National Hurricane Conference 371
National Information Service for Earthquake Engineering 190
National Institute of Mental Health 99, 134, 202
National Opinion Research Center 47, 95, 120, 202, 349, 362, 373
National Public Radio 358
National Research Council 22, 23, 122
National Science Foundation 39, 96, 99, 102, 137, 148, 153, 193, 201, 202, 356, 363, 373
Natural Hazards Observer 329
Natural Hazards Research and Applications Information Center 38, 145, 193, 201, 202, 206, 208, 209, 309, 356, 362

Natural Hazards Workshop 371
Naval War College, Newport, Rhode Island 290
Neal, D. 201, 203, 209
Neo-Kantian school of philosophy 41
New Glasgow, Nova Scotia 271, 276, 281, 283, 291
New Madrid (Missouri) earthquake fault 162
New Orleans, Louisiana 112
Nguyen, L. 29, 155
Nielsen, J. 365
Nijera Kori 340
Niobe 273, 274
Nishenko, S. 148
nongovernmental organizations 240, 309, 336, 340, 341, 343, 348, 379
Norris, F. 163
Northridge (California) earthquake 141, 158, 161, 162, 163, 164, 166, 168, 169, 171, 172, 176, 177, 178, 179, 180, 183, 187, 188, 193, 352, 363
Northwest (Canada) Rebellion 294
Novak, J. 334

O

Oak Ridge National Laboratory, Oak Ridge, Tennessee 202
observation in field research 115, 196, 206

Office of Civil Defense 96, 98, 99, 350
Office of Emergency Services, California 363
Office of Foreign Disaster Assistance, U.S. Agency for International Development 239
Ohio State University, The 96, 97, 127, 200, 350
Okamoto, T. 316
Old Colony 282
Old Kings Courthouse Museum 280
Oliver-Smith, A. 205, 238, 243
Omvik, A. 289
Organization of American States 239, 384
Organized Behavior in Disaster 300
Osborne, G. 284, 289, 290
Osmond, J. 281
Otway, H. 257
Øyen, E. 235, 237, 239, 244, 249

P

Pacific Earthquake Engineering Research Center 373
Palm, R. 157, 221, 222, 233
Patterson, P. 260
Payzant, J. 277
Payzant, L. 277
PC Praxis 312
PC Professional 314
Peacock, W. 31, 155, 242, 246, 248, 249, 351, 369
Pelanda, C. 249, 272, 321
Pennings, J. 139
Perrow, C. 39
Perry, R. 369
Petterson, J. 356
Petty, W. 315
Phalia Dighar, Bangladesh 339
Phillips, B. 30, 155, 203, 349, 352, 372
Picou, J. 356, 359
Pictou, Nova Scotia 287, 292
Pierce, J. 319
Piscitelli, A. 238
Ponting, J. 94
Poppe Tyson 313
Port Greville, Nova Scotia 290
post-traumatic stress disorder 163, 178, 308
Powell, J. 51
Prawzick, A. 267
President Bill Clinton 377
Presser, S. 359
Prince, S. 194, 200, 244, 271, 276, 277, 284, 290, 291, 296
Pring, G. 357
Project Impact 377
Providence, Rhode Island 272, 286
Przeworski, A. 235
Public Archives of Nova Scotia 274, 275, 278, 281, 291, 293, 296

Public Health Nurse 278
public information officers 359, 360
Public Records Office, London 280, 290
public-private partnerships 375
  and disasters 380
  legal constraints on 381
  needed reserach on 386
Punaro, T. 323
Puttner, C. 277

Q

QSR Nudist (computer software) 208
Qualitative Data Analysis 196
Qualitative Health Research 195
qualitative research 21, 22, 30, 155, 194, 195, 196, 197, 198, 199, 200, 201, 202, 203, 204, 205, 206, 207, 208, 209, 210, 365, 372
  compatibility with disaster research 204
  future of in disaster research 211
  history of in disaster research 202
  misconceptions of 200
  problems of in disaster research 207
  recent trends in 197

Quarantelli, E. 23, 25, 26, 27, 28, 30, 33, 37, 40, 45, 94, 95, 97, 100, 120, 125, 127, 131, 134, 145, 146, 200, 204, 267, 300, 301, 321, 349, 350, 352, 372, 373, 374
Quayle, D. (U.S. Vice President) 381
Quayle, M. 381
Quételet, A. 315, 316

R

R. J. Reynolds Tobacco Company 355
Raddall, T. 276, 295
Ragin, C. 235, 241, 243
random digit dialing (see also survey research) 160, 161, 162, 164, 165, 166, 173, 191
Rauch, M. 290
Rayner, J. 26, 51
Reagan, R. 138
reconnaissance, in field studies 60, 61, 118
research design in disaster studies 63
  selecting events for study 61, 107
research teams in comparative research 362
respondents, interview 66, 68, 74, 75, 76, 77, 78, 85, 87, 89, 91, 92, 93, 103,

110, 111, 117, 119, 123, 158, 159, 160, 161, 163, 164, 166, 167, 168, 169, 171, 172, 174, 178, 180, 183, 184, 187, 188, 189, 191, 196, 202, 203, 211, 341, 343, 344, 345, 346, 347, 357
Richards, L. 208
Richards, T. 208
Richardson, L. 196
Richardson, R. 269
Rickets, F. 293
Rio Grande River flood 91
risk communications 256
  amplification of risk in 253
  effects of conflict on 256
  perceptions of credibility in 255
Rockingham, Nova Scotia 298, 299
Romance of the Halifax Disaster, A 274
Rose, A. 229, 230
Ross, G. 277
Ross, J. 277
Rossi, P. 217
Rossman, G. 207
Roth, R. 226, 233
Royal Canadian Navy 290
Royal Naval College 285, 290
Rubin, C. 216, 217, 222, 225
Rubin, D. 270
Ruffman, A. 277, 289
Russel, T. 226, 227

Russell, L. 162, 169, 174
Rutherford, G. 269

S

Sage Publications 195
Saint John, New Brunswick 272, 284
Salvation Army 292
sampling 26, 51, 57, 58, 61, 63, 64, 65, 66, 67, 68, 69, 70, 71, 95, 105, 117, 138, 139, 141, 144, 159, 162, 165, 166, 167, 176, 178, 188, 312
  representativeness of samples in survey research 176
  snowball 269
Sandefjord Blad 273, 290
Sandefjord, Norway 281, 289, 297
Sawatsky, J. 269, 270
Scanlon, J. 32, 33, 155, 205, 267, 268, 269, 270, 271, 272, 301
Scarce, R. 355
Scawthorn, C. 218, 219
Schatzman, L. 195
School for the Deaf and Dumb (Nova Scotia) 280
Schrage, E. 316
Schwandt, T. 199
Schwartz, H. 195
SCOR Reinsurance 383
Scott, W. 92, 93
Shamgar-Handelman, L. 196

Shattered City 275
Shilling, J. 216, 218, 219, 222
Shoaf, K. 29, 155
Simile, C. 203, 369
Simon, J. 316
Slovic, P. 257
Smelser, N. 41
Smith, S. 40, 205, 238, 243, 273, 276
Social Darwinism 316
social problems, disasters as 152
Sociological Imagination, The 266, 268
Sorensen, J. 140
Spaeth, M. 160
Spencer, P. 163
Spradley, J. 195, 196, 203
Sproat, D. 268
St. Francis Xavier University, Antigonish, Nova Scotia 279
St. John Ambulance 293, 296, 300
St. Johns Memorial University (Newfoundland) 290
St. Marys College, Halifix 272
St. Marys University (Canada) 277, 292
St. Pauls Anglican Church, Halifax 272
St. Stephens Protestant Episcopal Church, New York 272
Stake, R. 196
Stallings, R. 30, 143, 151, 153, 319, 374
stand-by research capability 25
State Archives, Oslo (Norway) 292
State University of New York at Buffalo (New York) 373
Staten Island, New York 289
Statistical Bureau of Prussia 316
Steel Company of Canada 292
Stellarton, Nova Scotia 285
Stephenson, E. 193
Stephenson, J. 134
Stern 310
Stern, P. 256
strategic lawsuits against public participation 357, 358
Strauss, A. 101, 195, 196, 204
Streeter, C. 321, 330
Stretton, A. 267
subjects, securing cooperation from 72
subjects, selecting for research 65, 68
Summers, R. 245
survey research 21, 22, 29, 47, 52, 83, 94, 95, 134, 135, 155, 157, 158, 159, 160, 161, 162, 163, 164, 169, 171, 172, 173, 178,

179, 180, 187, 191, 192,
  214, 241, 247, 315, 330,
  331, 361, 372
control groups in 190
data archives 190
data-entry programs 190
definition of 160
methods of 162
recall of subjects in 183
responsiveness of subjects in
  178
telephone interviews in 30,
  140, 161, 166, 176
Survey Research Center,
  University of California,
  Los Angeles 162, 179
Survey, The 276
surveys
  cross-earthquake comparisons of 173
  postearthquake 169
  timing of after disasters 179
sustainability 148, 151
Swanson, G. 235
Swetnam, W. 298
Sykes, G. 135
Sztompka, P. 244

T

T-Online 312
Taeger, J. 316
Tampa Bay (Florida) fuel spill
  324
Taylor, C. 240
Taylor, J. 135

Taylor, V. 31, 94, 267, 277,
  374
Teune, H. 235, 239, 247
Texas A&M University 200
Texas City, Texas 26, 70
Third Reich 317
Thompson, J. 139
Tierney, K. 28, 37, 40, 157,
  200, 303, 359
Tonnessen, J. 289
Tooke, F. 277
Tornado in Worcester 90
tornadoes 26, 32, 50, 51, 60,
  63, 94, 108, 135, 145,
  318
Toronto Daily Star 273
Toronto Globe 276
tourist industry, study of 142
Tours, France 284
Town That Died, The 275
Townsend, P. 279
Transport Canada 292
Truro, Nova Scotia 271, 272,
  283, 284, 286, 295, 297
Turkey Point (Florida) Nuclear
  Power Plant 326
Turner, R. 158, 161, 162,
  169, 243

U

U.S. Agency for International
  Development 239, 240,
  381, 385
U.S. Army Corps of Engineers
  202, 224, 329

U.S. Coast Guard 324
U.S. Department of the Army 39, 202, 350
U.S. Office of Disaster Assistance 239, 240, 385
UNESCO 338
Ungerson, C. 243
University of California, Berkeley 190
University of California, Los Angeles 161, 162, 176, 179, 190
University of Colorado, Boulder 145, 201, 356, 362
University of Delaware 201, 270, 385
University of Denver 134, 135
University of Dhaka, Bangladesh 338
University of Maine, Orono 292
University of Maryland, College Park 349
University of North Texas 201, 202, 209
University of Oklahoma 47, 349
University of Southern California 162, 168, 193
University of Toronto 271
unobtrusive measures 206

V

Valentini, T. 257
Van Maanen, J. 196
van Schendel, W. 337
Victoria General Hospital 277, 285
von Steuben 280
von Winterfeldt, D. 255
vulnerability analysis 332

W

Wagner, J. 206
Wallace, A.F.C. 47, 51, 70, 90, 244
Wallace, W. 206
War Cry 292
Warheit, G. 97
Warner Robbins, Georgia 26
Warren, C. 365
Webb, E. 206
Webb, G. 359
Weber, Marc 309
Weber, Max 41, 42, 43
Weitzman, E. 208
Wenger, D. 200, 270
West, C. 229
West Side School, Halifax 291
Western Union 116, 294
White, G. 144, 201
Whittier Narrows (California) earthquake 158, 162, 163, 169, 171, 172, 175, 176, 177, 187, 190, 193
Whyte, W. 128

Wichita Falls, Texas, tornado 136
Wilkes Barre (Pennsylvania) flood 94
Wilkins, L. 260
Williams, H. 22, 47, 134
Windsor, Nova Scotia 282
Wolcott, H. 196
Wolensky, K. 324
Wolensky, R. 324
Wolfe, J. 163
Wolfville, Nova Scotia 279, 282
Worcester (Massachusetts) tornado 244
World Bank 147, 240, 339
World Health Organization 309
World Resources Institute 240
Worldmap of Natural Risks 318
Worth, S. 206
Wotherspoon, W. 268
Wright, J. 217, 242, 246
Wynne, B. 254

X

Xenia (Ohio) tornado 94

Y

Yellowstone National Park, forest fire in 318
Yezer, A. 31, 155, 216, 217, 222, 224
Yin, R. 140, 196
YMCA 272, 284

Z

Zurcher, L. 135

Made in the USA
Lexington, KY
23 January 2010